CISM Courses and Lectures

The series presents lecture notes, monographs, edited works and proceedings in the field of Mechanics, Engineering, Computer Science and Applied Mathematics.
Purpose of the series is to make known in the international scientific and technical community results obtained in some of the activities organized by CISM, the International Centre for Mechanical Sciences.

International Centre for Mechanical Sciences

Courses and Lectures Vol. 552

For further volumes:
www.springer.com/series/76

Tomasz Łodygowski · Alexis Rusinek
Editors

Constitutive Relations under Impact Loadings

Experiments, Theoretical and Numerical Aspects

Springer

Editors

Tomasz Łodygowski
Poznan University of Technology, Poland

Alexis Rusinek
National Engineering School of Metz, France

ISSN 0254-1971
ISBN 978-3-7091-1767-5 ISBN 978-3-7091-1768-2 (eBook)
DOI 10.1007/978-3-7091-1768-2
Springer Wien Heidelberg New York Dordrecht London

All contributions have been typeset by the authors
Printed in Italy

Printed on acid-free paper

Springer is part of Springer Science+Business Media (www.springer.com)

PREFACE

Testing the properties of matter, formulation of the constitutive equations and using them in computations are the main focus of the materials presented in this book. Today the industrial applications require better understanding of the material behavoiur in particular when the strong influence of loading type, which introduces temperature, strain rate dependence, fracture etc. are taken into account. We believe that understanding of these phenomena is of fundamental importance for responsible computations. In particular, using the well known commercial programs requires deep understanding of constitutive formulations and their restrictions.

The lectures are addressed to the users from industry who are always facing the crucial decisions in design, as well as, to the young scientists who work on a new models that describe and reflect the complexity of material behavior.

Finally, we would like to acknowledge the commitment of Professors: G. Gary (Ecole Polytechnique, Palaiseau, France), T. Łodygowski (Poznan University of Technology, Poznań, Poland), R. Pęcherski (Institute of Fundamental Technological Research, Warsaw, Poland), D. Rittel (Technion - Israel Institute of Technology, Haifa, Israel), A. Rusinek (National Engineering School of Metz, Metz, France), R. Zaera (University of Carlos III, Madrid-Leganes, Spain) in making the course possible and arranging the excellent lecturing. All the chapters of the book, being the results of these lectures, provide a basis for the future development of constitutive relations and their careful applications.

We would like to acknowledge, on behave of all the lectures and participants, an excellent organization of the course No. 396 by the CISM staff.

Tomasz Łodygowski

CONTENTS

CONTENTS

Testing With Bars From Dynamic to Quasi-static

Gérard Gary

Laboratoire de Mcanique des Solides, Ecole Polytechnique, Palaiseau, France

Abstract The numerical calculation of the dynamic loading of a structure includes a great number of steps in which various fundamental or engineering problems are involved. Most of them are addressed in the present course at CISM. In this paper, we discuss the testing of materials in order to model their behaviour.

Because of waves induced in the testing device by impulse loadings and short time measurements, the data analysis has to deal with transient effects. Using bars makes easier such an analysis. For this reason, Hopkinson bars are a very commonly used dynamic testing device.

Using the word "dynamic" means that "time" is considered as an active parameter in the evolution process. When dynamically testing a structure (a cylindrical specimen is a common example of such a structure) the effects of time appears in different ways.

There is not static equilibrium in the machine so that measurements at specimen ends cannot be simply deduced from measurements with sensors incorporated in the machine, as it is the case with quasi-static testing. Furthermore, most sensors (like force cells) have a limited high passing band.

Transient effects in the specimen induce waves and the non-homogeneity of mechanical parameters. Consequently, average or global measurements cannot be right away related to local ones.

Stresses cannot be simply related to forces measurements as inertia effects are also involved – the most known effect is the confinement induced by lateral inertia, especially important when testing a big specimen of brittle material.

Short testing times do no allow for isothermal testing – a metallic specimen can have a temperature increase up to 100°C during a SHBP test.

The behaviour of an elementary volume of the material can depend on the rate of change of basic mechanical parameters strain and/or stress. This last effect (strain rate sensitivity) is the (only) one that is expected to be measured, in most cases.

In (dynamic) mechanical testing it is then suitable to consider separately the global measurements made on the specimen (forces applied at a part of the specimen border and displacement measured at another - or the same - part) and the analysis of its mechanical evolution.

This is commonly done in the quasi-static side but is not always, for historical reasons, done in dynamic testing.

The above discussion does not answer the basic question of the boarder between quasi-static and dynamic testing. Theoretically, indeed, waves in solids are still present in quasi-static testing. The common criterion to evaluate this limit is to compare the time τ_e needed to reach equilibrium (say $< 5\%$ of non homogeneity of stresses and strains) to the measurement duration. Note that τ_e mostly depends on the specimen size and on the elastic speed of waves in the material and not on the measurement duration. In the classical SHPB literature, this problem is related to the "impedance matching problem", misunderstood in many publications, perfectly addressed in 1963 by Davies & Hunter. Based on this criterion, (too) many SHPB tests are considered as quasi-static ones.

1 Why using bars

Taking account of the transient response of the machine seems a difficult theoretical problem as far as a geometrically complex loading machine is considered. Two ways allow for avoiding this difficulty.

The first one is to make measurements directly at specimen faces. Even if modern optical devices can provide direct displacement measurements, force measurements need transducers in which mechanical waves are induced by the loading.

The second is to use a simple enough machine allowing for an analysis of transient effects. This is the case of Hopkinson bars.

In order to illustrate the problems encountered with a classical machine, we examine the case of a drop test.

We consider the simplified description of a drop test machine using, as an example, the compression test of an aluminium honey comb. As a first approximation, we assume that the force response of the specimen provides a constant value F_0 (fig. 1).

In this simulation, the force is measured in two different ways. We first consider a linear spring the shortening of which is measured by an instantaneous optical device. Secondly, the force is deduced from the deceleration of the known falling mass. The corresponding acceleration is measured with an accelerometer of finite size in its middle. For sake of simplicity, the test is modeled with the one-dimensional analysis. In this case, usual relations between jumps of stress, particle velocity and strain are used. See basic demonstration later or, in a more general frame, Achenbach (1978).

$$\Delta\nu = -c\Delta\varepsilon \quad \Delta\sigma = -\rho c\Delta v \tag{1}$$

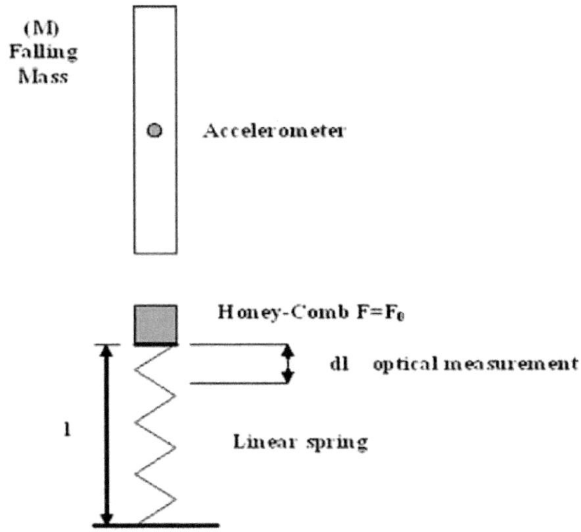

Fig. 1. One-dimensional scheme of a drop test.

At any time, the stress and the particle velocity can be calculated at any section of the falling mass and of the spring. Fig. 2 shows the calculated velocities at the accelerometer position and at the head of the spring.

Knowing the velocity of the accelerometer, its acceleration is obtained by derivation. To avoid obtaining an infinite acceleration, the speed considered is the average speed across the accelerometer and it depends on its size. If the falling mass is supposed to have this measured deceleration, a measured force is deduced.

From the velocity of the springhead, the displacement is obtained by integration and the relative variation of the length of the spring is known. If the force supported by the spring is supposed proportional to this displacement, another measured force is deduced.

Both forces, as they would be deduced from a quasi-static analysis of the test, are shown in fig. 3.

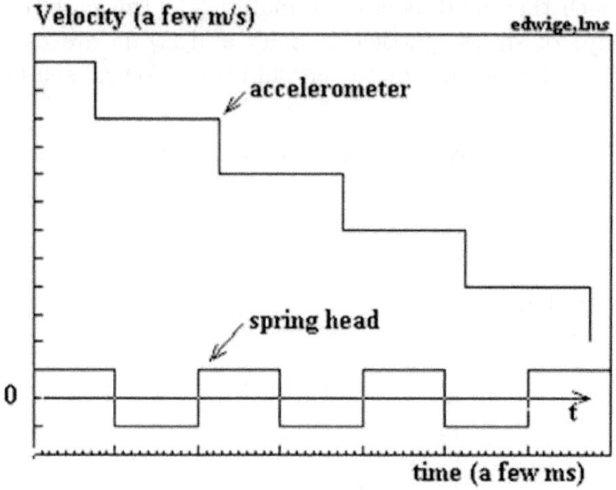

Fig. 2. Velocity at measuring points.

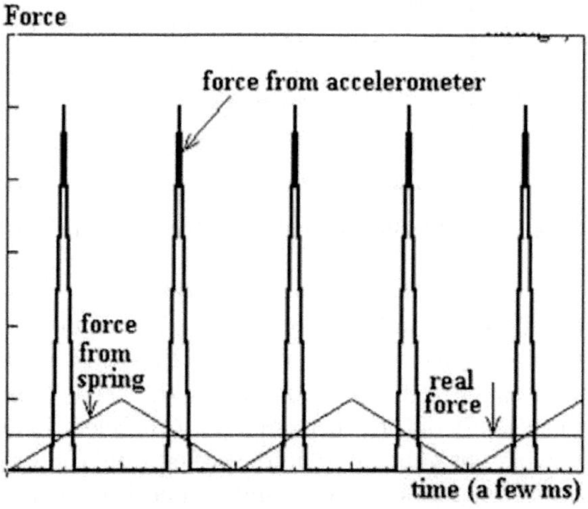

Fig. 3. "measured forces"

It is observed that the two methods give very different results. In both cases, the maximum force is over estimated, especially with the acceleration measurement. Nevertheless, it can be checked that the mean value of both force measurements corresponds to the exact one. It confirms that a quasi-static analysis of the test could be done provided that the duration of the test is great compared with the oscillation period of each system and that the passing band of the system is adapted.

This example purposely goes to extremes in order to show the importance of transient effects in dynamic testing. Such an analysis largely explains the difference observed between the two results of crash-tests done on the same structure (square tube) with an industrial drop test machine and, in our laboratory with a Hopkinson bar. Both results are shown together in fig. 4, with same scales on axes.

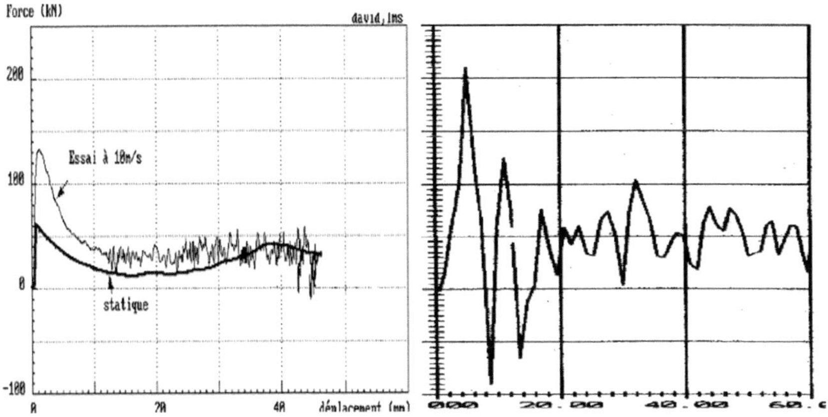

Fig. 4. Measurement with Hopkinson bars (left), falling mass (right)

Looking at fig. 4, one observes that, as expected, the dynamic response of the square tube shows a greater force than the static response of the same tube. The dynamic result also shows a good agreement with a numerical simulation of the test. The test duration of the Hopkinson bar system (6 m long aluminium input bar, diameter 80 mm) has been increased by using a deconvolution technique that will be described later.

This example shows how important are transient and inertia effects in dynamic testing machines and it also shows that a good account of them is taken with Hopkinson bars. There are various other dynamic testing techniques that will not be considered in this paper as we focus on the Hopkinson bars, or Kolsky apparatus.

2 Basic notions on 1-D waves

2.1 Dalembert's equation

The general form of the so called "Dalembert's equation" concerns one time variable t and one or more spatial variables x_1, x_2, ..., x_n, and a function $u = u(x_1, x_2, ...x_n, t)$, the values of which model some amplitude of a wave.

The wave equation for u is $\frac{\partial^2 u}{\partial t^2} = c^2 \nabla u$, where ∇ is the (spatial) Laplacian and where c is a fixed constant.

In solid mechanics, and within the Lagrangian formalism where variables are related to material points, the generic scalar function is the amplitude of the displacement vector at a point.

In the particular case of 1-D elastic bars, this displacement is a scalar where c is the speed of the 1-D wave. The wave equation can be then established in a very simple manner.

Let us consider the dynamic equilibrium a thin slice of a bar with a thickness dx between abscissa x and x+dx.

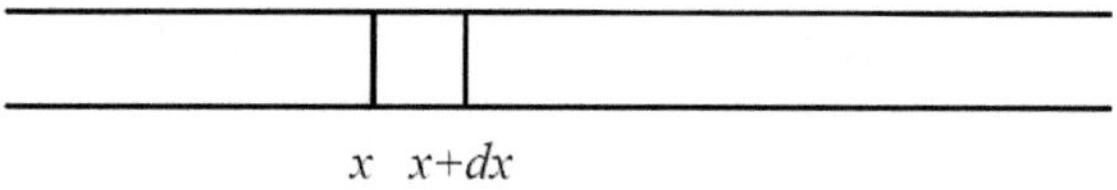

A is the area of the bar, ρ its density, σ the uniaxial stress, u the displacement, ε the uniaxial strain.

Dynamic equilibrium ($F = m\gamma$) for this slice reads:

$$A\sigma(x + dx) - A\sigma(x) = \rho A dx \frac{\partial^2 u}{\partial t^2}$$

Using elasticity 1-D ($\sigma = E\varepsilon$) and the definition of 1-D strain ($\varepsilon = \partial u/\partial x$) one obtains:

$$AE\frac{\partial \varepsilon}{\partial x} dx = \rho A dx \frac{\partial^2 u}{\partial t^2},$$

and the Dalembert's 1-D equation:

$$\frac{\partial^2 u}{\partial t^2} = c^2 \frac{\partial^2 u}{\partial x^2} \text{ with } c^2 = \frac{E}{\rho}$$

Both terms support to be differentiated with respect to x, so that Dalembert's equation is also verified by the uniaxial strain (and by the linearly linked uniaxial stress). In mechanical applications the equation of waves more often considers the strain.

The general solution of this equation is known to be in the form:

$$u = f(x - ct) + g(x + ct)$$

where $f$and g are arbitrary functions.

The variable $x - ct$ means that the value of u at point x and time t is the same as at an other position x' and an other time t'such as

$x - ct = x' - ct'$ or $x - x' = c(t - t')$

The value u is observed at a distance ($x - x'$) after a delay $(x - x')/c$. It defines the propagation of a signal in one direction at speed c.

The variable $x + ct$ defines, in a similar way, the propagation in the opposite direction.

2.2 Relations between strain, stress and particle velocity

These relations can be defined as a particular case of "jump equations" established for material waves.

An easier and more comprehensive way to derive them is to recall the general form of a single wave propagating in the positive direction (conventionally defined) and its derivatives with respect to space and time.

$$u = f(x - ct) \quad \varepsilon = \frac{\partial u}{\partial x} = f'(x - ct) \quad v = \frac{\partial u}{\partial t} = -cf'(x - ct)$$

For a wave propagating in the positive direction, on has then.

$v = -c\varepsilon$ and (from $E = \rho c^2$) $\sigma = -\rho c v$

On has to keep in mind that these relations are valid in a Galilean referential. Considering for instance the shock of a striker on a bar needs to carefully account for the initial speed of the striker. In terms of jumps, equations (2) are recovered.

$$\Delta\nu = -c\Delta\varepsilon \quad \Delta\sigma = -\rho c\Delta v \tag{2}$$

Equations (2), together with the propagation relations, are the foundations of Hopkinson bar measurements, where forces and displacements are computed at bar end on the base of strain measurements.

3 Real bars: dispersion relations

3.1 Case of elastic bars

A philosophical point

From an epistemological point of view, it is interesting to observe that, in the general 3-D elastic case, one can find two solutions **only** of the Dalembert's equation corresponding to the so called **P** and **S** waves. By reference to seismology, **P** is for "premières" (first arrived, speed c_p) and **S** for "secondes"(speed c_s).

$$c_p = \sqrt{\frac{E(1-\nu)}{\rho(1+\nu)(1-2\nu)}} \quad c_s = \sqrt{\frac{E}{2\rho(1+\nu))}}$$

For a Young's modulus equal to 200 GPa, a Poisson's ratio 0.3 and a density 7850 kg/m3 (values for steel), c_p =5860 m/s, c_s =3130 m/s, while the speed of the 1-D wave is 5047 m/s.

Note that for torsion bars, the speed of the torsion wave is c_s.

The 1-D wave (that theoretically does not exist, as no bar is purely 1-D) has a speed in between 3-D speeds and is a theoretical concept that describes very well the wave signals observed in bars. Such a "wave" could be also considered as a mixture of 3-D waves traveling at speeds $c_p$and c_s.

Dispersion

Knowing that the 1-D model is not perfect, as a careful observation of waves might show it at once, it is worth looking at a more complete model.

The main point is that, due to a positive value of Poisson's ratio, the compression goes with a diameter increase inducing lateral inertia effects.

On the contrary, shear waves (torsion waves in torsion bars), which do not induce volume change, are non dispersive.

The wave dispersion effects (due to radial expansion) on the longitudinal elastic wave propagating in cylindrical bars have been studied experimentally by Davies (1948). On the basis of the longitudinal wave solution for an infinite cylindrical elastic bar given by Pochhammer's (1876) and Chree (1889), a dispersion correction has been proposed and verified by the experimental data. Even though the Pochhammer-Chree solution is not the exact one for a finite bar, it is found easily applicable and sufficiently accurate - see Davies (1948).

In the Pochhammer-Chree's longitudinal wave analysis, it is assumed that the wave is harmonic as follows.

$$\mathrm{u}(r,z,t) = \frac{1}{2\pi}\int_{-\infty}^{+\infty} \bar{\mathrm{u}}(r,\omega)e^{i[\xi(\omega)z-\omega t]}d\omega \tag{3}$$

where u(r,z,t) is the displacement vector.

The complete solution of the governing equations with the boundary condition on the external surface of the bars leads to a frequency equation that gives a relation between the wave number ξ and the pulsation ω.

$$f(\xi) = (2\alpha/r_0)(\beta^2+\xi^2)J_1(\alpha.r_0)J_1(\beta.r_0) - (\beta^2-\xi^2)^2 J_0(\alpha.r_0)J_1(\beta.r_0) - 4\xi^2\alpha.\beta.J_1(\alpha.r_0)J_0(\beta.r_0) = 0 \tag{4}$$

with $\alpha^2 = \frac{\rho\omega^2}{\lambda+2\mu} - \xi^2; \beta^2 = \frac{\rho\omega^2}{\mu} - \xi^2$; and J_0, J_1 are the Bessel functions, λ and μ are the elastic coefficients, r_0 is the radius of the bar.

It is important to note that this equation has an infinite set of solutions which are called the propagating modes. As they do not define a basis of a vector space, (they are not orthogonal to each other) it cannot be decided, from a single measurement, how the energy is split into the different modes.

A very nice point is that, at low frequencies, only one mode has a real celerity. The frequency at which could appear the second mode is around $0.22c/a$ (for steel, with a radius equal to 10 mm,$f_{cut} = 105$ kHz). It can be checked that such a frequency is generally above the spectrum of a standard test (even without pulse shaper).

It is then of common use with Hopkinson bars to consider the first mode solution of equation (4). This use has been carefully validated by Tyas & Watson (2001).

This harmonic wave solution has been studied numerically by Bancroft (1941). Bancroft's data is given in the form where the phase velocity variation C/C_0 is a function of Poisson's coefficient ν and of the ratio between radius and wave length r_0/λ for the non dimensional interest.

$$C/C_0 = F(r_0/\lambda, v), \tag{5}$$

with $C = \omega/\xi$ and $\lambda = 2\pi/\xi$

The previous works - Follansbee & Franz (1983), Gorham (1983), Gong et al.(1990), Lifshitz & Leber (1994)- use this data to correct the wave dispersion in bars. Following Yew & Chen's works (1978), they calculate harmonic components in frequency domain of signals by Fast Fourier Transform (FFT) and find the phase difference for each component from (Eqn. 4). The corrected signals in time domain can be recovered from the corrected frequency components.

Their correction procedure in term of variation of phase velocity can be re-written as follows. From the knowledge of the dispersion relation between wave number ξ and frequency ω given by the solution of (4) or (5), one can calculate, from a measured wave $u_z^m(t)$, the wave $u_z^i(t)$ reached at a distance Δz. Using the components in z of the formula (2) at the surface of the bar, one can write $u_z^m(t)$ and $u_z^i(t)$.

$$\mathrm{u}_z^m(t) = \frac{1}{2\pi}\int_{-\infty}^{+\infty} \bar{\mathrm{u}}_z(r_0,\omega)e^{i[\xi(\omega)z_0-\omega t]}d\omega$$

$$\mathrm{u}_z^i(t) = \frac{1}{2\pi}\int_{-\infty}^{+\infty} \bar{\mathrm{u}}_z(r_0,\omega)e^{i[\xi(\omega)(z_0\Delta z)-\omega t]}d\omega \tag{6}$$

The wave shifting procedure can be then performed numerically by the FFT.

$$u_z^i(t) = FFT^{-1}\left[e^{i\xi(\omega)\Delta z}FFT\left[u_z^m(t)\right]\right] \tag{7}$$

Accordingly, the correction accuracy depends only on that of the dispersion relation $\xi(\omega)$. The use of Bancroft's data must be done carefully as one has to take care of the following points. First, as the value is given only in form of a table for a certain Poisson's coefficient and for certain values of the ratio between radius and wave length r_0/λ , which is known discretely, an interpolation is needed. Second, equation (5) only gives an implicit relation for correction, between wave number ξ and frequency ω. It is then recommended to calculate directly from (4) the dispersion relation, in the form of wave's number ξ as function of ω. for a given Poisson's ratio and Young's modulus.

The application of formula (7) with modern computers is instantaneous and should be systematically used. Using "pulse-shapers" to reduce the high frequencies in the recorded signals (and consequently reducing the need of the correction) also reduces the rising time of the signals and then the duration of the commonly found "constant plateau".

It is observed that the dispersion changes the slope of the signal in the early stage of the test and has a sensitive effect on the average stress-strain curve in the range of small strains, as shown by Gary et al. (1991) and Zhao and Gary (1996).

A good way to test the quality of the dispersion relation is to check whether the oscillations induced by the dispersion disappear at the impact side where the signal can be transported, as shown in fig. 5.

The two "ears" are due to two plastic rings used to guide the striker. (The non-nul slope is not explained)

Another way is to calculate the stress at a free end, which must be zero, as shown in fig. 6.

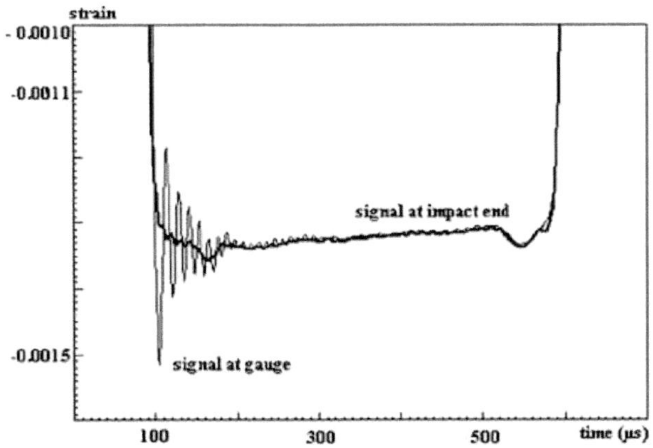

Fig. 5. Incident signal transported at striker side. (striker 1.3 m and input bar 3 m made of steel, diameter 20 mm. Incident gauge in the middle, striker speed 13.3 m/s)

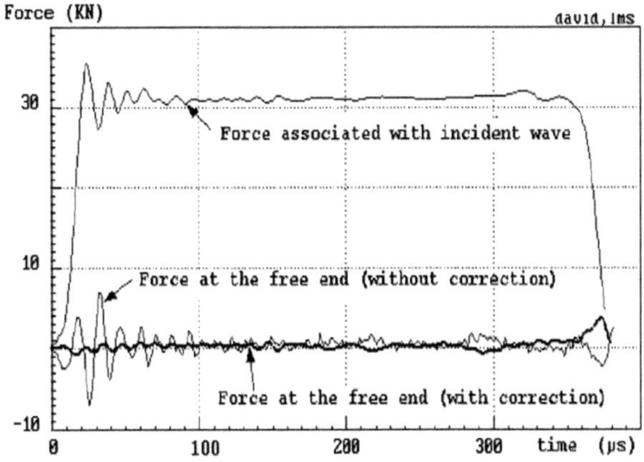

Fig. 6. The forces at a free end.

3.2 Case of visco-elastic bars

For viscoelastic bars, the argument ξ in equation (4) is a complex number so that a solution cannot be expressed in a simplified form such as (5). Consequently, the computation of the dispersion relation needs the functions $\lambda^*(\omega)$ and $\mu^*(\omega)$ that describe the viscoelastic behaviour of the bars.

The material properties of standard viscoelastic materials are often represented with a rheological model. The simplest one (called Zenner model) needs 3 parameters to be identified (first 3 left elements in figure 7).

They could be identified in different ways, more often when dealing with bars by an inverse approach based on the measurement of the wave propagation on the bar itself. This is a safe method as, unlike for metals, material properties depend on various external factors.

Zhao and Gary (1995) have used an inverse method based on such an idea. They found that the Zenner model was not complete enough, and had to use a more complex model with 9 parameters (figure 7). Furthermore, following many authors, they assumed a constant Poisson's ratio.

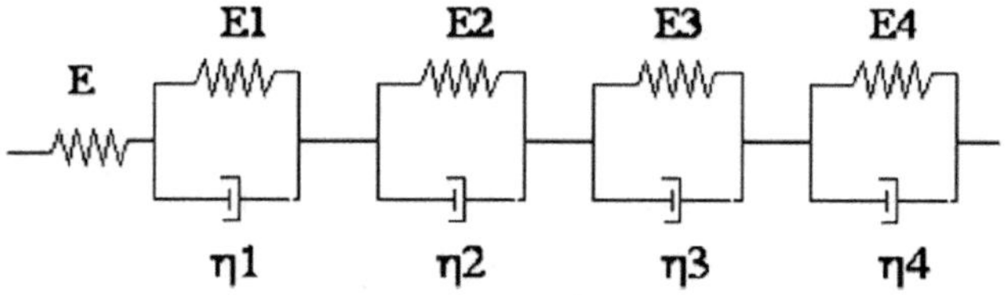

Fig. 7. A linear viscoelastic model

The same experimental set-up as in an impulse test is applied, where the waves are recorded at two different points in the bar. Using the wave at a recording point as input data, the parameters are determined in comparing the predicted wave and the recorded wave at another point (like in fig. 8)

It is also possible to avoid to solve equation (4). Dispersion may indeed be experimentally determined by comparing the wave Fourier components measured on two points on the rod – Blanc (1971), Lundberg and Blanc (1988), Gorham (1983), Bacon (1998). More recently, Hillström & al. (2000) developed a multi-point method using least squares. Another method, very accurate, is also proposed by Othman & al. (2001) based on a one-point measurement method using a spectral analysis of the resonant frequencies of the rod. An recent extension and improvement of this method proposed by Collet et al (2012) is presented in chapter **7.**

In order to illustrate the excellent quality of the dispersion relations generally obtained, an original record and two other records at respective distances of 4 m and 8 m of the original one, compared with their predictions, are shown in fig. 8.

By comparison with the elastic case (see fig. 4) it is observed that the dispersion correction is more important for common size viscoelastic bars and could not be neglected.

The importance of the correction in a real test situation is illustrated by showing both forces calculated in a test without specimen where the output force must obviously be equal to the input one (fig. 9 and 10).

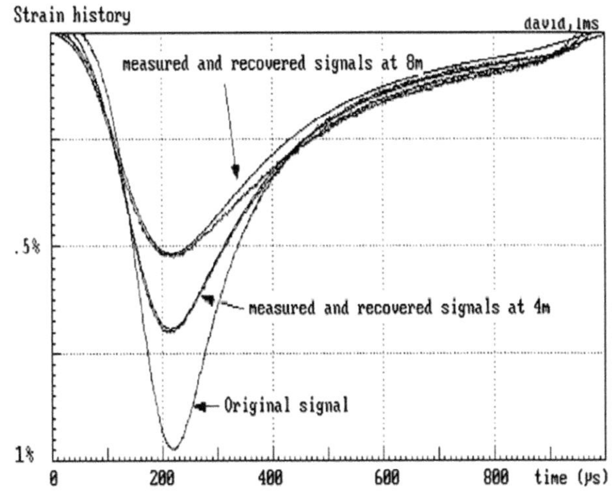

Fig. 8. Test of the dispersion relation.

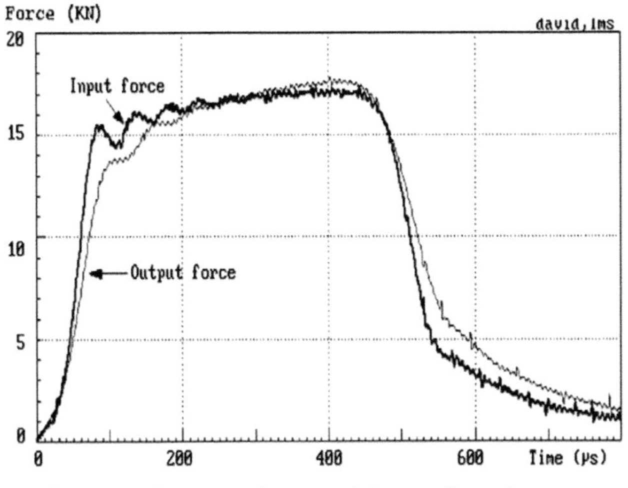

Fig. 9. Input and output forces **without** dipersion correction.

For **both viscoelastic and elastic bars**, a precise dispersion correction **does not suppress** oscillations of the incident force in the case of sharp loading pulses. These oscillations are indeed the physical consequence of the wave dispersion.

At very high speeds of loading with short specimens, oscillations can also be seen in the output signal

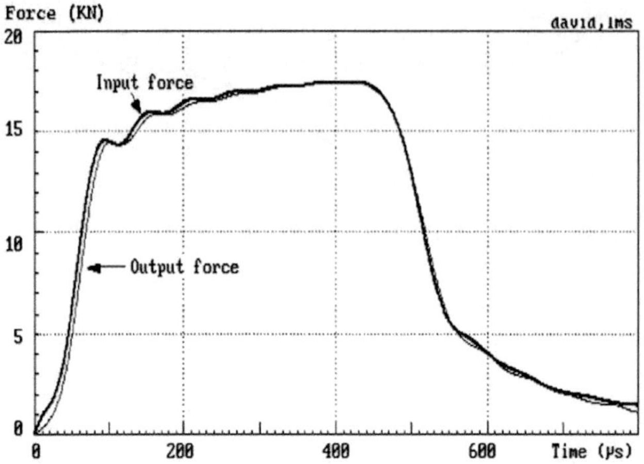

Fig. 10. Input and output forces **with** dipersion correction.

3.3 Planerity of waves

The planarity of waves is extensively studied by Davies for the first mode. The solution of equation (4) is well known and illustrated in his paper for particular values of the frequency, as shown in figure 11 - from Davies (1948).

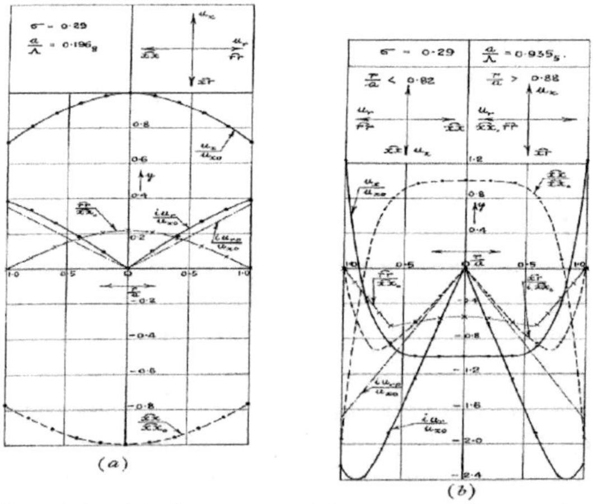

FIGURE 19. The variation of the displacements and the stresses over the cross-section of a bar of radius a. $\widehat{xx}$ = longitudinal stress; $\widehat{xr}$ = shearing stress; $\widehat{rr}$ = radial stress; u_x = longitudinal displacements; u_r = radial displacement; u_{rp} = radial displacement in distortionless bar.

Fig. 11. Stresses and strains function of radius and frequency.

In this figure one sees, among other variables, the relative variation of the stress along the radius (xx/xx_0). It shows that, at a relative high frequency (in 19b, 0.935 corresponds to a non realistic value of the frequency for our standard steel bar: 280 kHz), the stress at the surface of the bar is negative when the average stress is positive. At a more realistic frequency (figure 19a – 94 kHz for our bar) one can see that the average stress is more or less 10% higher than the stress at the surface. The strain is not proportional to the strain, but those observations are also valid for the average strain. On has then to keep in mind that, depending on the frequency:

a) - The strain measured at the surface is not proportional to the average strain.

b) - The axial strain and the orthoradial strain are not proportional to each other.

c) - The axial stress is not proportional to the axial strain.

d) - The average axial speed is not proportional to the axial strain.

Point (b) is well illustrated by Tyas and Watson (2001).

At lower frequencies, all the measurements become close to be proportional to each other.

A few authors – Safford (1992), Merle and Zhao (2006) have checked the validity of usual hypothesis commonly used. They proved that it is not worth taking care when frequencies f are under $2c/a$ (where a is the radius of the bar, equivalent for steel, to $a/\lambda < 0.2$). For the standard 20 mm steel bar, one observes that frequencies are in the good range.

4 Direct impact test

Recall here that, with non strictly 1_D bars.

The average strain is not proportional to the strain measured at the surface of the bar. The average stress is not proportional to the average strain

The dispersion of the waves induces significant changes in rising times of the signals.

The displacement at bar end is over estimated as the punching of the bar by the specimen is neglected (this point will be addressed later).

In the following, the basic understanding of SHB is recalled based on the 1-D propagation theory.

A convenient way to have a good understanding of waves propagation and reflections in a system of bars is to use the “Lagrange” representation together with formulas (2). In a space-time (abscissa-ordinate) diagram, the fact that the celerity of waves is constant induces that a wave event follows a straight line with an absolute slope c (figure 12).

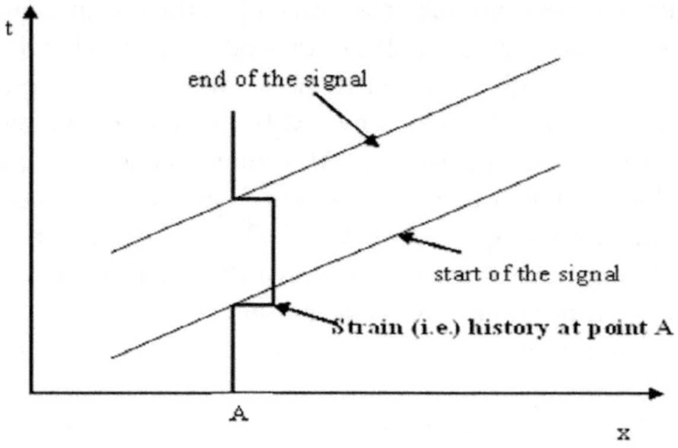

Fig. 12. A typical Lagrange diagram. The two sloping lines show the path of the beginning and the end of a square signal in the space-time diagram. At a given point A, the time story show what could have been recorded by a gauge at this point.

Direct impact test: 1 force and 1 displacement measurements.
A force is applied at the end of a bar of length l starting at time $t = 0$. It produces a strain-time signal which is measured by a gauge at a (short) distance a of the bar end. The corresponding signal $\varepsilon_i(t)$ starts at time $t_s = a/c$. The beginning of the signal reaches the other end (say output) of the bar at a time $t_e = l/c$.

A given boundary condition is applied at the output end of the bar. In order to satisfy this (unknown) boundary condition, a new wave $\varepsilon_r(t)$ in the opposite direction is produced which starts at time t_e.

This reflected wave reaches point A at time $t_r = l/c + (l - a)/c$.

From this time t_r, the strain measured by the gauge is the sum of the initial wave and of the unknown reflected one.

Consequently, the measurement duration based on the record of the initial wave is $\Delta t_m = t_r - t_s = 2(l - a)/c$.

Say the strain measured at the gauge is $\varepsilon_m(t)$.

The measurement duration is limited by the overlapping of waves.

During this time Δt_m ,

The strain at bar (input) end is $\varepsilon(t) = \varepsilon_m(t - a/c)$

The force at input bar is $F = E_b S_b \varepsilon$

The speed at input bar is $v = -c\varepsilon$

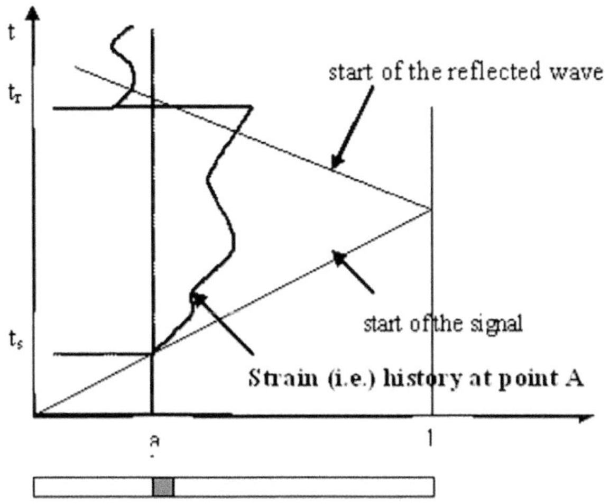

Fig. 13. A Lagrange diagram for an impact test.

If

- one side of the specimen is fixed at the input end,
- the speed of the other side of the specimen can be measured,
- the equilibrium of the specimen is assumed,

Then,

The force–(relative) displacement relation of the specimen can be computed.

5 SH(P)B. 2 forces and 2 displacements measurements

The specimen is sandwiched between two bars. One force and one speed measurement are made on each bar-end in contact with the specimen. The output bars works as for the direct impact. It is not the case for the input bar where, due to the loading, there exists a wave prior to the loading of the specimen itself.

Torsion bars and direct tension bars will not be explicitly addressed here. Their analysis is based on the same knowledge of basic wave propagation theory, but they need to solve some specific technical problems concerning the loading of the system and the specimen clamping.

In this chapter, some problems related to the measurement side (that of forces and speeds) will be addressed. Anyway, due to the very common use of Kolsky's formulas, this historical case is briefly recalled first, where problems dealing with the specimen behaviour have to be also introduced.

5.1 The special case of Kolsky bars

This very special case is interesting. For historical reasons, the gauges are equally distant from the specimen and both bars are identical. Considering the values of the strain at specimen ends, forces and displacements at both specimen ends are given by formulas (7,8)

$$v_i = -c_b(\varepsilon_i - \varepsilon_r) \quad v_o = -c_b(\varepsilon_t) \tag{8}$$

$$F_i = A_b E_b(\varepsilon_i + \varepsilon_r) \quad F_o = A_b E_b \varepsilon_t \tag{9}$$

Where v_i, v_o, F_i, F_o are input and output speeds and input and output forces at specimen faces, respectively. *Ab, Eb, cb* are area, Young's modulus and celerity of waves in (identical) bars, respectively.

ε_i, ε_r, ε_t are incident, reflected and transmitted waves, respectively, computed at specimen faces.

The measurement finishes here. In particular, the "Kolsky" formulas that will be discussed later are resulting of more assumptions, mainly: identical bars and specimen equilibrium. This second assumption is always an approximation.

The situation is indeed here exactly the same as any measurement where the machine basically provides forces and displacements. Consequently it is possible, without any lost in the quality of measurements, to use different bars.

In Kolsky's standard processing, the special assumption of equilibrium is used in the case of identical (same material, same diameter) bars. This yields - along with the superposition principle and formulas (7,8)- the well known relation:

$$\varepsilon_i(t) + \varepsilon_r(t) = \varepsilon_t(t) \tag{10}$$

Consequently, assuming the homogeneity of stresses and strains within the specimen, the incident wave does not explicitly appear in the simplified formulas:

$$\varepsilon_s(t) = \int_0^t \dot{\varepsilon}_s(\tau)d\tau = -\frac{2c_b}{l_s}\int_0 \varepsilon_r(\tau)d\tau \quad \sigma_s(t) = \frac{A_b}{A_s}E_b\varepsilon_t(t) \tag{11}$$

where A_s is the area of the specimen.

These simplified formulas are **both** based on equilibrium. When equilibrium is only badly verified, some researchers suggest the use an average formula for the stress (sometimes called the "three waves formula"). First, one must keep in mind that the average force could give a worth response than one of other input force or output force. Second, the simplified formula for the strain could also produce an imperfect result.

5.2 Overlapping of waves

The best way to address easily this problem is to make use of Lagrange diagram. It already has been seen for the direct impact test which corresponds, with the Kolsky apparatus, to the output bar. When equilibrium is not assumed and has to be checked, a measure of both input and output forces is needed. In this case, the best position for the input gauge appears to be in the middle of the input bar. Consequently the length of the output bar is such that the length between the gauge and its end is half of the input bar. If the bars do not have the same impedance, one must take care of propagation time and not of length.

Theoretically, the length of the striker could be as long as half of the input bar. In real situations one must account that the duration of the input wave is greater than that of the theoretical "square" one. The rising time of the signal has indeed to be added. For a given striker, this time is related to the imperfect geometrical matching of both striker and input bar ends. This mismatch can be due to imperfect surfacing and/or imperfect alignment but it can be considered equivalent to an imperfection of a constant thickness. A realistic interpretation of the rising time is that of the time needed for the bars to flatten the geometric mismatch.

In the case of a standard steel SHPB 20mm, the rising time is around 20μs with a striker speed of 10 m/s. It should be around 100μs at a striker speed of 2 m/s.

If the minimum loading speed is 2 m/s, these consideration lead to the standard optimal SHPB configuration, with:

Input bar length = $2l$

Output bar length = $l + 5d$ (5 diameters form the end for the gauge position)

Striker length = l *-0.25 m* (theoretical loading duration decreased by 100 s)

Note that for "pulse shaper" users, the rising time will increase and depend on the "pulse shaper" geometry and material.

In the case of viscoelastic bars and a viscoelastic striker, the analysis is a little different. With a striker and an input bar having the same impedance, the loading duration (in the 1-D analysis) is infinite, due to a tail. It is not easy to decide when the amplitude of the tail is small enough. It is then recommended, with viscoelastic strikers and bars, to use a striker with a smaller (by around 10%) diameter than the bars. Bussac et al. (2008) have indeed demonstrated that the incident pulse has, in this case, a finite duration.

5.3 Wave shifting. A precise method for SHPB.

The 1-D analysis of the waves implicitly takes account of the Saint-Venant principle: a certain distance is needed between the end of the bar and the strain gauge to insure the homogeneity of the strain across the bar (typically 5 diameters).

One needs then to use a wave theory to deduce the strain (as it would be is this point was not an end) at the end a bar. This operation is called shifting.

It is clear that the input force is proportional to the sum of the incident and reflected waves and that a relative imperfect shifting in time would induce an error, especially at the beginning of the loading. Some authors have proposed to use some geometrical trick at the beginning of the signals to define a "foot" for the curves and determine the shifting process on the knowledge of these points: we call that the "foot shifting" technique. Acceptable results could be obtained for the input force, with this method.

If we extend the method to the output wave, it induces a wrong evaluation, at least, of the strain, if it neglects the times needed by the loading to go across the specimen.

The shifting process must then be precise. A method, introduced by Zhao & Gary (1996),can be used. It is based on the transient simulation of an initial elastic behaviour of the specimen. The method also allows for a correction of specimen geometry and a measurement of the Young's modulus of the specimen material.

The incident wave at the input specimen face been known (after the shifting process), reflected and transmitted waves can be computed – depending on specimen dimensions, bar dimensions and mechanical properties, specimen Young's modulus. The only unknown is the last one. Using a try and error method, one rapidly finds the Young's modulus that gives shapes of both simulated waves similar to that of transmitted and reflected waves as they are known (after the shifting process) at the input and output specimen faces.

Note that this operation does not work if the dispersion is not taken into account, even with elastic bars. The reason is that the elastic response of the specimen concerns the first instants of the loading where the initial slope is strongly affected by the dispersion.

Some distance in time is still often observed between real and simulated waves. This distance can be due to an imperfect knowledge of the speed of waves or, more often, to some geometrical imperfection of the specimen. For an optimized stress-strain curve, real waves are shifted to simulated ones.

This very short shift (equivalent of a few cm in distance) is not affected by the dispersion. An illustration of the method is presented in figures 14 and 15.

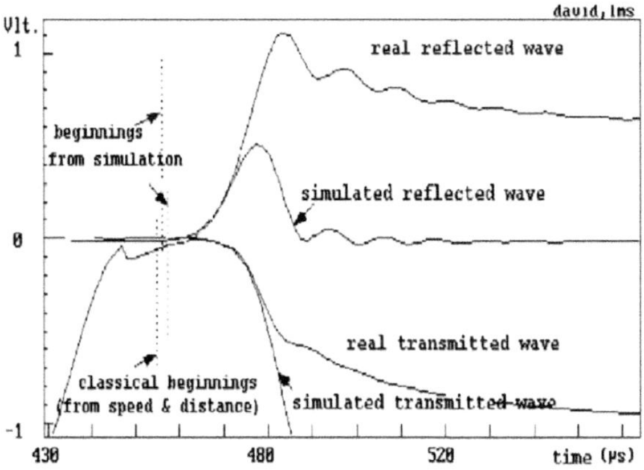

Fig. 14. Optimal choice of Young's modulus for transmitted and reflected waves.

The precise way stress and strain are computed will be addressed later. Remind that the best results are derived from the force and displacement time histories at the specimen boundaries without artificial time shifts.

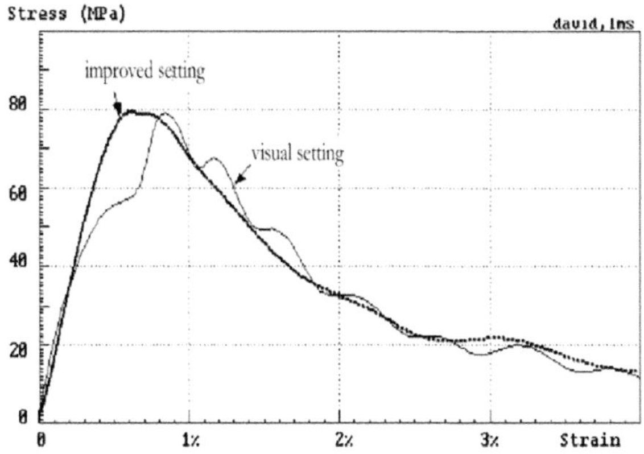

Fig. 15. an example of possible effect of the method. Case of a concrete specimen.

Recall that the method does not work correctly without the dispersion correction and that it relies on an initial non negligible elastic behaviour of the specimen.

It still perfectly works with (slightly) viscoelastic bars (as Nylon or PMMA bars) but not with a purely viscous specimen.

5.4 Punching of the bar-end by the specimen

After using this method during years, we have observed that the expected modulus of the specimen was found when the specimen diameter was close to that of the bars. The measured value was lower than the expected one in most situations when the specimen diameter was smaller than that of the bars. Furthermore, for a given specimen diameter, the discrepancy was decreasing with specimen length.

This problem, due to the local punching of the bar end by the specimen, has been addressed by Safa & Gary (2010). They have shown that this effect is like that of an added spring at the end of the bars: the displacement due to punching is proportional to the applied (and measured) force. A simple formula has been established that gives the value of the displacement due to punching. Results are summarised in figure 16 next page.

Note that the 1-D elastic behaviour of punching allows for taking account of its contribution in the elastic simulation process. Consequently, the Young's modulus of any material specimen can be directly estimated by means of the elastic simulation process.
It is clear that the relative influence of the displacement due to punching increases when the length of the specimen decreases. This is especially important at high or very high strain-rates that need shorter specimen (as the incident striker-speed is limited by the care of the gauges) when the specimen diameter is under the half of that of the bar.

For a real test made at 10000/s, with a specimen 3 mm thick, diameter 6.5mm (steel bars, 12 mm, striker speed 25m/s), the equivalent in strain due to punching, at the plastic plateau, is around 14%. On the other side, at low strain rates, when the total measured strain is small, the effect of punching is still important as illustrated in figure 17.

Summary of main results for SHPB

We consider a classical SHPB apparatus where input and output bars are identical with diameter D and with ρ, ν and E the density, Poisson ratio and Young's modulus, respectively, of their material. The geometrical characteristics of the specimen are d, l_s and S_s, corresponding to its diameter, length and cross sectional area, respectively.

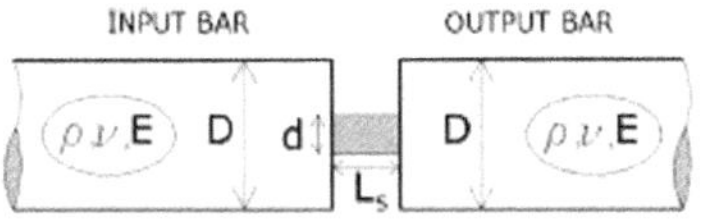

At any time throughout the experiment, the strain of the sample is obtained as

$$\epsilon(t) = \epsilon_{SHPB}(t) - \epsilon_{punch}(t)$$

where

- $\epsilon_{punch}(t) = 2K_p \dfrac{\sigma_{SHPB}(t) S_s}{l_s}$
- $\epsilon_{SHPB}(t)$ and $\sigma_{SHPB}(t)$ are the strain and stress, respectively, obtained by standard SHPB formulas.
- $K_p = \dfrac{16}{3\pi^2} \dfrac{1-\nu^2}{dE} H_p(\dfrac{d}{D})$

$$H_p(x) = 2 - (x + \frac{1}{x})E(x) - (x - \frac{1}{x})K(x)$$

$$E(x) = \int_0^{\frac{\pi}{2}} \sqrt{1 - x^2 \sin^2\theta}\, d\theta, \quad K(x) = \int_0^{\frac{\pi}{2}} \frac{d\theta}{\sqrt{1 - x^2 \sin^2\theta}}$$

Tabulated results for the function $H_p(x)$ are given in the table underneath.
Values for $0.5 < x \leq 1$ are obtained by linear interpolation.

x	0.10	0.15	0.20	0.25	0.30	0.35	0.40	0.45	0.50	0.60	0.70	0.80	0.90	1.00
$H_p(x)$	1.765	1.648	1.531	1.416	1.301	1.188	1.076	0.967	0.860	0.688	0.516	0.344	0.172	0

Fig. 16. Summary of punching data for SHPB.

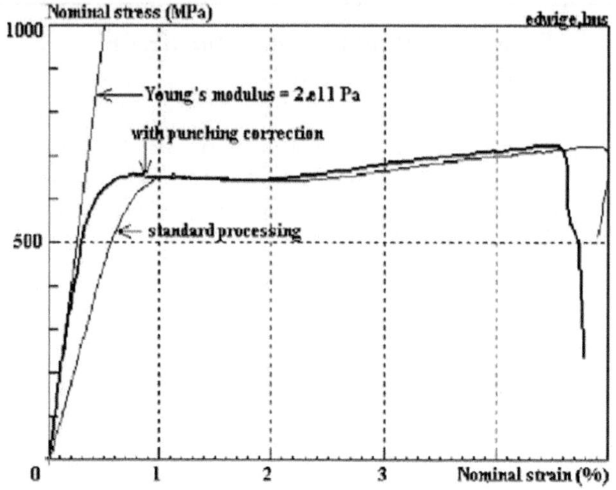

Fig. 17. Influence of the punching correction. (20 mm steel bar, striker speed 2.79 m/s, strain-rate around 100/s).

5.5 Force matching. Critical case of the input side.

This problem has nothing to do with a so called "impedance matching" problem (discussed later).

In most cases, especially when testing metallic specimens, input and output forces are almost equal. This is not always the case, as shown later. Anyway, in the following analysis, we use this assumption to make clearer the "force matching" problem.

Considering formula (9), it appears that a correct measurement of the output force is only possible when the measured output strain is itself significant. It can be considered that strain measurements are done with a precision around 5.10^{-6} when using classical strain-gauges. With semiconductor gauges, it can be improved by a factor 50. Considering that an acceptable relative precision of 2% is suitable for any measurement, the maximum strain measured should be more than $2{,}5.10^{-4}$. Using formula (9) allows for the calculation of the minimum suitable output force.

In the example of a steel bar (20 mm diameter), this force is around 15 kN. The specimen area being at maximum equal to the bar diameter, the corresponding stress is 50 MPa. For weaker materials, it is necessary to use semiconductor gauges or a better amplifying system (or a bar with a low Young's modulus, as it is the case of most viscoelastic bars).

Let us then assume that a material, with a maximum stress of 50 MPa, is tested with the steel bar, very carefully. Considering now formula (9) for the input force, it appears that a precise measurement is not possible when the amplitude of the incident wave and the opposite of the amplitude of the reflected wave are too close.

Keeping this example, if we wish to obtain high strain rates, we use a small specimen length. For sake of simplicity, we assume that the tested material is not strongly rate sensitive. A simple simulation, with a specimen 5 mm long, shows that the striker speed must be 15 m/s to obtain an average strain rate close to 3000/s. In this case, the absolute amplitude of the reflected wave only differs by 15% from the amplitude of the incident wave. A precise calculation of the input force with formula (9) is then difficult.

If we consider that an acceptable result is obtained when both waves differ by more than 20%, the maximum strain rate giving an acceptable result is, for this example, around 400/s. It can be concluded that materials with a plateau stress under 50 MPa cannot be safely tested with 20 mm steel bars at strain-rates over 400/s.

More generally speaking, weak materials need the use of soft bars.

This problem should be called the "force-matching" problem as it has nothing to do with the specimen impedance, but only with the product of the Young's modulus of the bar by its area to be compared with the maximum force to be measured.

Optimizing specimen dimensions. In order to insure good measurements on the input side (more often of the force) the specimen has to be designed depending on the required strain-rate.The problem is also to choose the appropriate striker speed to obtain the strain rate. One can use a method presented by Gary (2001), which is based on simple assumptions (one-dimensional wave analysis, specimen equilibrium, evaluation of the material stress: say σ_y). The amplitude of the incident wave is deduced from the striker speed V (eq. 11a). The amplitude of the transmitted wave is deduced from the specimen section S_s and σ_y (eq. 11b), and the amplitude of the reflected wave is deduced from the equilibrium (eq. 11c).

$$\varepsilon_i = \frac{V}{2c} \text{ (11a)}, \quad \varepsilon_t = \frac{S_s \sigma_y}{S_b E_b} \text{ (11b)},$$

$$\varepsilon_r = -\varepsilon_i + \varepsilon_t \text{ (11c)}, \quad \dot{\varepsilon}_s = \frac{2c\varepsilon_r}{l_s} \text{ (11d)}$$

where Sb and Eb are the section and the Young's modulus of the bars, respectively.

The average strain rate is now given by formula (11d), where c is the wave speed in the bars and ls is the specimen length. It is then possible to make sure that the relative amplitude of the reflected wave is small enough in comparison with the relative amplitude of the incident wave. If we look at formulas (8) and (9), it can be seen that the best situation, i.e., that giving the most accurate force and speed results, is that occurring when the amplitudes of the transmitted and reflected waves are fairly similar. These formulas make it possible to easily determine optimized specimen dimensions and striker speeds.

5.6 Impedance matching

The impedance matching problem, has been addressed by Davies and Hunter (1963). It mainly deals with the homogeneity of stresses and strains in the specimen, during a Hopkinson bar test, and especially at early instants.

During this loading phase, input and output forces are alternatively bigger than the other. An example is shown in fig 18, from Klepaczko (2007) - figure 3.11a, p 206.The gap between both forces decreases when the speed of waves in the specimen increases and when its length decreases. It also decreases when the amplitude in force (in stress in the present figure) of the first step decreases.

Here is the "impedance" matching problem that describes the conditions to make this step small enough. Furthermore, this step is softened if the rising time of the loading is increased. This point is one reason for using pulse shapers.

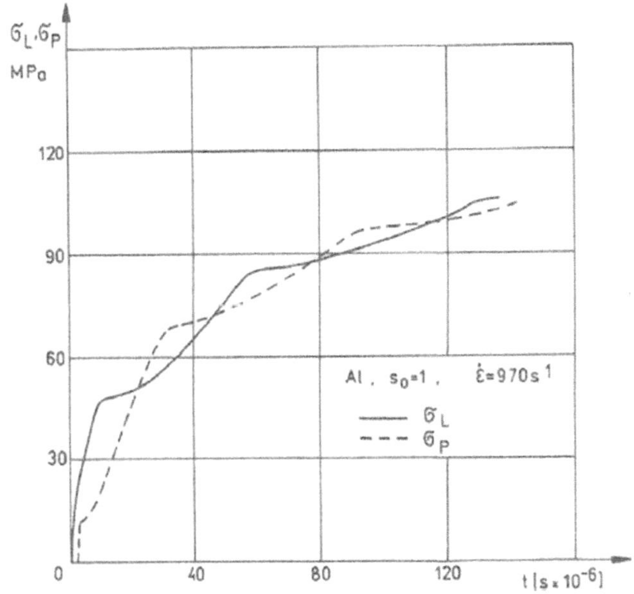

Fig. 18. Elastic input and output stresses at early instant of the loading - from Klepaczko (2007).

5.7 Holding the specimen: dealing with one-point measurements

Measurements with bars are always analyzed within the 1-D hypothesis (even with dispersion and punching corrections as both are only expressed as a function of the axial coordinate of the bar). In other words, force and speed measurements, on each side, are made at a **single point**. This **basic hypothesis** does not appear when Kolsky's formulas are used for a direct stress-strain relation. It is then important to keep in mind this aspect, especially in situations where holding the specimen leads to the use of devices that induce an impedance change along the wave propagation path.

Some examples are given here.

When compression bars are used to test a material specimen, the specimen diameter must be smaller than the diameter of the bar to ensure a quasi-uniaxial state of stress. When very weak material are tested, even with viscoelastic bars, the quality of the measurement is limited by force matching problems described above.

In order to overcome this limitation, one could use at bar ends a device with a larger diameter and then be able to hold a specimen with a much larger specimen. Such a technique is illustrated in figures 19 and 20.

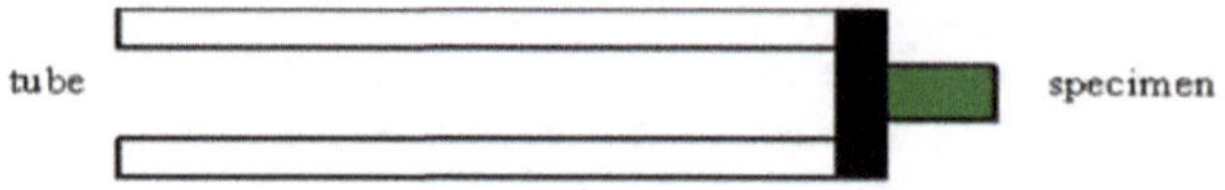

Fig. 19. Impedance change between the bar (tube) and the specimen.

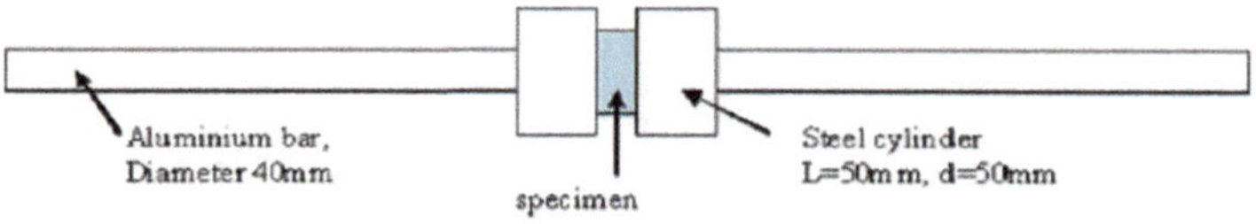

Fig. 20. Special device for a larger specimen.

The first question here is to decide at which point is made the measurement: end of the bar or specimen face?

In order to illustrate this problem, a brass specimen has been tested in the situation shown in fig. 20 and in a classical situation with the same (aluminium) bars. The corresponding recorded waves are shown in fig. 21

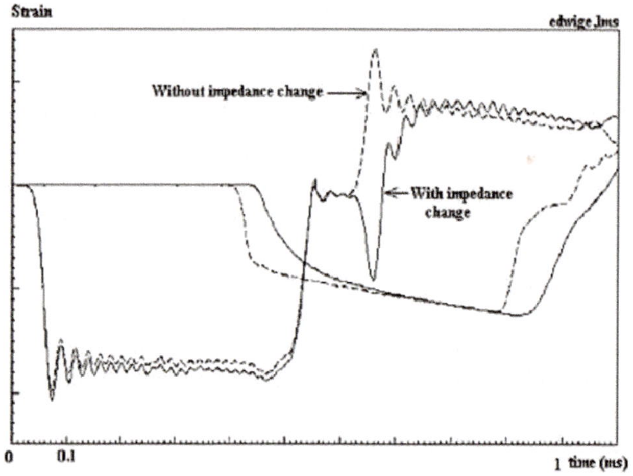

Fig. 21. Illustration of the influence of a matching device – waves. It is observed that the strongest influence appears in the reflected wave, and that the transmitted wave is smoothed (this is not a general results as it depends on each impedance ratio device/bar).

A standard processing of the waves will then give wrong results when matching devices are used, as shown in figures 22 and 23.

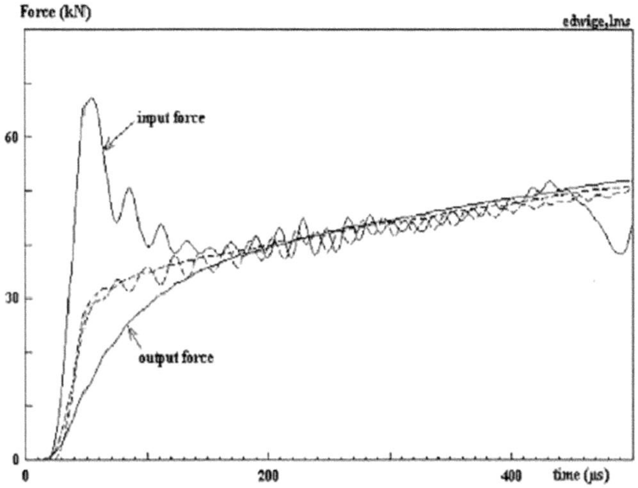

Fig. 22. Apparent equilibrium with and without devices.

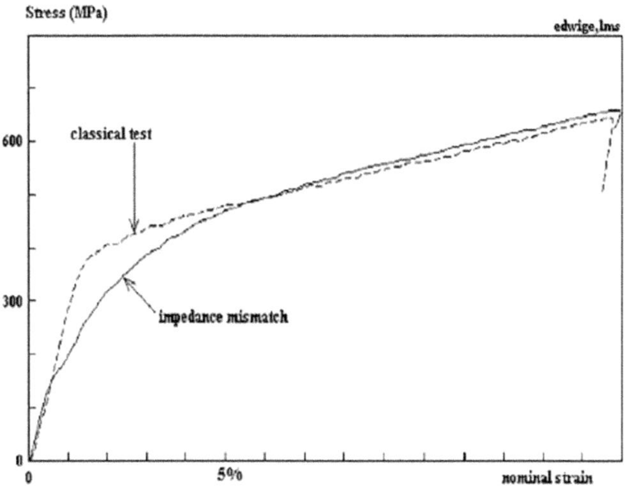

Fig. 23. Apparent stress-strain relation with and without devices.

Looking at the stress-strain diagram shows a difference that is not so large, in particular for larger strains. The main effect is a smoothing of the stress-strain relation that prevents for an acceptable evaluation of the elastic limit of the tested material.

When it is not possible to avoid the use of a matching device, an impedance correction is possible. A simple way is to use a one dimensional analysis. Knowing the waves at the end of the bar (in the part with a constant impedance), the waves at specimen faces can be calculated, assuming the 1-D elastic behaviour of the devices of known impedance. The forces and the displacements applied to the specimen faces are then deduced.

5.8 Undetermined specimen ends

A good example is found in a work of Gary and Nowacki (1994). They proposed a device to convert compression in shear. The idea is summarised in fig. 24.

The (say) input hollow cylinder has the same impedance than the input bar when the output cylinder has the same impedance as the output bar. The problem here is that the beginning and the end of the specimen are not defined. Consequently one has to consider that the shear-displacement curve is filtered with a cut-off frequency depending on the length of the shear zone.

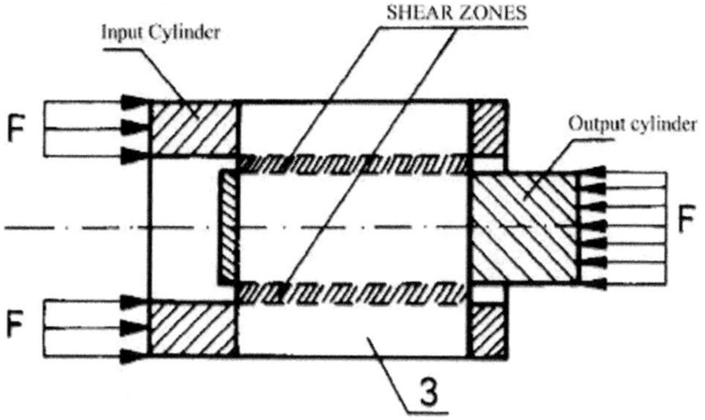

Fig. 24. Conversion compression to shear.

6 Specimen behaviour

The Split Hopkinson pressure bar arrangement can give very accurate measurements of forces and velocities at both sample faces if the data processing is carefully performed. Relating material properties to measurements of forces and velocities at the two specimen faces is a completely independent problem.

For this reason, we have distinguished the problems of measuring accuracy of SHPB and material property identification from the measured data.

6.1 Axial homogeneity

Due to the presence of waves in dynamic experiments, both the stress and strain fields within a specimen are seldom uniform. A dynamic material test should be designed such as to minimize this inherent non-uniformity, a condition which is typically associated with "quasi-static equilibrium". However, when testing purely elastic materials such as fiber reinforced composites or low impedance materials, the validity of this assumption needs to be checked with care. Note that there are situations where force equilibrium can exist without strain homogeneity as it is often the case with structural materials like honey-comb or foam.

Before computers became generally available, the assumption of quasi-static equilibrium of the specimen had a special importance from a data processing point of view.

The historical point is that, in the 50's, signals could not be recorded. The idea was to associate, in a "memory scope" (the image was stored in an analogic way) the "strain" and the "stress" signals and directly observe a curve proportional to the stress-strain curve of the material. First, the strain-rate was analogically integrated with respect to time. Then, the simplest way to synchronize both signals was to produce them at the same time. This was the reason to use both bars of same length with each gauge in the middle.

With the use of modern numerical acquisition, the output gage can now be positioned closer to the specimen interface; consequently, the output bar can be shorter than the input one (and **there is no good reason** to choose its length equal to that of the input bar). The use of "equilibrium" condition was also crucial at that time as the input force could not be directly measured.

Optimal design with equilibrium. It is worth noting here that this equilibrium condition could be used with different input and output bars leading to slightly more complex formulas. In the same way, other formulas could be derived using any pair of waves (among three).

Let's consider the idea of using only the incident and transmitted waves. For sake of simplicity we suppose that the two bars are still identical but the results could be generalized to a SHPB made of two different bars.

Assuming equilibrium, formula 3 can also be written:

$$\varepsilon_s(t) = \int_0^t \dot{\varepsilon}_s(\tau)d\tau = -\frac{2c_b}{l_s}\int_0 (\varepsilon_i(\tau) - \varepsilon_t(\tau))d\tau \quad \sigma_s(t) = \frac{A_b}{A_s}E_b\varepsilon_t(t)$$

where A_b is the area of the bar, E_b its Young's modulus and A_s the area of the specimen.

It can be seen that only the incident and the transmitted waves are now involved in the processing formulas. Finding an optimal position for the gauges, close to the input side of the bars where the waves arrive first, (see the case of direct impact) will then lead to an increased measured strain. It is then easy to find the optimal design close to the one shown in figure 25. The bars have now to have the same length to allow for the same duration of the incident and the transmitted waves. Consequently, the striker has to be as long (or longer) as the remaining distance between the gauges and the end of the bars.

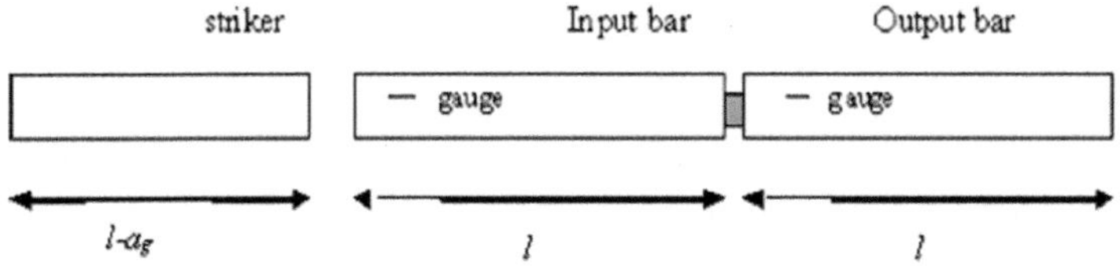

Fig. 25. With equilibrium, a configuration for an increased duration measurement.

Evaluation of specimen behaviour. The evidence of non equilibrium is found in early instants of the loading, when the first wave in the specimen has not yet reached its end. At this instant the input force exists when the output force is still null. Knowing that specimen equilibrium is never achieved exactly, we seek the best of the commonly used stress-strain curve estimates in a SHPB experiment. This question is studied by Mohr et al. (2010). The time shift of the waves is found to play a critical role as far as the accuracy is concerned.

More specifically, it is found that the omission of artificial time shifts (as illustrated in fig 26) provides the best stress-strain curve estimates. In other words, once the force and displacement histories are known at the specimen boundaries, accurate estimates of the stress-strain curves should be made without further shifting the signals on the time axis.

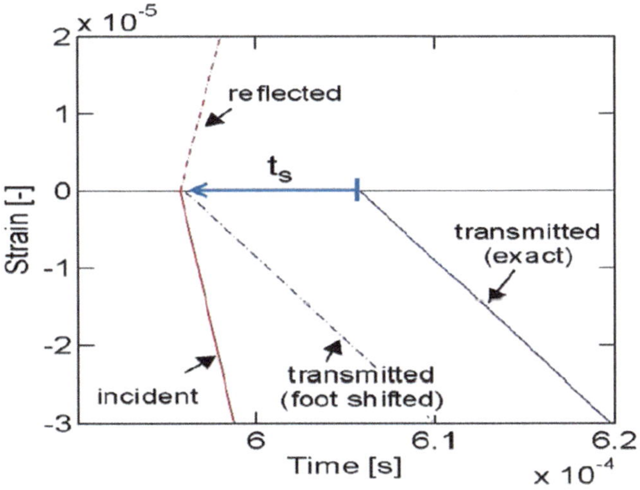

Fig. 26. Exact theoretical waves as shifted at specimen faces. The dashed blue line shows the strain history of the transmitted wave after shifting the beginning of this wave in time (so-called "foot shifting") such that all strain histories begin simultaneously.

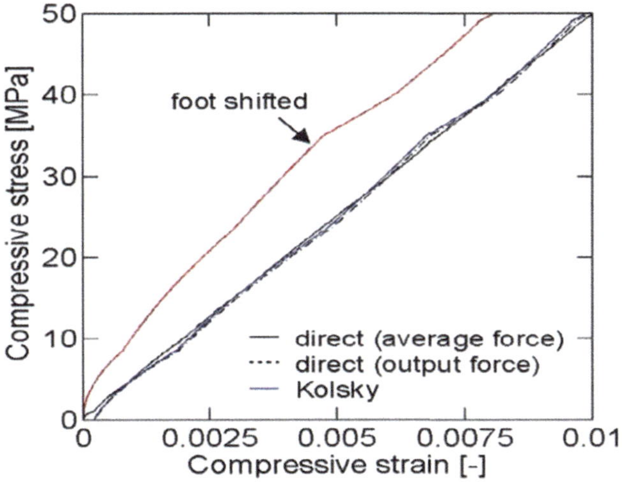

Fig. 27. Plot of the estimated stress strain curves for a dynamic compression experiment on PMMA (steel bars 20 mm).

Estimations evaluated by Mohr et al. (2010) are based on exact explicit calculation done in the frequency domain. Results confirm that the error (for all formulas) increase with the impedance mismatch between the specimen and the bars. It is emphasized that, in standard testing situations, exact shifting (illustrated in fig 26) is much more important than strain or force approximation (with 2 or 3 waves).

The estimates which are based on the force and displacement time histories at the specimen boundaries without artificial time shifts, provide the most accurate stress-strain curve. Unless accurate input force measurements are available, the combination of the average strain with the output force based stress estimate is recommended for standard SHPB experiments. An illustration is given in fig. 27.

Recall that in the elastic range and in standard SHPB testing, when the specimen diameter is smaller than half of that of the bars, the incidence of the **punching correction** on the quality of the results **is the most important**.

6.2 Radial homogeneity. Friction and inertia

We must keep in mind that, alike in all testing situations, axial stress and strain homogeneity (considered up to now) is not the only one to be considered. In quasi-static testing, it is possible to directly measure the strain in a "gauge section" of the specimen where 1-D assumptions are well verified. In SHPB, when using a cylindrical specimen, it is more difficult to avoid friction at specimen faces. This friction prevents the expected radial expansion quantified by Poisson's ratio.

The consequence is that the triaxial stress-state is not uniform along the axial loading. It induces an average over-stress in the common situation of elastoplastic materials.

This radial expansion is also prevented by inertia effects that induce lateral confinement.

An empirical formula has been established by Malinowski and Klepaczko, (1986) that takes account of both effects. This formula is valid for purely plastic materials but gives a first order correction that applies to many other materials.

$$\sigma - \sigma_0 = \frac{\mu\sigma}{3s} + \frac{\rho d^2}{12}(s^2 - \frac{3}{16})(\dot{\varepsilon}^2 + \ddot{\varepsilon}) + \frac{3\rho d^2}{64}\ddot{\varepsilon}$$

where, σ_0 : ideal value, σ : actual value (true stresses)

$l$and $d$actual values of length and specimen diameter and $s = l/d$

ρ , density assumed to be constant

μ friction Coulomb coefficient (usually <0.1).

This formula indicates that the inertia effect is much smaller than that of friction for common metal testing (bar diameter < 20 mm). It becomes significant for bigger specimens (>30mm) like found for rock or concrete testing when big specimen sizes are required for a good average material representation. It is seen in this case that some artificial strain rate effects can be attributed to bi-axial loading induced by inertial confinement.

In the case of testing where a quasi-static radial confinement is added - see Gary and Bailly (1998), and Forquin et al. (2010) - inertia effects are prevented and do not operate anymore.

6.3 Adiabatic heating

The duration of a dynamic compression test with SHPB is almost always under 1 ms. For many materials, it induces large strains (depending on strain-rate) that produce work which is (partly ?) converted into heat. In case of metal or polymer testing, it can induce a specimen temperature increase in the range up to 100°C.

It is then important to account for this temperature increase for an improved evaluation of the material behaviour. This question has been investigated by various authors who have proposed techniques for high speed temperature measurements (more often based on infrared specimen emission). See for instance Rittel (1999) and Negreanu et al. (2009).

6.4 Inverse methods

On the basis of force and velocities measurements at specimen faces, an approach of the specimen behaviour based on an inverse method is theoretically possible, as shown by Rota (1994), as these four values are superabundant measurements.

3-D dynamic numerical simulation could allow, in this case, accounting for friction, radial inertia and possibly temperature increase. If an appropriate form of the material behaviour with some parameters to be determined is known, using a part of data (two velocities, for example) as the input data, another part of data (the two forces) associated with the given parameters can be calculated. The best set of parameter which gives the calculated forces well in agreement with the measured ones can theoretically be found.

In order to further investigate this approach, we shall limit the analysis to axial 1-D effects by means of a fast 1-D transient simulation based on the Sokolovsky (1948) and Malvern (1951) approach.

The uniaxial governing equation and the constitutive law are written in equations (12)

$$
\begin{aligned}
\frac{\partial\sigma(x,t)}{\partial x} &= \rho\frac{\partial \mathrm{v}(x,t)}{\partial t} \\
\frac{\partial\varepsilon(x,t)}{\partial t} &= \frac{\partial \mathrm{v}(x,t)}{\partial x} \qquad (12) \\
f(\sigma,\varepsilon,\dot{\sigma},\dot{\varepsilon},...) &= 0
\end{aligned}
$$

where $\sigma, \varepsilon, \mathrm{v}$ are the stress, the strain, and the particle velocity in the specimen. ρ is the mass density.

The boundary conditions at the two faces of the specimen are given as follows:

$$
\sigma(x,t) - \frac{E_B S_B}{C_B S_S}\mathrm{v}(x,t) = 2\frac{S_B}{S_S}E_B\varepsilon_i(t) \quad \text{at the input side}
$$

$$
\sigma(x,t) + \frac{E_B S_B}{C_B S_S}\mathrm{v}(x,t) = 0 \quad \text{at the output side} \qquad (13)
$$

E_B, C_B, S_B, S_s denote the Young's modulus, wave velocity, cross-sectional area of the bar and the section of the specimen.

Once the specimen behaviour is assumed, the direct problem described by equations (12, 13) corresponds to an uni-dimensional SHPB simulation. As mentioned above, the simulation would be just the preliminary stage of the inverse problem which consists in finding the parameters of the model. The inverse methods need fast calculation procedures to allow for a great number of direct calculations with different sets of parameters if necessary. For this purpose, the specimen behaviour should be numerically easy to calculate. This is why it is chosen to use a Sokolovsky-Malvern model type for the specimen, (14).

$$
\begin{aligned}
\frac{\partial\varepsilon}{\partial t} &= \frac{1}{E}\frac{\partial\sigma}{\partial t} \quad if \quad \sigma \le \sigma_s \\
\frac{\partial\varepsilon}{\partial t} &= \frac{1}{E}\frac{\partial\sigma}{\partial t} + g(\sigma,\varepsilon) \quad if \quad \sigma > \sigma_s
\end{aligned} \qquad (14)
$$

The characteristic network in this case is composed of families of straight lines so that the numerical integration of equation (13, 14) by the method of characteristics is very efficient. In comparing the given behaviour with the average stress strain curve obtained from three simulated waves, the efficiency of classical SHPB analysis for this type of materials can be evaluated.

It is noted that the Sokolovsky-Malvern constitutive model is a quite general rate-sensitive one. Though it has been introduced initially for metals, it can be also used to describe the non-metallic materials if the function $g(\sigma,\varepsilon)$ is correctly chosen –see Critescu (1967).

In order to have a general idea of the efficiency of the classical SHPB analysis for this type of constitutive models, a popular rheological model (Fig. 28) is examined here, which is of Sokolovsky-Malvern type with the $g(\sigma, \varepsilon)$ expressed as follows,

$$g(\sigma, \varepsilon) = \frac{(1 + \frac{E_v}{E})\sigma - E_v\varepsilon}{\eta} \tag{15}$$

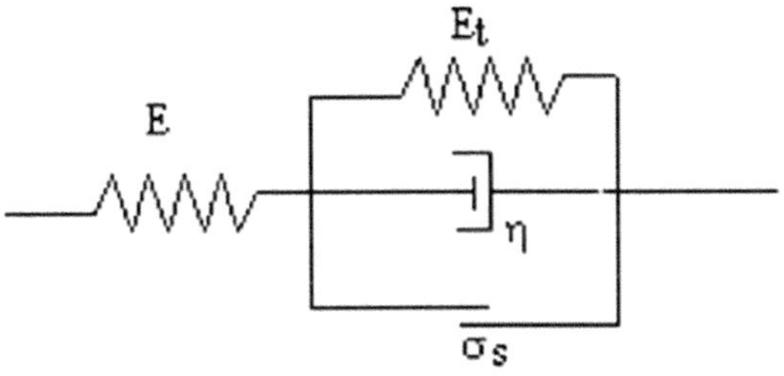

Fig. 28. Rheological model for 1-D standard elasto-visco-plasticity.

A number of simulations were done for different sets of parameters. It has been found, like for previous results on metals, that the classical average stress-strain curve is in good agreement with the curve given by SHPB standard analysis at relative important strains. However, in the range of small strains, it is acceptable when the viscosity is low even if there is no axial uniformity. On the other hand, when the viscosity is high, the average (SHPB) curve is quite far from the given behaviour.

An illustration of the method is presented in two particular cases: for salt in fig. (29) which exhibits a strong viscoplastic behaviour and for concrete fig. (30) which breaks, showing a negative strain hardening, at small strains. In both cases, equilibrium conditions are obviously not satisfied.

The chosen model with the set of parameters which gives the best agreement can be considered as a representative model of the specimen in this test. As a result, the stress and strain fields in the specimen are evaluated so that a stress-strain curve is found. The assumption on the uniformity of stress and strain fields is then not needed in such a method. The only hypothesis used is that specimen tested manifests the same properties everywhere. Furthermore, it is possible to give a stress strain curve at a constant strain rate via the model and the set of parameters identified.

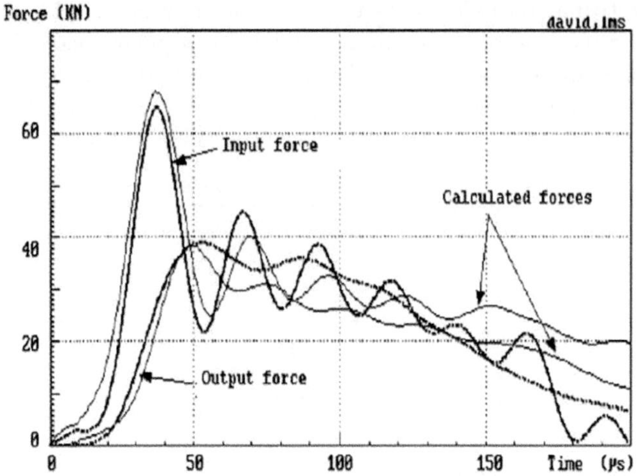

Fig. 29. Simulated and measured forces for rock-salt.

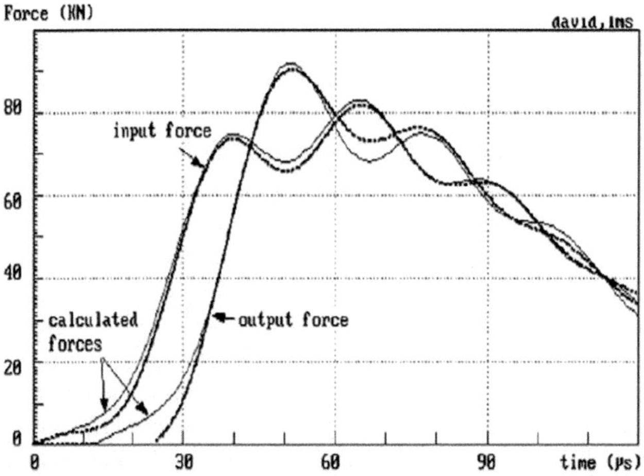

Fig. 30 Simulated and measured forces for concrete.

This stress-strain curve can then be compared with the one obtained with the classical SHPB analysis. For the test on salt, the two curves are quite far from each other in the range of small strains. The inverse calculation is then in this case the only way to obtain an accurate result for this type of materials. On the other hand, for the test of concrete, the average stress-strain curve is rather close to the curve derived from the model, as a kind of averaging operates between input and output forces. The inverse calculation could then be avoided.

7 Measurement of the dispersion relations

A method is briefly presented here that provides a very accurate estimation of the dispersion relations. One reason is that it is a one point method; the other is that it is based on simple formulas and an easy identification of the wave speed and damping at a given finite set of frequencies.

Furthermore, it allows for a subsequent mathematical analysis which includes noise reduction and provides a rheological model with a number of elements that is in accordance with the complexity of the material.

7.1 Impact test method. Theory

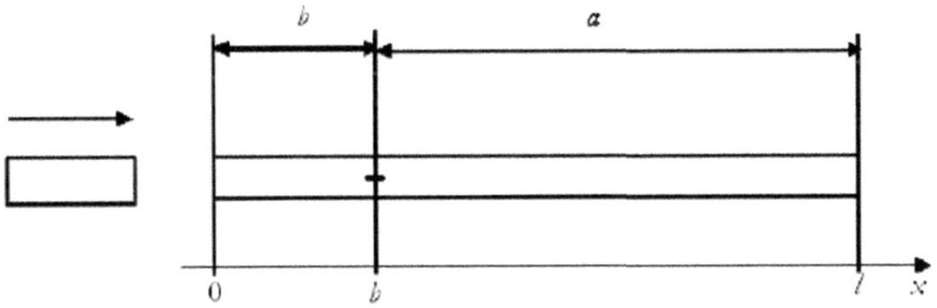

Fig. 31 Impact test with a uniform viscoelastic bar specimen.

At one end, x=0, the bar is impacted axially by a striker that separates from the bar after the generation of a compressive primary pulse in the bar. The other end of the bar, $x = l$, is free. The strain $\varepsilon_b(t) = \varepsilon(b, t)$ is recorded at a distance a from the free end and b from the impacted end ($a + b = l$). The primary pulse should be shorter than $2a$ so that there is no overlap in the measured strain of this pulse and the first pulse reflected from the free end. However, it should be much longer than the diameter of the bar so that approximate 1D conditions prevail - Hillström et al (2000).

In the frequency domain, the strain in the bar can be expressed as:

$$\hat{\varepsilon}(x, \omega) = A(\omega)\, e^{-\mathrm{i}\xi(\omega)\, x} + B(\omega)\, e^{\mathrm{i}\xi(\omega)\, x}, \tag{16}$$

where

$$\xi^2(\omega) = \rho\omega^2/E(\omega), \xi(\omega) = k(\omega) - \mathrm{i}\alpha(\omega) \tag{17}$$

Here, $\hat{\varepsilon}(x, \omega)$ is the Fourier transform of $\varepsilon(x, t)$, and $A(\omega)$ and $B(\omega)$ are complex amplitudes of waves travelling in the directions of increasing

and decreasing x, respectively, $k(\omega)$ is the wave number and $\alpha(\omega)$ is the damping coefficient.
In order to determine an expression for the recorded strain $\hat{\varepsilon}_b^1(\omega) = \hat{\varepsilon}(b,\omega)$ associated with the primary pulse alone, the amplitudes A and B are first determined from Eq. (16) and the boundary conditions $\hat{\varepsilon}(0,\omega) = \hat{\varepsilon}_0(\omega)$ and $B(\omega) = 0$ for a semi-infinite bar $x \geq 0$, where $\hat{\varepsilon}_0(\omega)$ is the strain at the impacted end. With these amplitudes and $x = b$ inserted, Eq. (16) gives

$$\hat{\varepsilon}_b^1(\omega) = e^{-i\xi b}\hat{\varepsilon}_0(\omega) \tag{18}$$

The recorded strain $\hat{\varepsilon}_b^1(\omega) = \hat{\varepsilon}(b,\omega)$ associated with the complete train of pulses is determined similarly. With amplitudes A and B, and $x = b$ and $l - b = a$ inserted, Eq. (16) gives

$$\hat{\varepsilon}_b^\infty(\omega) = \frac{\sin(\xi a)}{\sin(\xi l)}\hat{\varepsilon}_0(\omega). \tag{19}$$

Dividing the members of Eq. (19) by those of Eq. (18) eliminates the strain $\hat{\varepsilon}_0(\omega)$ at the impacted end which normally cannot be measured. The elimination of this strain makes the difference between the method used here and that used by Othman et al. (2001). Substituting ξ from the second of Eqs. (17) into the result, one gets the ratio $\hat{\varepsilon}_b^\infty(\omega)/\hat{\varepsilon}_b^1(\omega)$ which gives

$$\left|\hat{\varepsilon}_b^\infty(\omega)\right|^2 = \left|\hat{\varepsilon}_b^1(\omega)\right|^2 e^{2\alpha b}\frac{\sin^2(ka) + \sinh^2(\alpha a)}{\sin^2(kl) + \sinh^2(\alpha l)}. \tag{20}$$

Resonance occurs at the angular frequencies $\omega = \omega_m$, m=1, 2,... which correspond to the wave numbers, wavelengths and phase velocities

$$k_m = \frac{m\pi}{l}, \lambda_m = \frac{2\pi}{k_m} = \frac{2l}{m}, c_m = \frac{\omega_m}{k_m} = \frac{\omega_m l}{m\pi}, \tag{21}$$

respectively.
It is assumed that within the *m:th* resonance peak $\alpha = \alpha_m\omega/\omega_m$can be taken as directly proportional to angular frequency and $c(\omega) = \omega/k(\omega) = c_m$ can be taken as constant. By use of the third of Eqs. (21) we then obtain the relation between wave number and angular frequency within the resonance peak. Inserting these expressions for $\alpha(\omega)$ and $k(\omega)$ into Eq. (20) we get $(m = 1, 2, \dots)$

$$\left|\hat{\varepsilon}_b^\infty(\omega)\right|^2 = \left|\hat{\varepsilon}_b^1(\omega)\right|^2 e^{2\alpha_m b\omega/\omega_m}\frac{\sin^2(k_m a\omega/\omega_m) + \sinh^2(\alpha_m a\omega/\omega_m)}{\sin^2(k_m l\omega/\omega_m) + \sinh^2(\alpha_m l\omega/\omega_m)} \tag{22}$$

within the *m:th* resonance peak. In this relation, the spectra $\left|\hat{\varepsilon}_b^\infty(\omega)\right|^2$ and $\left|\hat{\varepsilon}_b^1(\omega)\right|^2$ can be determined experimentally.

For each resonance m=1,2,..., the resonance frequency ω_m and damping coefficient α_m can be estimated by minimizing the difference between the resonance peaks represented by the left and right members of Eq. (16). From ω_m and m, the complex modulus E_m at each resonance frequency can be finally obtained as

$$E_m = \rho \left(\frac{\omega_m}{\xi_m} \right)^2, \xi_m = \frac{m\pi}{l} - \mathrm{i}\alpha_m, m = 1, 2, \ldots.$$

7.2 Experimental set-up and procedure

Impact tests were carried out with a Nylon bar specimen of length 3045 mm, diameter 10.2 mm, and density 1149 kg/m. The striker had length 174 mm and the same diameter 10.2 mm and material as the bar, and its impact velocity was 3.7 m/s. The strain was recorded with 500 kHz sampling frequency, and the signal vanished completely before the end of the recording window (fig. 32 and 33)

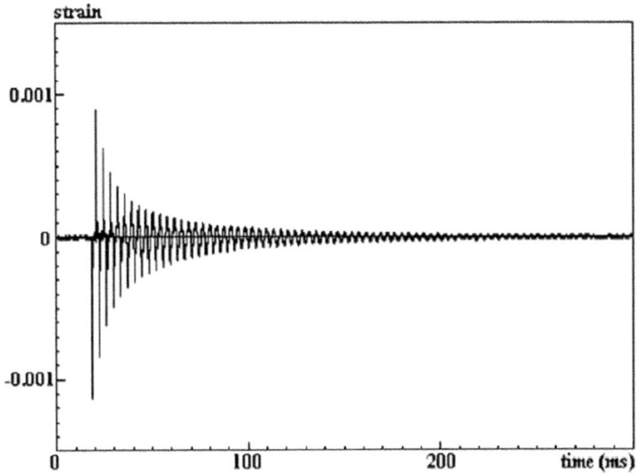

Fig. 32 Recorded strain in the Nylon bar specimen. Long-time record of strain pulses.

The spectrum $|\hat{\varepsilon}_b^\infty|$, computed from the long-time strain record, is shown in Fig. 34, 35 and 36. It is easily checked that resonance peaks are clearly identified. In figures 35 and 36, the experimental peak is shown with the theoretical one identified by the minimization process described above (in figure 35, the superposition is perfect at the scale of the drawing).

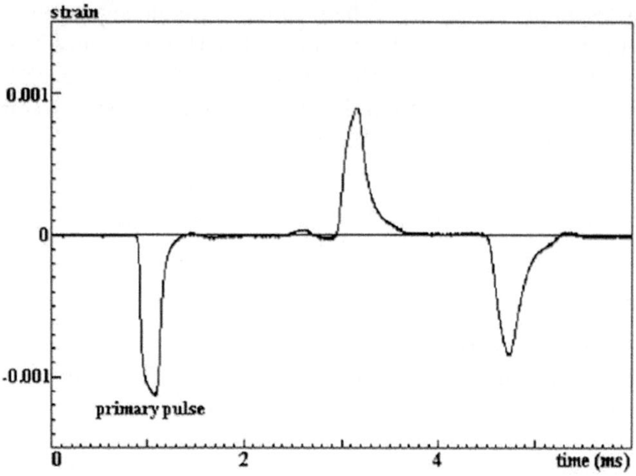

Fig. 33 Primary strain pulse followed by strain pulses which have undergone one and two free-end reflections

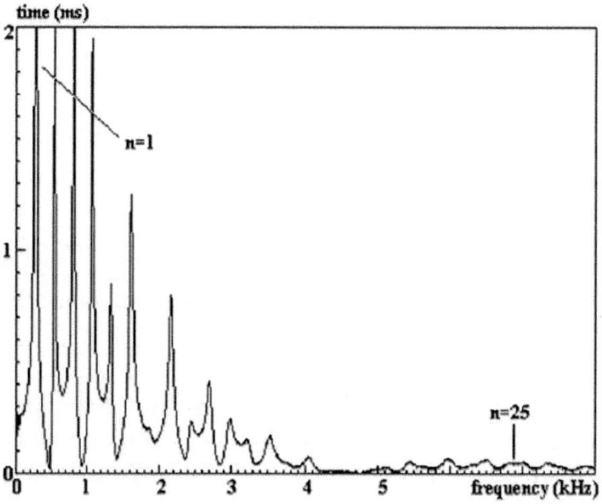

Fig. 34 Experimental spectrum

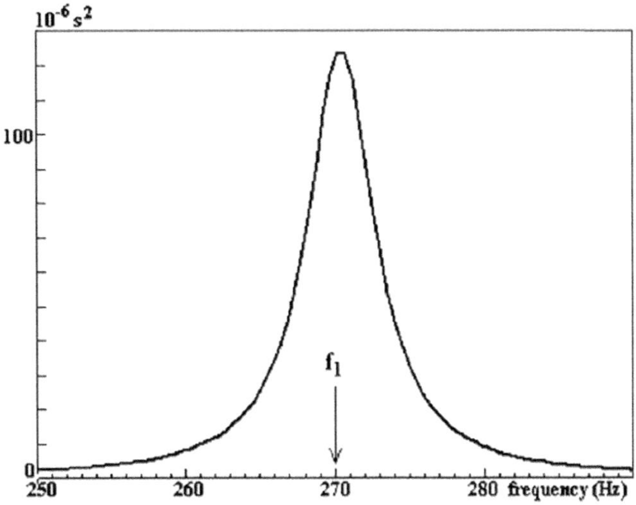

Fig. 35 Spectrum, zoom on f=270 Hz

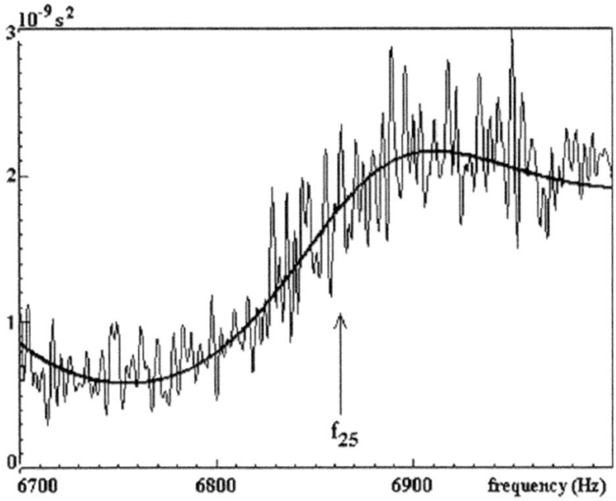

Fig. 36 Spectrum, zoom on f=6860 Hz

Corresponding to the peaks, a set of experimental points is obtained which provides, for each frequency, a value for wave speed and damping (circles in figures 37, 38)

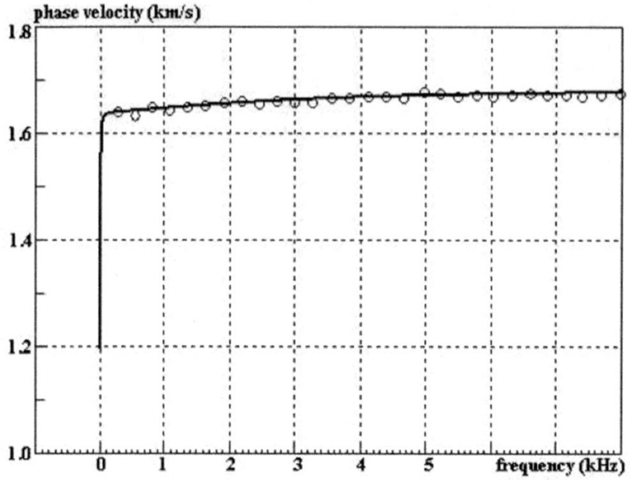

Fig. 37 Wave speed

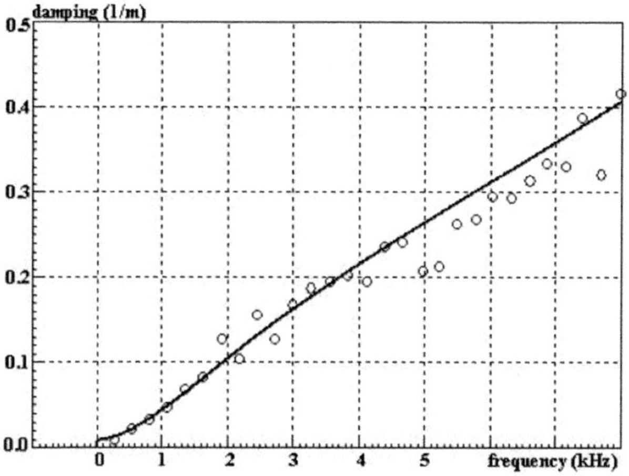

Fig. 38 Damping

By averaging, the results in figures 37 and 38 immediately provide the dispersion relations that could be used right away for the wave shifting.

Using the mathematical analysis presented by Collet et al. (2012), an optimal rheological model type is built and identified with 7 springs and dashpots. Note that the number of elements is not an assumption - as was done in Othman et al. (2001) - but is a result of the method. Knowing the corresponding Young's modulus, wave speed and damping as functions of frequency are then built. In the present example, they correspond to black continuous curves in figures 37 and 38.

8 Long-time measurements

Basic measuring techniques involving the use of bars require the knowledge of the two elementary waves which propagate in opposite directions. Once they have been characterised, they can be shifted to the appropriate cross-sections (the bar specimen interfaces for example) and all the mechanical values required can be calculated. The SHPB technique involves the independent measurement of each wave.

Some authors - Lundberg & Henchoz (1977), Yanagihara (1978), Zhao & Gary (1997), Jacquelin & Hamelin (2001) - have carried out two measurements on each bar. The pioneers in this respect were Lundberg & Henchoz (1977) and Yanagihara (1978), who independently developed a wave separation technique based on one-dimensional wave propagation theory. Their methods didn't take into account the wave dispersion. Some attempts have been made to develop methods also accounting for the wave dispersion, starting with Zhao & Gary (1997). However, the solutions proposed so far are sensitive to noise. A wave separation technique based on multiple measurements has been proposed by Othman & al. (2001) and Bussac & al. (2002). In line with Jacquelin and Hamelin (2003), this method will be subsequently called the BCGO method. It is based on the Maximum Likelihood approach and involves performing multiple strain measurements on each bar. This method, which is not noise-sensitive, requires an extra velocity measurement at very low strain rates. Jacquelin & Hamelin (2003) have developed an alternative three-point wave separation technique which is also insensitive to noise but the gauges are cemented to specific points and the force is calculated at one bar end (which is the only point where it can be calculated). The BCGO method is illustrated with the analysis of Split Hopkinson Bar tests, and an extended range of strain-rates (from 10^{-1} to 5 10^3 s-1) for the study the rate sensitivity of an aluminum alloy.

8.1 Wave separation: the BCGO method

In this section, the BCGO method is briefly presented.

Let us consider an elastic or viscoelastic bar with length l. In the case of single mode propagating longitudinal waves, the stress, strain, displacement and velocity are expressed in terms of the Fourier transforms as follows:

$$\begin{aligned}
\tilde{\varepsilon}(x,\omega) &= A(\omega)\ e^{-i\xi(\omega)x} + B(\omega)\ e^{i\xi(\omega)x} \\
\tilde{\sigma}(x,\omega) &= E^*(\omega)\ \left(A(\omega)\ e^{-i\xi(\omega)x} + B(\omega)\ e^{i\xi(\omega)x}\right) \\
\tilde{v}(x,\omega) &= -\omega\left(A(\omega)\ e^{-i\xi(\omega)x} - B(\omega)\ e^{i\xi(\omega)x}\right)\Big/\xi(\omega) \\
\tilde{u}(x,\omega) &= i\left(A(\omega)\ e^{-i\xi(\omega)x} - B(\omega)\ e^{i\xi(\omega)x}\right)\Big/\xi(\omega)
\end{aligned} \tag{23}$$

where $A(\omega)$ and $B(\omega)$ are the Fourier components of the forward and downward waves, respectively, at the origin of the bar, $E^*(\omega)$ is the complex Young's modulus and $\xi(\omega) = k(\omega) + i\alpha(\omega)$ is the complex wave number. Eqs. (23) show that the strain, stress, displacement and particle velocity can be obtained at any point on the bar if the following four parameters are known: $\xi(\omega)$, $E^*(\omega)$, $A(\omega)$ and $B(\omega)$.

The two parameters $E^*(\omega)$ and $\xi(\omega)$ depend only on the bar characteristics (its geometry and material). They only need to be determined once. Here we used the method presented in chapter 7.

In what follows, $E^*(\omega)$ and $\xi(\omega)$ are therefore assumed to be known. $A(\omega)$ and $B(\omega)$ are calculated based on the data obtained by performing three strain measurements and one velocity measurement. We express the fact that the signals recorded are noisy by writing that they are the sum of the exact value of the strain (or the velocity) and an unknown noise. The statistical distribution of the noise is assumed to be Gaussian. Consequently, the Maximum Likelihood Method can be used to estimate the two functions $A(\omega)$ and $B(\omega)$. This consists in writing that the signals measured correspond to the most probable event. Our problem is therefore equivalent to the minimization of a functional: this minimization yields an explicit formula for $A(\omega)$ and $B(\omega)$ as a function of the material and geometric parameters and the measured quantities - see Bussac & al. (2002).

The elementary functions $A(\omega)$ and $B(\omega)$ are then computed for each test. Using eqs. (23), one can now determine the force and the velocity at each point on the bar, especially at the bar ends. By applying the BCGO method to each of the bars, it is then possible to assess the force and the velocity at the two bar/specimen interfaces.

8.2 Experimental set-up

As an example, it is proposed to explore the strain-rate sensitivity of aluminium. The behaviour of this material is explored under a large range of strain rate conditions: quasi-static, medium and high strain rates.

In the high strain-rate tests, the classical time domain approach corresponding to the Split Hopkinson Bar was used. In the quasi-static and medium strain-rate tests, the new method involving the use of extra sensors on the SHPB was adopted, as explained above. The corresponding apparatus is a modified SHPB system loaded by an actuator, of the kind first introduced by Zhao & Gary (1997). In the quasi-static and medium strain-rate tests, low velocity loading was required. The kinetic energy of a striker would not suffice to induce large strains in the specimen, and longer loading durations were required. The bar system was therefore loaded using a hydraulic actuator and selecting the speed as required. Low loading speeds (of less than 0.1 m/s) can be monitored and automatically kept constant throughout the test. At higher speeds of up to 5 m/s, the requisite value is maintained approximately constant during the test. The new apparatus is called the "slow bars" apparatus. In the present case, three strain gauges and an optical displacement extensometer were used on each bar. The derivative of the displacement was then calculated numerically to obtain the velocity. A simplified scheme is presented in Fig. 39 and pictures of the set-up are given in figures 40a and 40b.

Aluminium bars 40mm in diameter and 3m in length were used in these tests on the strain-sensitivity of the aluminium.

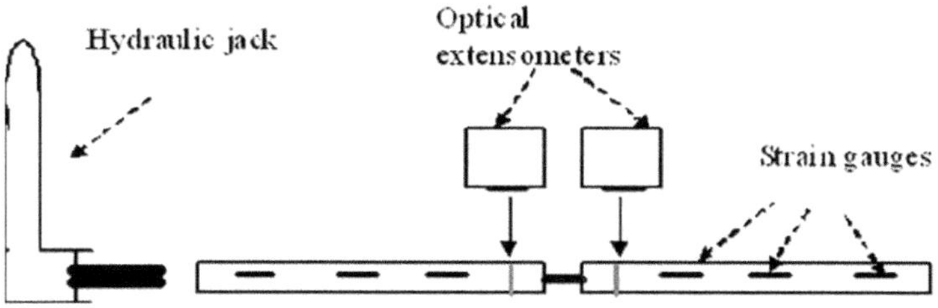

Fig. 39 Simplified scheme of the "slow bar" set-up

Both the Hopkinson bar and the slow bars give force and displacement measurements at the two bar/specimen interfaces. To determine the behaviour of the material, we assume the stress and strain to be homogeneous in the specimens. This assumption was systematically checked in the case of the slow bars and SHPB by making sure that the forces measured in the bars on each side of the specimen were practically equal. As was to be expected, it was observed that the smaller the loading rate, the more exactly this condition (called equilibrium) was fulfilled.

Fig. 40a View of the slow bar rig **Fig. 40b** Hydraulic actuator

8.3 A bar-bar test check

To check the consistency of the complete system (as well as testing the accuracy of the BCGO method), a bar/bar test was carried out, in which the two bars were put in contact without placing a specimen between them. The force and the displacement were computed at the bar/bar interface in two independent ways, using the measurements obtained on each bar separately and checking whether the two results obtained were equal. In the present example, the velocity of the hydraulic jack was set at 1.5m/s.

The forces and displacements calculated at bar ends are compared in Fig. 41 and Fig. 42, respectively. The results of computations made on each bar were almost equal, as was to be expected (with the bars in contact).

8.4 Aluminium characterisation

In the slow bar tests, the specimens used were 6mm in length and 6mm in diameter. The velocity of the hydraulic jack ranged from 2.10^{-4} to 2.5m/s. The strain rate ranged approximately from 10^{-1}on to 400/s. In Hopkinson bar tests, the specimen geometry and the striker impact velocity have to be adapted to the strain rates, which ranges approximately from 150 to 5000/s. In each test, the assumption that the force equilibrium conditions were satisfied between the two bar/specimen interfaces was checked. An example is given in Fig. 43 in the case of slow bars. The forces dropped suddenly at approximately 0.032s because the specimen broke at that point.

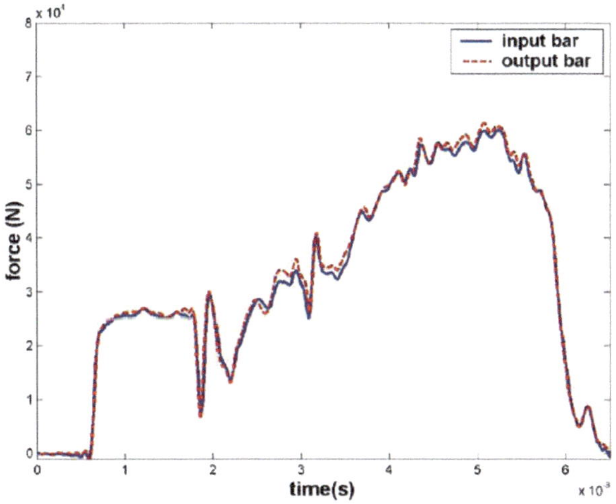

Fig. 41 Forces at the bar/bar interface.

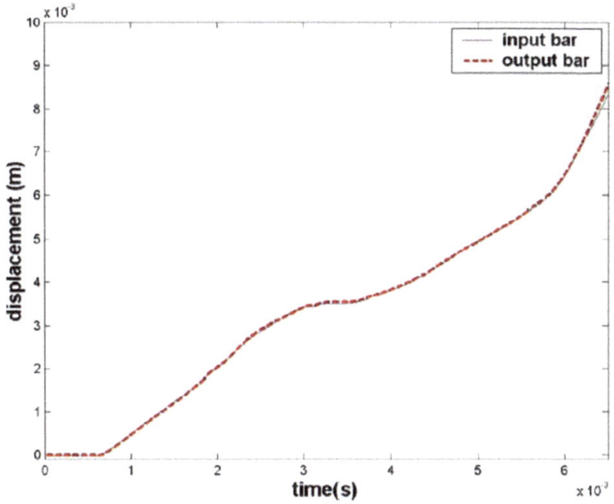

Fig. 42 Displacements at the bar/bar interface.

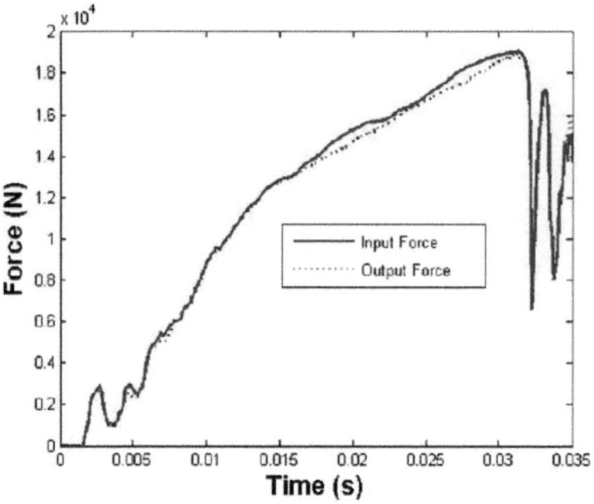

Fig. 43 Specimen equilibrium.

Displacement data were also calculated at each of the specimen faces, as shown in Fig. 44.

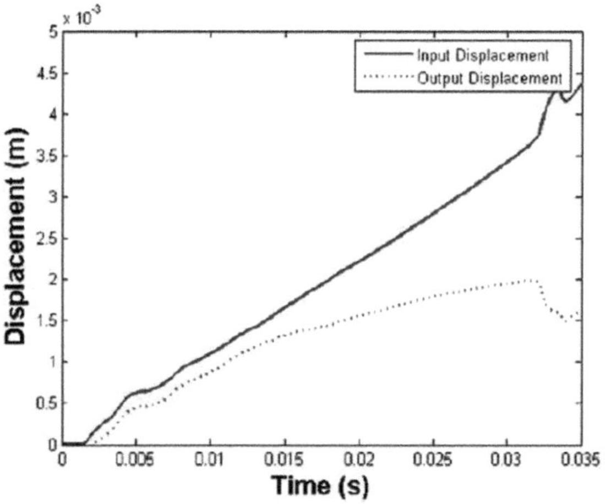

Fig. 44 Displacements at specimen faces.

Assuming (as well as checking) that the equilibrium conditions were satisfied, the stresses, strains and strain rates were obtained from these measurements. It can be noted that the duration of the tests increased considerably when using slow bars, reaching several seconds, in comparison with the usual duration of $500\mu s$ in the case of the classical Hopkinson bar tests carried out on the same bars.

The results of the aluminum test show that this material is only slightly sensitive to the strain rate. In Fig. 45, the changes in the stress corresponding to a 10% strain level were plotted versus the strain rate corresponding to the same strain level. The stress increased by approximately 15% when the strain rate increased from 10^{-1} to 5000/s. Some slow bar tests and Hopkinson bar tests were also both carried out at strain rates in the 150 to 400/s range. The mean difference between the results obtained with these two methods was less than 20MPa.

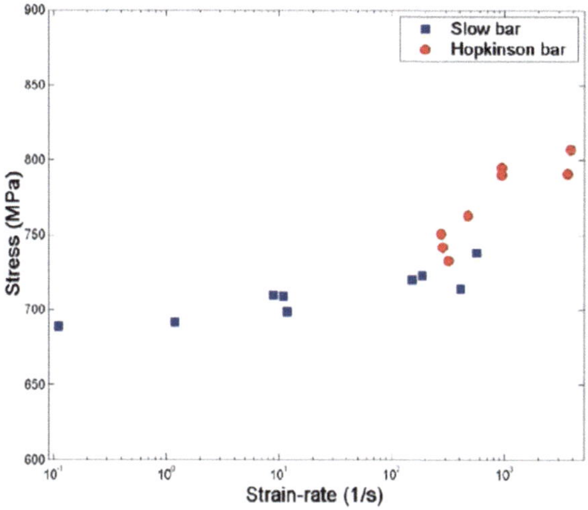

Fig. 45 Strain-rate sensitivity of aluminum.

8.5 A powerful tool

An easy interrupted test.

A simple way to make an interrupted test is to ensure that, after the first loading, the speed of the output bar at specimen side is larger than the speed of the intput bar at specimen side. This test also needs, of course, that the output bar is free of moving, that provides an extra "strain" measurement at its end (null strain). The total strain amplitude is then limited by the striker length (provided that the specimen behaviour is estimated as it also governs the strain rate). This will happen, when the transmitted wave will be greater, in amplitude, than half of the incident strain (with bars of the same impedance).

A test have been designed for this purpose, with aluminium bars (diameter 40mm, in input bar 1.8 m, output bar 1.3 m, striker 0.8 m – speed 10 m/s).

The corresponding wave records are shown in fig. 46 and 47. One extra gauge has been added on each bar and the record duration has been increased in order to check the validity of the "interrupted test" assumption.

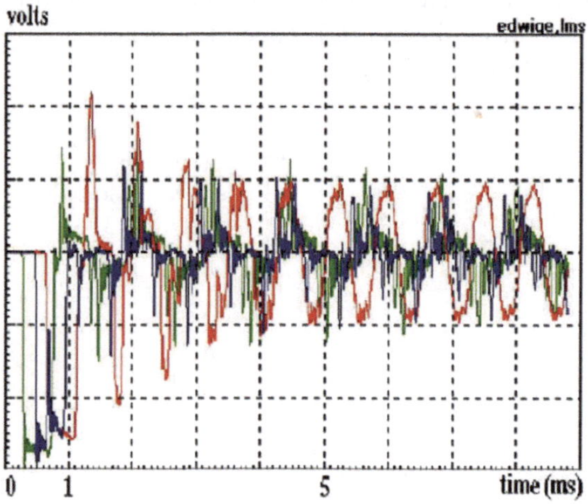

Fig. 46 Recorded waves.

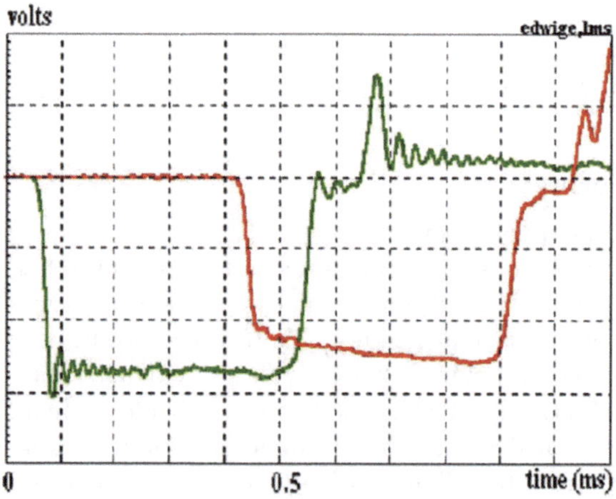

Fig. 47 Recorded waves (early instants).

Using the BCGO method, displacements computed at bar ends (specimen side) prove that the test was indeed an "interrupted test" as the specimen could not be reloaded, as shown in figs. 48 and 49.

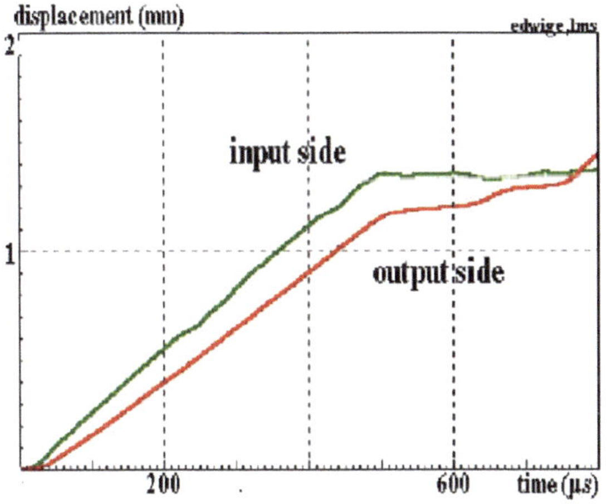

Fig. 48 Displacements at bar ends (specimen side) – short time

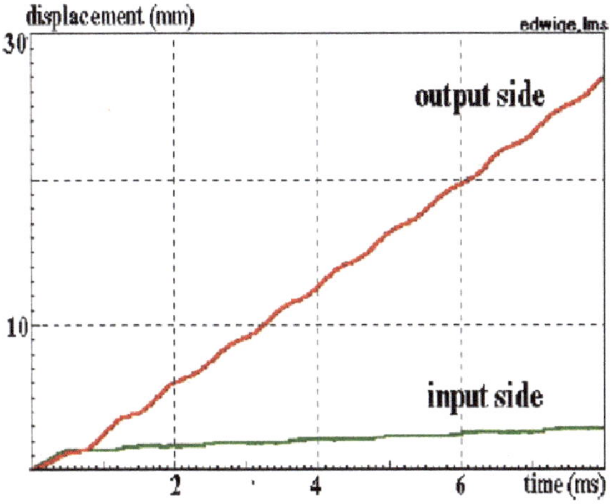

Fig. 49 Displacements at bar ends (specimen side) – long time

The BCGO method is valid at quasi-static strain-rates.
As the method works in the frequency domain and as the size of numerical records is limited, the data acquisition frequency used at very low strain

rates is no more than 1 kHz. Considering that the corresponding time base corresponds to 5 m of an elastic wave propagation, one could fear that the description of waves in bars is insufficient.

This is not the case, as demonstrated in fig. 54 where a test at a loading speed of 5mm/s is shown. The specimen is a water cylinder contained in a pressure vessel in which the bars act as pistons.

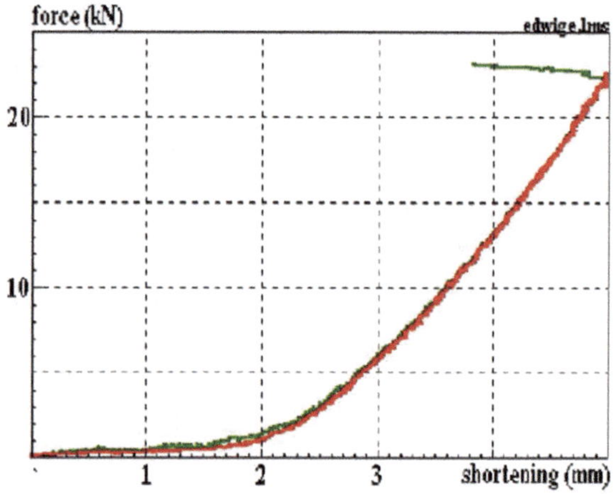

Fig. 50 Comparison of the BCGO method with a quasi-static analysis

For the quasi-static analysis, the displacement of each bar is corrected with the mean strain.

$$\varepsilon_m = (\varepsilon_1 + \varepsilon_2 + \varepsilon_3)/3 \quad d_e = d_{meas} + \varepsilon_m d(x_{meas}, specimen)$$

The force is simply deduced from the mean strain.

$$F = S_b E_b \varepsilon_m$$

9 Conclusion

Most problems related to dynamic testing have been addressed here, in particular those dealing with bars that are very commonly used at high strain rates, from 200/s to 5000/s.

Higher strain-rates have not been studied as they generally involve very expensive equipments.

On the other strain-rate side, some suggestions have been proposed to extend the use of bars at medium strain-rates (1/ to 200/s) where no industrial experimental method is easily available.

Recall that dynamic material testing needs to keep in mind a few points:

- Direct measurements of specimen mechanical properties like strains and stresses are almost impossible.

- Forces and displacements must be done (or computed) as close as possible to specimen faces.

- Going from global measurements to mechanical properties needs to evaluate dynamic effects (stresses and strains homogeneity, lateral inertia, temperature increase) in order to quantify the subsequent approximation.

Some modern optical methods which allow for direct measurements during high rates of testing, like image correlation for a more local estimation of the strain and infra-red temperature measurements, are welcome when available, in order to confirm and complete the validity of this difficult process that leads, from global forces and displacements measurements, to some knowledge of the material behaviour.

Bibliography

Achenbach, J.D.(1978), Wave propagation in elastic solids., North-Holland publishing company.

Bacon, C., An experimental method for considering dispersion and attenuation in a viscoelastic Hopkinson bar, Experimental Mechanics, 36, n 4 (1998) 242-249.

Bancroft, D., (1941) The velocity of longitudinal waves in cylindrical bars. Phys. Rev. 59, 588-593.

Blanc, R. H., (1971). Détermination de l'équation de comportement des corps visco-élastiques linéaires par une méthode d'impulsion. Ph. D. Thesis, Université d'Aix-Marseille, published in part in Problémes de la Rhéologie (W.K. Nowacki, editor), 65-85. IPPT PAN, Warsaw

Bussac M.N., Collet P., Gary G., Othman R., "An optimisation method for separating and rebuilding one-dimensional dispersive waves from multipoint measurements. Application to elastic or viscoelastic bars," J. Mech. Phys. Solids, 50, 321-350 (2002).

Bussac, M.,N.,Collet, P., Gary, G., Lundberg, B. Mousavi, S., (2008) Viscoelastic impact between a cylindrical striker and a long cylindrical bar Int J. of Impact Engineering. Vol 35 (5), 2008, 226-239

Collet P., Lundberg B., Gary G., Mohr D., (2012) Complex Modulus Estimation Respecting Causality: Application to Viscoelastic Bars. DYMAT, placeFreiburg, *accepted for publication*

Collet P., Lundberg B., Gary G., (2012) Noise-corrected Estimation of Complex Modulus in Accord With Causality and Thermodynamics: Application to an Impact test. J. of App. Mechanics, *accepted for publication*

Cristescu, N. (1967). Dynamic Plasticity. North-Holland, Amsterdam.

Davies, R.M. (1948), A critical study of Hopkinson pressure bar. Phil. Trans. Roy. Soc., A240, 375-457.

Davies E.D.H., Hunter S.C., (1963), The dynamic compression testing of solids by the method of the split Hopkinson pressure bar. *J. Mech. Phys. Solids*, **11**, 155-179.

Chree, C. (1889) The equations of an isotropic elastic solid in polar and cylindrical coordinates, their solutions and applications. Cambridge Phil. Soc., Trans. 14, 250-369.

Follansbee, P.S. and Franz, C. (1983) Wave propagation in the split Hopkinson pressure bar. J. Engng. Mater. Tech., 105, 61-66.

Forquin,P., Safa, K., Gary, G. (2010) Influence of free water on the quasistatic and dynamic strength of concrete in confined compression tests, Cement and Concrete Research, 40, 2, February 2010, 321–333

Gary, G., Klepaczko, J., R., Zhao, H. (1991) Correction de dispersion pour l'analyse des petites dformations aux barres de Hopkinson, Journal de Physique IV, 1(1991), C.3, 403-410.

Gary, G., Nowacki, W.,K.(1994). Essai de cisaillement plan appliqu des tles minces, *Journal de Physique IV*, **4**(1994), C8, 65-70

Gary.,G. and Bailly.,P., (1998), Behaviour of quasi-brittle material at high strain rate. Experiment and modelling, European Journal of Mechanics - A/Solids, 17, 3, (1998), 403–420.

Gary, G. (2001), Some aspects of dynamic testing with wave-guides, New Experimental Methods in Material Dynamics and Impact, Trends in Mechanics of Materials, Volume 3, W.K.Nowacki and J.R.Klepaczko (Eds), INB ZTUREK, P.O.Box 374, 00-950 Warsaw 1, Poland, 2001, 179-222

Gong, J.C. Malvern, L.E. and Jenkins, D.A. (1990) Dispersion investigation in the split Hopkinson pressure bar. J. Engng. Mater. Tech., 112, 309-314.

Gorham, D.A. (1983) A numerical method for the correction of dispersion in pressure bar signals. J. Phys. E: Sci. Instrum. 16, 477-479

Hillström, L., Mossberg,M; and Lundberg, B. (2000) Identification of complex modulus from measured strains on an axially impacted bar using least squares. Journal of Sound and Vibration, 230(4), 689-707.

Jacquelin, E. and Hamelin P., (2001) "Block-bar device for energy absorption analysis," Mech. Systems Signal Processing, 15, 603-617 (2001).

Jacquelin, E. and Hamelin, P., "Force recovered from three recorded strains," Int. J. Solids Structures, 40, 73-88 (2003).

Klepaczko, J., R., (2007) Introduction to experimental techniques for mate-

rials testing at high strain rates. Strength of materials under impact, Institute of Aviation ed. ,Al. Krakowska 110/114, 02-256 WARSAW, POLAND
Lifshitz J.M., Leber H., (1994), Data processing in the split Hopkinson pressure bar tests. Int. J. Impact Engng., 15, 723-733.
Lundberg, B. and Henchoz, A.,(1977) "Analysis of elastic waves from two-point strain measurement," Exper. Mech., 17, 213-218 (1977).
Lundberg, B. and Blanc, R. H. (1988) Determination of mechanical material properties from the two-point response of an impacted linearly viscoelastic rod specimen, J. Sound and Vibration, 126, 97-108.
Malinowski, J.,.Z., Klepaczko, J.,R., (1986) A unified analytic and numerical approach to specimen behaviour in the Split-Hopkinson pressure bar, International Journal of Mechanical Sciences, 28, 6, (1986), 381–391
Malvern L.E., (1951), Propagation of longitudinal waves of plastic deformation. J. Appl. Mech., 18, 203-208.
Merle. R., & Zhao., H. (2006), On the errors associated with the use of large diameter SHPB, correction for radially non-uniform distribution of stress and particle velocity in SHPB testing, International Journal of Impact Engineering 32 (2006) 1964–1980.
Mohr, D., Gary, G.; Lundberg, B.; (2010) Evaluation of Stress-strain Curve Estimates in Dynamic Experiments, IE-D-08-00215R1, International Journal of Impact Engineering 37, 2010, 161-169
Negreanu, I., Gary, G., Mohr, D., Enhanced infrared radiation method for temperature measurement in dynamic experiments, DYMAT 2009, EDP Sciences, 2009, 179-183
Othman R., Blanc R.H., Bussac M.N., Collet P., Gary G., (2001), A spectral method for wave dispersion analysis. Application to an aluminium rod. Proc. 4th Int. Symp. on Impact Engng, Kumamoto, Japan.
Othman R., Bussac M.N., Collet P., Gary G., "Sparation et reconstruction des ondes dans les barres lastiques et viscolastiques partir de mesures Redondantes," Comptes Rendus de l'Acadmie des Sciences, Srie IIb, 329, 369-376 (2001).
Pochhammer L., (1876), Uber die fortpflanzungsgeschwindigkeiten kleiner schwingungen in einem unbergrenzten isotropen kreiszylinder. J. fr die Reine und Angewandte Mathematik, 81, 324-336.
Rittel, D., (1999) On the conversion of plastic work to heat during high strain rate deformation of glassy polymers, Mechanics of Materials, 31, 2, February 1999, 131–139.
placeRota, L., (1994) An inverse approach of identification of dynamic constitutive equations, International symposium on inverse problems, Ed. Balkema
Safa, K.; Gary, G.(2010); Displacement correction for punching at a dynam-

ically loaded bar end, IE-1835, International Journal of Impact Engineering 37, 2010, 371-384

Safford. N., A., (1992) Laterial testing up to 10^5 s^{-1}using a miniaturized Hopkinson bar with dispersion corrections, Proc. 2nd Intl. Symp. On Intense Dynamic Loading and its Effects, Sichuan University Press, Chengdu, China, (1992), 378

Sokolovsky V.V, (1948), The propagation of elastic-viscous-plastic wave in bars. Prikl. Mat. Mekh., 12, 261280.

Tyas A., Watson A. (2001) An investigation of frequency domain dispersion of pressure signals. *Int. J. Impact Engng.*, **25**, 87-101

Yanagihara, N., (1978) "New measuring method of impact force," Bull. Jpn. Soc. Mech. Engng., 21, 1085-1088 (1978).

Yew E.H. and Chen C.S., (1978) Experimental study of dispersive waves in beam and rod using FFT, J Appl. Mech., 45, 940-942.

Zhao, H., and Gary, G., (1995) A three dimensional longitudinal wave propagation in an infinite linear viscoelastic cylindrical bar. Application to experimental techniques, *J. Mech. Phys. Solids,* **43**(1995), 1335-1348

Zhao, H., Gary, G., (1996) On the use of SHPB technique to determine the dynamic behavior of the materials in the range of small strains, Int. J. Solid. & Structure. 33(1996), 3363-3375.

Zhao, H and Gary, G., (1997) "A new method for the separation of waves. Application to the SHPB technique for an unlimited measuring duration," J.Mech.Phys.Solids, 45, 1185-1202 (1997).

Dynamic testing of materials: Selected topics

Daniel Rittel
Faculty of Mechanical Engineering, Technion, Haifa, Israel

Abstract Dynamic testing of materials is a vast subject involving a large variety of techniques according to the investigated properties. Consequently, one cannot cover it extensively and the present chapter will address three selected topics that were discussed during the lecture series given at CISM. For those subjects, the main experimental tool is the Split Hopkinson Pressure Bar (SHPB, Kolsky apparatus), used in its two or one bar version. The SHPB is extensively covered in another chapter so that we will assume that the reader is reasonably familiar with this technique, and will focus here on selected applications, noting that the SHPB is, before anything else, an experimental setup that allows for determination of the boundary conditions (load, displacements) applied to a structure (a specimen being in that case a special case of structure). The following topics will be addressed:
Dynamic fracture of materials: Here we will present the one-point impact technique and its applications to both fracture mechanics testing, and also to dynamic tensile testing of quasi-brittle materials.
Pressure sensitivity of materials: We will describe the basic technique and its application to metallic and polymeric materials, with selected results.
Thermomechanical couplings: We will address here the effect itself, followed by various experimental techniques to measure transient temperature changes in impacted solids, followed by selected results.

As alluded before, such a chapter cannot be extensive, and is based mostly on the author's experience. However, we will outline throughout the text relevant references that will allow the reader to expand his knowledge on the field.

1 Dynamic fracture testing

Dynamic fracture mechanics addresses the evolution of cracks in structures subjected to transient loadings. It can be broadly divided into two main domains, namely stationary and propagating cracks. The interested reader can refer to Freund's book on the subject to get familiar with the analytical aspects of the subject (1990). We will focus here on stationary cracks only, since from an engineering point of view, the designer needs to have a criterion and the matching material property to predict whether the crack will propagate or not (Cox et al., 2005). The simplest (one parameter) way to analyze the dynamic stationary crack problem is analogous to the quasi-static case. Namely, the crack-tip stress field can be described by:

$$\sigma_{ij}(t) = \frac{K_\alpha(t)}{\sqrt{2\pi r}} f_{ij}(\theta) \, . \tag{1}$$

Where *(r,θ)* are the coordinates of a point from the crack-tip located at the origin, σ is the stress tensor, f is a tabulated geometric function and t stands for time. The indices i and j vary from 1 to 3. K has the usual meaning of the stress intensity factor (SIF), with a=I, II or III according to the loading mode. Note that the only difference with the quasi-static case lies in the fact that both the stress tensor and the stress intensity factor are now functions of time. Eqn. (1) is valid as long as the crack remains stationary. Consequently, a fracture criterion can be written as:

$$K_I(t) = K_I^d. \tag{2}$$

Where K_I is the stress intensity factor assuming mode I for the sake of simplicity, and its critical value K_I^d is called the *dynamic initiation (fracture) toughness.*

This section deals with the measurement of the dynamic fracture toughness, leaving aside considerations as to whether this is a true material property, a point on which there is no wide agreement yet.

To measure the dynamic fracture toughness, one needs to select an appropriate specimen and the corresponding experimental setup. While several variations can be found in the literature (see e.g. (Jiang and Vecchio, 2009)), it seems like the simplest configuration that can be used for testing materials of a limited ductility is the one-point impact technique (Giovanola, 1986). From a practical point of view, the specimen is a rectangular beam into which a sharp crack has been introduced. Belenky et al. (2010) investigated the dynamic fracture of transparent nanograined alumina, using the one-point impact technique. To produce sharp cracks, these authors used micro-indentation (Vickers) to obtain initially 4 cracks at each corner of the pyramidal indentation. Subsequent careful controlled bending allowed a pair of cracks to grow through the thickness of the specimen to a controlled depth, as shown in Fig. 1.

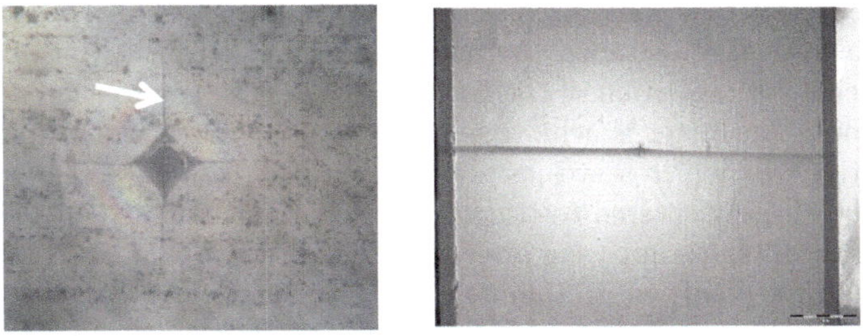

Fig.1. Left: Pyramidal indentation with cracks at each corner (arrowed). Right: The fully grown sharp crack.

In the experiment, the specimen is loaded dynamically at its center (Fig. 2) by means of a striker-instrumented bar system (Hopkinson bar). The bar is brought in contact with the specimen that is loaded by a stress pulse. As a result, wave propagation causes fracture of the specimen in a predominantly bending mode, even if the latter is unsupported, as a result of its inertial resistance. All that is needed in such experiments is a precise knowledge of the fracture time, more precisely the time at which the crack starts to propagate. To this extent, a thin line of conductive paint can be silk-screened in the immediate vicinity of the crack, on both sides of the specimen (Figs. 2).

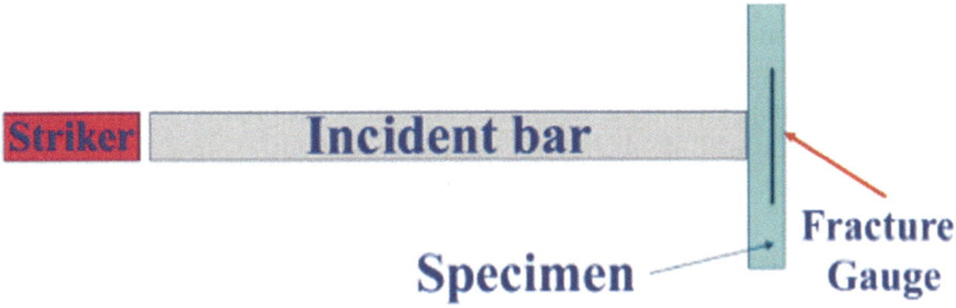

Fig.2.One-point impact testing of the dynamic fracture toughness.

Fracture time is recorded and synchronized with the bar signals, when time origin can be chosen for example as the time at which the stress wave first impinges on the specimen. The next part of the experiment consists of modeling the fracture mechanics of the cracked beam, while the crack is stationary so that the time evolution of the (mode I) stress intensity factor is calculated numerically. Finally, the value of the stress intensity factor at fracture time will be considered as the dynamic initiation fracture toughness of the investigated material. As noted long ago by Kalthoff et al. (1977), the precise determination of the fracture time is the key to successful measurements. The latter is achieved, as mentioned, via the fracture of a thin conductive wire which serves as a circuit breaker. However, the wire needs not to be entirely severed to trigger the signal, and a noticeable change in the trigger voltage is apparent after some 200ns, which for any realistic crack velocity corresponds to an infinitesimal advance. Since two such fracture gauges are silk-screened on each side of the cracked specimen, care should be paid to make sure that the two signals are identical in terms of timing. Lack of similarity may indicate that one side of the crack propagated earlier than the second one, which may invalidate the measurements. Moreover, while this device is not absolutely needed, a high speed camera recording is always a valuable complement in terms of validation of the measured fracture times. However, one should also keep in mind that such experiments induce stress intensity rates of the order of $10^6\ MPa\sqrt{m/s}$ so that accurate timing is quite central. An example of raw experimental signals is shown in Fig. 3.

The next point to be addressed is that of the accurate determination of the boundary conditions. In such an experiment, the boundary condition is neither of the prescribed displacement nor prescribed load type. Yet, as illustrated in Fig. 3, when the acoustic impedance of the tested specimen is such that most of the incident signal is actually reflected with a minor change that corresponds to the part of the signal that actually loaded the specimen, one finds that the incident force (sum of the incident and reflected signals) is by far less accurate than the incident displacement (difference of those signals). Consequently, one may opt for a prescribed displacement or velocity type of boundary condition as preferential, from an experimental point of view. This is the *only boundary condition* applied in the numerical model. The nature of the contact can vary from a simple distributed load to a more Hertz-like distribution, and here no recommendation is made. The above mentioned boundary condition is actually the only one to be introduced in the model, thus illustrating the relative advantage of the one point-contact method over other methods like the 3 or 4 point methods, in which special care must be paid both experimentally and numerically to the potential loss of contact between the specimen and the supports, as pointed out by Kalthoff et al. (1977). Moreover, one may observe that in such tests, low ductility materials fracture shortly after impact in a much shorter time than that required for the signal to travel across the specimen and reach the supports. Such an observation was reported some time ago by Rittel and Maigre (1993), which led to the conclusion that the supports are not needed in those experiments, so that the one-point configuration was subsequently adopted.

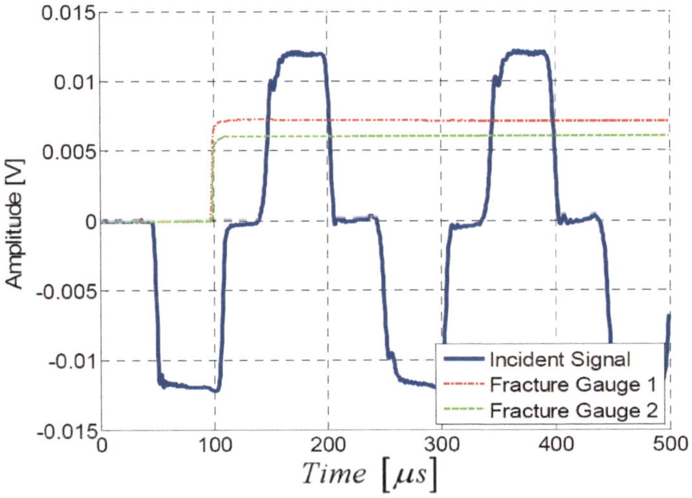

Fig. 3. Raw experimental signals. Note the slight difference between the incident (first) and reflected (second) signals. The fracture gauges gave identical fracture times in this experiment. Signals are not synchronized.

We will now survey some results and applications of the one-point impact testing method. Weisbrod and Rittel (Rittel and Weisbrod, 2001; Weisbrod and Rittel, 2000) applied this methodology to the testing of short beam specimens made of tungsten base heavy alloys. The challenge in this work was to test short (25 mm long) beams extracted from 25 mm diameter cylindrical rods, into which the initial fatigue pre-cracks were oriented along the longitudinal axis of the rod. To validate the overall approach, 0.2 mm strain gauges were carefully cemented close to the crack-tip so that the local strain component could be recorded during the experiment, as long as the gauge did not get destroyed by the propagating crack, which could also be interpreted as the onset of final fracture. Next, the corresponding stress intensity factor was calculated from the strain-gauge reading. The comparison of measured and calculated (using the approach described above) stress intensity factors revealed a high degree of similarity (Fig. 4) which not only serves to validate the overall hybrid experimental-numerical approach, but also reveals that the use of 1 term fracture mechanics concepts is sufficiently accurate in the present case. One should note that instead of calculating the stress intensity factor, which in the present case is required by the highly compact specimen geometry, the latter can be determined by analytical means as proposed, e.g. by Rokach et al. (Rokach, 1998; Rokach and Labedzki, 2009). A summary of the experimentally determined values of the dynamic initiation toughness for this heavy alloy, together with the quasi-static fracture toughness are shown in Fig. 5.

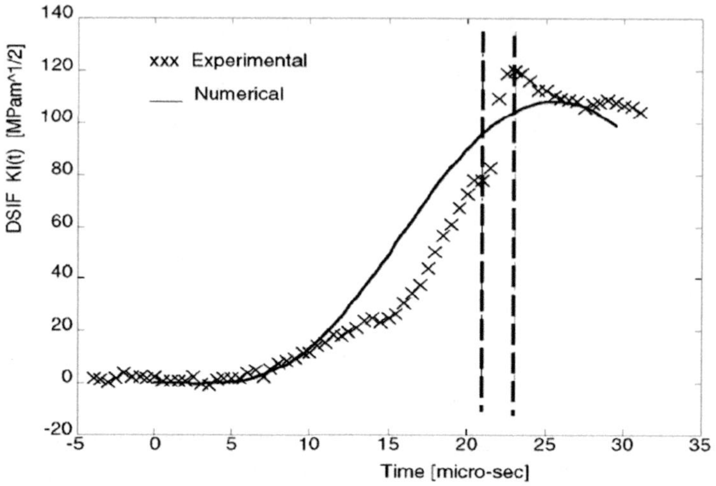

Fig. 4. A comparison of experimentally and numerically determined stress intensity factors (Weisbrod and Rittel, 2000). A strain gauge was cemented in the vicinity of the crack to measure the SIF. The numerical calculation used the measured displacement as a boundary condition for the finite element model of the cracked beam.

We will now address a different, yet closely related issue, namely that of the *dynamic tensile strength of brittle materials.* It is well-known that the determination of the static strength of brittle materials (e.g. ceramics) is a delicate experimental procedure since slight specimen misalignments with the loading train will induce bending and thus early fracture in nominally tensile test. Consequently, alternative procedures have been developed, which are listed in Belenky and Rittel's paper (2011), with a mention of their respective advantages or disadvantages. It so happens that, again, the simplest possible test that can be carried out is the 3-point bending tests, in which the measured strength is called "tensile rupture strength". While the stress distribution in the bent bar is not uniform, fracture that occurs from the tensile side of the beam can be viewed as the attainment of a critical local tensile stress. If we consider now the dynamic determination of this property, it appears that the subject has been barely investigated, or that the specimens used for this purpose all have serious limitations. One noticeable exception is the recent work of Delvare et al. (2010) who used dynamic 3 point-bending technique to determine the dynamic tensile strength of brittle materials. One should however note that in those tests, the specimen size is relatively large, a fact that can pose some problems when testing industrial ceramics that are usually manufactured as small tiles rather than large plates. But the second and more important point is that, similar to previous observations, fracture occurs most likely before the load pulse has reached the supports, in other words, fracture is purely inertial and of the one-point impact type. At about the same period, Belenky and Rittel (2011) proposed a simple adaptation of their approach for dynamic initiation toughness testing to measure the dynamic tensile strength of brittle materials. The technique remains essentially identical with the modification that this time, the specimen is not *pre-cracked.* Like before, a conductive gauge is silk-screened now on the tensile side of the beam to signal the onset of rupture. Like before, numerical simulations are used, aimed this time at calculating the tensile stress operating on the tensile side of the beam. In addition, a high speed camera can be used to corroborate the fracture gauge readings in terms of timing. Besides its simplicity, the main advantage of the proposed method is that large sample sizes can now be tested in a relatively short time, to provide statistical reliability to the obtained results. Fig. 6 shows the experimental setup used for testing commercial alumina bars.

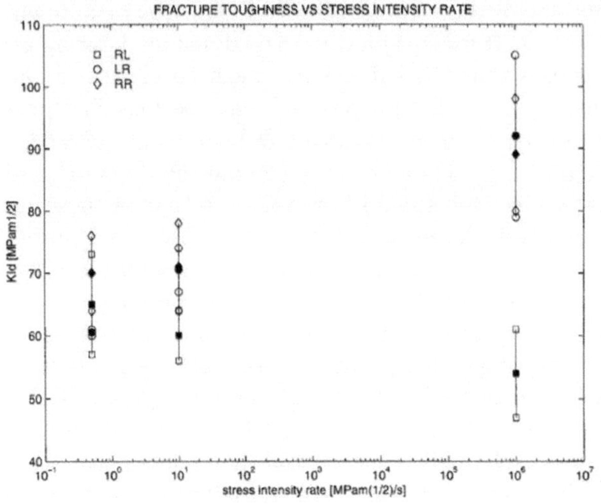

Fig. 5. A comparison the dynamic fracture toughness as a function of the strain rate for tungsten-base heavy alloys in 3 different orientations of the bar (Rittel and Weisbrod, 2001).

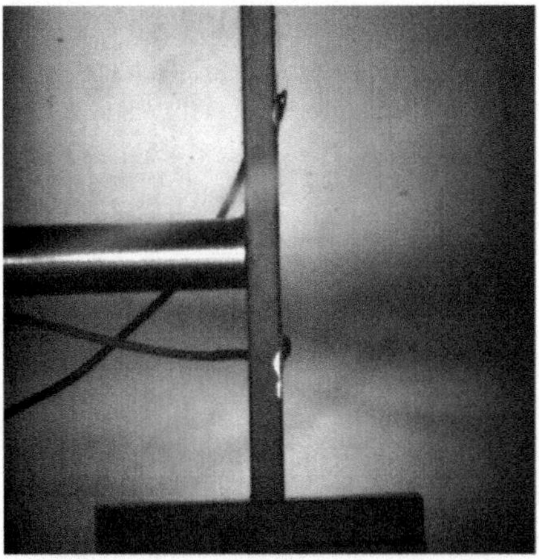

Fig. 6.Dynamic one-point impact strength testing of a commercial alumina beam. The specimen is in contact with the incident steel bar, and a conductive fracture gauge is silk-screened on its back face, to which two leads are attached as a circuit breaker.

Here too, it was observed that while the static tensile strength values are not much scattered, a large degree of scattering is obtained in the dynamic measurements. However, the distributions of the results can be compared showing beyond any doubt that the dynamic strength of commercial alumina, in this case, is significantly higher than its quasi-static counterpart. The reasons for this observation and the observed scatter were rationalized in terms of microstructure. Namely, commercial alumina contains some initial porosity, whose value and scatter do not evolve during the test, as opposed to microvoids in a ductile deforming material. Quantitative characterization of the porosity was extracted from the fracture surfaces of the specimens, and it was concluded that the population of disclosed pores in the quasi-static specimens is significantly larger than that in the dynamic case. While modeling of this finding may require heavy computational resources, it was argued that the fact that an initiating quasi-static crack "sees" a large population of microvoids causes overall material weakening as opposed to a dynamically initiating crack that "sees" much less of those voids. Leaving those considerations aside, one should nevertheless mention the fact that the one-point impact technique was applied to a systematic characterization of the influence of the material's microstructure on the one hand (Belenky and Rittel, 2012a), and separately to that of the influence of the initial surface condition of the specimen (Belenky and Rittel, 2012b). The above statement is brought here to illustrate the point that because of its simplicity, the proposed technique is well suited to in-depth studies of brittle materials.

2 Pressure sensitivity of materials

The sensitivity of the mechanical and failure properties of materials is a very vast subject that has been mostly explored for the quasi-static range of loadings. However, the occurrence of relatively large pressures is almost unavoidable in many dynamic situations, such as impact and perforation of structures. Moreover, strain localization situations may or may not be affected by local hydrostatic pressures. Finally, materials themselves, some of which are traditionally known to be pressure sensitive (e.g. polymers) are also likely to be affected in the dynamic regime, but this remains to be determined. We will therefore briefly review two techniques here, showing some characteristic outcomes.

2.1 Confinement by means of a metal jacket

Confining a cylindrical compression specimen in a tightly-fit cylindrical metallic jacket is not a new technique per-se. The classical experiments of that sort are carried out with sufficiently thick jackets so that they remain elastic throughout the whole test. This setup has the definite advantage that relatively high hydrostatic pressures can be reached in the specimen, but at the same time, the problem is that the pressure itself is not constant and it increases with the compressive load applied on the cylinder. If one specific point is of interest, such as a failure stress or strain, the technique is definitely valuable. If, on the other hand, one is interested to apply a constant confining pressure, an alternative solution must be sought, which may include hydraulic confinement with active control of the applied fluid pressure. Such setups are usually cumbersome and have this limitation that they cannot be used in dynamic impact experiments where the time scale is such that the controlling unit will not respond in real time. Consequently, a simple alternative is to confine the specimen in a thin metallic jacket that will yield at the desired stress level. Moreover, if the jacket's metal has a low strain-hardening character, the applied pressure will remain relatively constant. Plastic yielding of a thick-walled cylinder has been extensively studied (see e.g. (Kachanov, 1974)), so that simple solutions are available to describe the evolution of the elastic-plastic boundary as a function of the applied pressure. Note that the thickness of the jacket must be determined not only to provoke controlled yielding, but also to ensure that the transient elastic-plastic phase is of a limited character so that the stage of a constant confining pressure is reached as soon as possible. A graphic representation of the basic assumptions and confining setup is shown in Fig. 7 and 8 respectively (Hanina et al., 2007).

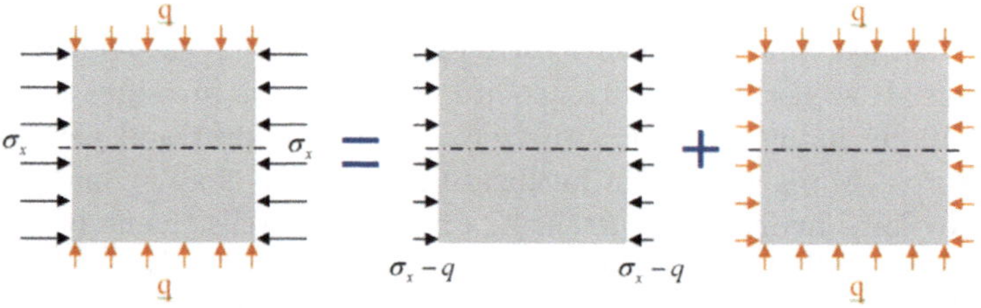

Fig. 7. Stress state of a uniaxial compression cylindrical specimen enclosed in a jacket. The symbol q indicates the hydrostatic pressure component.

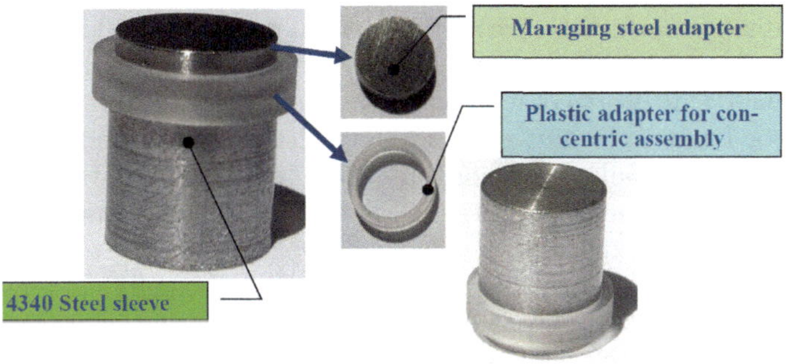

Fig. 7. The components of the confining system. The specimen (not shown) is tightly inserted into a 4340 steel jacket. On the upper part, a hard steel adapter is positioned and centered via a plastic ring adapter, so that the vertical load is solely applied to the specimen and not to the jacket.

Therefore, a preliminary step in the testing is the determination of the flow properties of the jacket material, either static or dynamic. The experimental setup can be of the kind shown in Fig. 7, where, once the specimen has been inserted into the jacket (close machining tolerances are required for this step), a small cylindrical piece, made of hard steel is inserted and guided onto the assembly to ensure that only the specimen (and not the whole assembly) is loaded. Note that the selection of the jacket material dictates the level of confining pressure that can be reached. It must be specified that while the method is not highly accurate as long as elastic straining is imposed, the flow stress at constant pressure can be determined accurately once the jacket has entirely yielded. As mentioned above, the confining pressure can be determined analytically, however, additional accuracy can be gained if friction at the specimen- jacket interface is taken into account. This can be done numerically, for example.

The determination of the confining pressure for pressure-sensitive and insensitive materials can be found in Rittel et al. (2008) . At this stage, it should be noted that the very same setup can be used in quasi-static tests as well as in dynamic ones, using a Kolsky apparatus (Split Hopkinson Pressure bar) for instance (1949). Several studies were carried out with this setup to characterize the pressure sensitivity of some alloys (Hanina et al., 2007) as well as polymeric materials (Rittel and Brill, 2008; Rittel and Dorogoy, 2008). For polymers, simple rate-dependent Drucker-Prager types of constitutive equations (Kachanov, 1974) could be fitted over a wide range of strain rates. Since then additional work has been carried out using this confining methodology coupled to numerical calculations, which has proved to be sufficiently mature for routine characterization of materials. Additional results worth of interest can be mentioned at this stage. The first one can be found in Hanina et al. (2007). These authors studied a dynamic shear localization failure mechanism called "adiabatic shear banding" and its relationship to the hydrostatic pressure. One main outcome of this study was that, while as expected, the plastic flow of the investigated metal is not influenced by the hydrostatic pressure component, the failure strain at which adiabatic shear developed was definitely increasing with the level of applied confinement. The relevance of this observation to impact or penetration situations is obvious and it can be noted that this is the first and only reported observation, so far, of this failure-pressure dependence. Following this result, it became of interest to characterize the response of a brittle polymer (PMMA), tested under confinement at high strain-rates. In this context, one must mention the results of Chen and Ravichandran (1997). These authors studied a glassy ceramic (MACOR) under confinement at high strain rate and discovered a failure-mode transition, related to the level of applied confinement, in which the material pulverizes in specific directions instead of shattering as expected. While such transition cannot strictly be qualified of a brittle-ductile one, it nevertheless shows that confinement has implications on the level of absorbable energy by brittle materials.

At this point, the conclusion of this paragraph is that a technique, that is quite mature, is available for the routine characterization of the pressure sensitivity of materials and associated failure mechanisms under dynamic loading. However, the range of confining pressures remains limited since jacket materials, all of which metallic, have only a limited spectrum of yield strengths, which once attained do not allow for further increase of the pressure. In that respect, it is felt that the present technique can be considered as a useful extension of the thick elastic jacket confinement which puts virtually no limitations on the confining pressure, albeit variable during the test.

2.2 The Shear Compression Disk (SCD)

The shear compression disk (SCD) was designed to allow for investigations of the influence of confining pressure and stress triaxiality on the mechanical response of various materials (Dorogoy et al., 2011; Karp et al., 2013). The specimen and experimental setup are shown in Fig. 8. The basic idea is quite simple and consists of confining a disk-like specimen by means of a thick steel die in which the conical geometry of the disk and the die create a state of confinement that is directly proportional to the depth of insertion of the disk. The latter has a pair of circular grooves of slightly different diameters on its upper and lower face respectively. Therefore, the first step of a test (preset) consists of inserting the disk to a predetermined depth in its conical guide to create a predefined state of hydrostatic stress (and triaxiality). This is usually carried out using an LVDT that allows for a precise measurement of the insertion depth. At this stage, the disk is held in place by the strong friction that develops with the die's walls. The setup is now ready for the second stage, namely quasi-static or impact loading of the specimen. It is important to note at that stage that the specimen's geometry, overall setup and approach allow both quasi-static and dynamic transient testing in a seamless manner, thus making this specimen quite attractive for systematic investigations of combined high strain-rates and confining pressures for example.

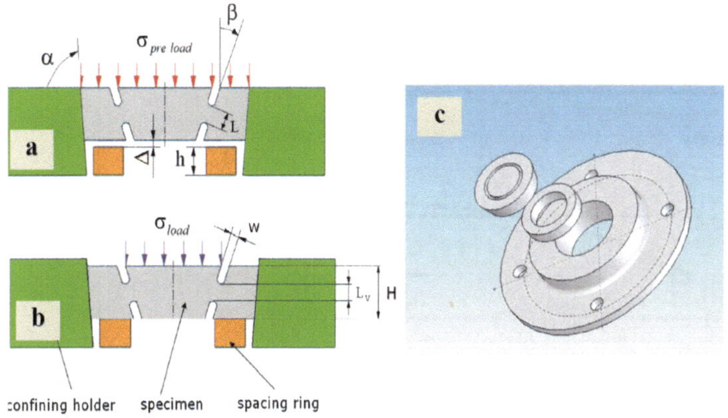

Fig. 8. The Shear Compression Disk. The specimen consists of a disk into which a pair of grooves are machined on each opposite surface at a given angle (b). Figure (a) describes the initial pre-load step to achieve confinement through insertion to depth D into a hard steel die. Figure (b) describes the actual experimental step for which pressure is applied only to the central part of the disk. Figure (c) shows details of the specimen, spacer and confining steel die.

However, just like with other equivalent specimens, the reduction of the measured loads and displacements into equivalent stress-strain curves requires numerical modeling of the experiment, much like is done for another specimen, the Shear Compression Specimen (Dorogoy et al., 2011; Rittel et al., 2002). As mentioned earlier, the hydrostatic pressure (or stress triaxiality) must also be determined by preliminary numerical simulations. A very attractive feature of the SCD, is the observation that the strain distribution is rather uniform in the gauge section which is a sort of prerequisite for a valid test. But the main advantage of the SCD lies in the fact that the confining pressure (or stress triaxiality) is both homogeneous in the gauge section but most of all is nearly constant. In that respect, the SCD can be compared to other experimental techniques in which confinement is applied by means of an elastic jacket for instance, which causes a continuously variable confining pressure as discussed in a previous section. Fig. 9 shows calculations of the triaxiality level using a finite element model of the specimen. Note that the angle b is of 20, which can be varied as a function of the desired level of triaxiality.

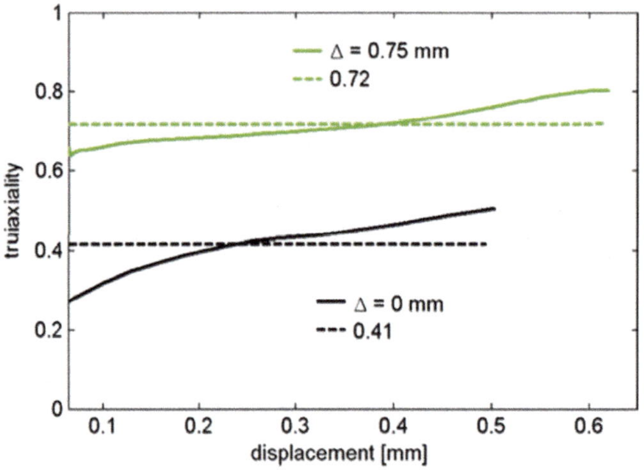

Fig. 9. Calculated levels of triaxiality for two different heights of initial insertion D. The triaxiality is calculated as a function of the prescribed experimental displacement following the pre-insertion phase. Note that the level of triaxiality is reasonably constant throughout the test, and its average value is shown by the dashed lines.

As of today, the specimen and the overall approach have been devised and validated, thus opening new perspectives in the field, noting that while the stress triaxiality has been markedly improved over the traditional value of 1/3 (uniaxial tests), it does not exceed values of 1 in the current tests. Therefore, while one can reasonably expect to observe trends in the dependence of the flow and failure properties, the quest for significantly higher triaxialities such as those encountered in penetration problems is still open.

3 Thermomechanical couplings

Thermomechanical coupling expresses the fact that a coupling exists between the mechanical energy invested in deforming a structure and thermal energy fluctuations during the process. In other words, the subject addresses the conversion of mechanical energy into heat. The reader is referred to the classical textbook of Boley and Wiener concerning theoretical aspects of the subject (Boley and Weiner, 1960).

The starting point is therefore the coupled heat equation which writes as:

$$k\nabla^2 T - \alpha\left(3\lambda + 2\mu\right) T_0 \dot{\varepsilon}_e^{kk} + \beta_{diff} \sigma_{ij} \dot{\varepsilon}_{ij}^p = \rho C_p \dot{T}. \tag{3}$$

Where k is the heat conductance, α is the linear thermal expansion coefficient, m and l are Lamé coefficients, T_0 is a reference temperature, and C_p is the heat capacity. The strain tensor components ε_{ij} are divided into elastic and inelastic (plastic). The coefficient β, often referred to as the Taylor-Quinney (T-Q) coefficient (Taylor and Quinney, 1934) is the central part if this paragraph, and it expresses the part of the mechanical *power* that acts as a thermal source. Two heat sources are noticeable: The first one relates to elastic volume changes and its contribution is usually minor in terms of resulting thermal evolutions. The second, on the other hand, consists of the fraction of dissipated plastic power that acts as a heat source, and the latter is quite significant. If adiabatic conditions prevail , the differential Taylor-Quinney coefficient becomes a simple ratio of the thermal to mechanical *powers*. Alternatively, one may evaluate the ratio of the thermal to mechanical *energies* as in Taylor and Quinney's paper. In this case, one will consider an integral factor, given by (with proper integration limits):

$$\beta_{int} = \frac{\rho C_p \Delta T}{\int \sigma_{ij} \dot{\varepsilon}_{ij}^p}. \tag{4}$$

The meaning and distinction between the differential and integral factors are discussed in Rittel (1999). From thereon, we will focus our attention integral T-Q factor, with emphasis on transient thermal measurements. We will not elaborate here on the many theoretical aspects of the thermomechanical energy storage and its important complement, the so-called stored energy of cold work (Bever et al., 1973), except for mentioning that those factors are intimately related to evolutions of the material's microstructure (e.g. dislocations density and arrangement, twinning, in crystalline solids).

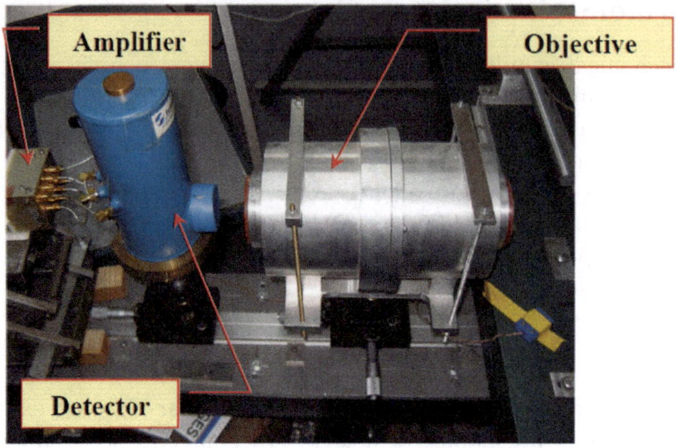

Fig. 10. Infrared detector setup (Dynamic Fracture Laboratory, Technion). The detector (Fermionics, Ca) consists of 8 liquid nitrogen cooled in-line pixels (HgCdTe), encased in a Dewar with a KCl window facing the specimen in the Split Hopkinson bar. The objective (double Schwartzchild) has a magnification ratio of 1:1, and each pixel is 35mm^2. The detected signals are amplified by an 8 channel amplifier.

Looking at Equation (4), it appears that in order to measure β_{int}, one must *simultaneously* measure the evolution of the stress, strain and *temperature* as well. The reader is already acquainted with the Kolsky apparatus and its many variants as one of the most basic and reliable tools to measure dynamic stress-strain curves of solids, so that the problem amounts to devising a robust transient thermal measurement. While fine thermocouples have been successfully used with polymeric materials (Bendersky, 1953; Bloomquist and Sheffield, 1980; Rabin and Rittel, 1999; Regev and Rittel, 2008; Rittel, 1998), this technique is not suitable for metallic materials. One would naturally consider thermal cameras that provide a full-field thermal map of the object's surface. However, such cameras are too slow for transient measurements which are carried out at typical frequencies of the order of the MHz. A viable alternative consists of infrared thermal detectors, whether comprising a single detector (pixel) or an array of detectors (usually 8 in line). Such detectors have been utilized a few decades ago, with a relatively coarse pixel size (of the order of the mm), see e.g. (Brockenbrough et al., 1988; Duffy and Chi, 1992; Hartley et al., 1985; Hartley et al., 1987; Marchand and Duffy, 1988). The interested reader will find in the above-listed papers a description of the detector and the associated optics that serve to map a point on the specimen's surface onto the detector. A typical setup, located in the Dynamic Fracture Laboratory (Technion) is shown in Fig. 10.

The detector itself provides a voltage reading that must be converted into a temperature. One can proceed from first principles, and given the emissivity of the investigated surface, convert the emitted radiation into a voltage. One should note here that the emissivity itself is not a constant, as it reflects the surface's condition. Consequently, while the specimen deforms, its surface condition may change due to roughening (e.g. "orange-peel") or even some light oxidation. Previous research has shown that for moderate strains (of the order of 0.2), surface alterations are not dramatic so that the emissivity is not altered (Mason et al., 1994), although this point deserves additional consideration. An alternative to first principle calibration consists of probing a hot specimen whose temperature is monitored, sampling its thermal signal and correlating the two. While being somewhat simplistic, this method is routinely used in high strain rate thermomechanical experiments (Belenky et al., 2010; Regev and Rittel, 2008; Rittel et al., 2007; Rittel et al., 2006; Rittel et al., 2009). Here the key experimental point is that calibrations must be routinely performed and consistence must be ascertained. In other words, calibrations must be performed several times before the actual test, and also after the test to make sure that the setup has not changed. Extreme care must be paid to calibrate the IR system in experimental conditions that closely mimic the actual experiment, including the experimentalist's location, as a source of IR radiation for him/herself! This preparation stage may necessitate several ad-hoc adjustments. As an example, let us mention recent experiments carried out on pure copper. When the calibration specimen is preheated in the oven, then placed in front of the detector and allowed to cool, the surface of the specimen covers with a very thin layer of oxide which affects the surface emissivity. This is unavoidable. Consequently, before performing actual tests with copper specimens, those were preheated in the oven for a very short time and allowed to cool down in an attempt to create an initial surface condition that is similar to that of the calibration specimen. The same concern could definitely apply to machining marks of the specimen. A polished and an as-machined specimen do not have the same surface emissivity and this should be kept in mind. A typical example of calibration curves to convert the measured infrared signal into the specimen's surface temperature is shown in Fig. 11. In this example, repeated experiments were carried out with polycrystalline tantalum specimens that were used either in the as-machined form or after being pre-strained quasi-statically. Here, one can note that as long as the temperature does not exceed some 80C, the difference in signal intensity as a result of the surface texture of the specimen is not very pronounced. Since in most experiments, the measured temperature does not significantly exceed the above-mentioned limit, an average value that is determined from the 2 curves shown in the figure is reasonably reliable. Note that for each specimen condition, several experiments are carried out to ensure repeatability of the results.

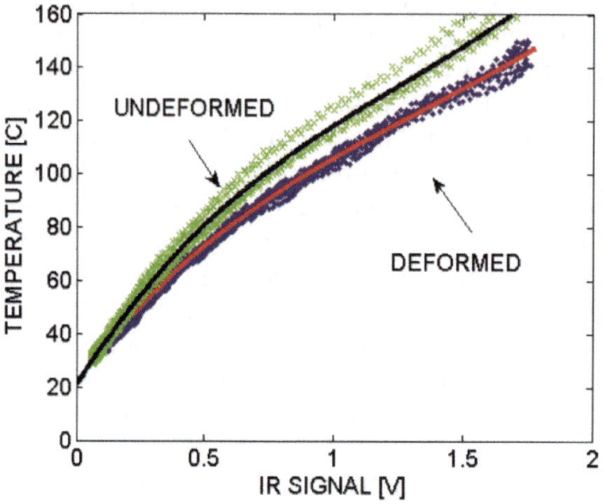

Fig. 11. Typical calibration curves obtained for polycrystalline tantalum specimens. The specimens were tested in the as-machined and deformed condition. Note that for temperatures of up to 80C, the curves are rather similar, so that surface texture is not very influential at that stage.

Another interesting point that is seldom addressed is that of the quality of focusing of the optics on the specimen's surface. Focusing is carried out with visible light, and for cylindrical specimens, accurate focusing may be a real challenge. The question which arises here is whether the quality of the focus affects the measurement, and if yes to which extent. This issue was discussed in Regev and Rittel (2008), who pointed out and verified experimentally that IR monitoring is not concerned with obtaining an accurate and sharp image of the investigated object, but rather collecting the IR emission from its surface. In other words, collecting the IR-energy is not related to the quality of the focus, which translates practically to the fact that the measurement is not very sensitive to the quality of the focusing, which is of tremendous help to the practitioner. Another practical point to be cared about is that of the field of view. The latter is related the optical setup, aperture and distance between the specimen and the detector, specimen dimensions, all by simple geometrical considerations. However, it happens quite frequently that for specimens of small thickness (or after a severe deformation) the Hopkinson bars themselves appear in the field of view of the detector, which may in turn cause an apparent reduction in the level of the thermal signal that could be misinterpreted as a temperature drop, a phenomenon that is not physical in terms of thermomechanical coupling. Indeed, while temperature will drop as a result of heat transfer, this will occur after much longer time intervals than those that characterize Hopkinson bar experiments.

Another important consideration is related to the specimen's surface temperature. Here, one could wonder whether the latter is a true representation of the specimen's temperature or simply a local measurement when temperature gradients are present in the specimen. Rabin and Rittel (2000) addressed this issue and provided non-dimensional graphic solutions to the transient heat equation that illustrate the extent of thermal gradients in a dynamically loaded specimen, thereby providing guidelines for experimental design, including determination of the specimen's dimensions. For metallic specimens of a few millimeter diameter, as those typical of a Hopkinson bar test, those gradients are negligible, meaning the under adiabatic conditions, the surface temperature is representative of the bulk temperature of the specimen. However, care must be exercised when other types of materials are considered.

Next we will briefly review some recent results obtained with the above-described measurement technique, with a discussion of their novelty and relevance.

A first interesting result concerns the magnitude of the temperature rise during impact and plastic flow, as well as the rate at which the temperature buildup occurs. If for instance, one assumes a constant value of the integral T-Q coefficient, one can expect a significant temperature rise to develop in the specimen. This practice is almost universal in the many analytical/numerical works dedicated to the development of dynamic shear instabilities, i.e. adiabatic shear bands, for which thermal softening is a dominant factor in the localization process (Zener and Hollomon, 1944). Fig. 12 shows a simultaneous record of the stress-strain-temperature in a Ti6Al4V specimen deformed at high strain rate. An array of 3 pixels was used in this experiment. Those results show clearly that the temperature does not increase linearly with the strain (the material has little strain-hardening). More striking is the fact that the maximal temperature rise is overall quite modest, of the order of a few tens of degrees. Such an extent cannot be thought of causing a significant thermal softening in this material. A more careful look at the thermal evolution also shows that the significant temperature rise occurs close to final failure, whereas during the homogeneous deformation phase, this thermal excursion is almost insignificant. Those observations can be wrapped up in the observation that instead of being a constant (e.g. 0.9), the β_{int} factor is in fact varying the with the amount of plastic strain, a fact that was already noted by Hodowany et al. (2000). It should be noted here that the trends shown for Ti6Al4V were also observed for other metallic alloys. From a practical point of view, such results can help in fine-tuning numerical models of dynamic events, which otherwise may predict temperatures that are significantly higher than those observed.

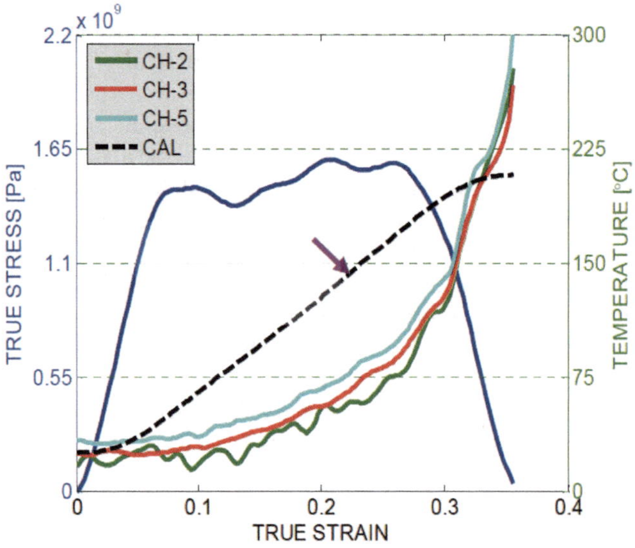

Fig. 12. Stress-strain-temperature measurement of a Ti6Al-4V specimen subjected to high strain rate impact (Rittel and Wang, 2008). 3 different channels are used for the temperature measurement. The dashed line indicates the calculated temperature rise that should be measured in the case of a full thermomechanical conversion. Note that until the stress starts dropping, the overall temperature rise is quite modest. Similarly, the measured and calculated temperature rise are quite different, which indicates that for this material, assuming β_{int}=1 is quite inaccurate.

Another interesting issue is related to the fact that it is universally reported that β_{int} cannot exceed a value of 1, according to the first law of thermodynamics. This is to be taken as a first principle. However, one may take a second look at equation (4) in which the thermal energy is divided by the mechanical one. The thermal energy term is *implicitly* considered to be *solely generated by mechanical work.* However, whatever may be its origins, the T-Q integral coefficient remains defined by eqn. (4). Extraneous heat sources may exist in the system that are not purely mechanical. Consider for example phase transformations, in which the enthalpy of transformation (*DH*, the latent heat) may be released in the system for an exothermal transformation. In this case, one should rewrite the integral T-Q coefficient as:

$$\beta_{int} = \frac{\rho C_p \Delta T + \Delta H}{\int \sigma_{ij} \dot{\varepsilon}^p_{ij}}. \tag{5}$$

The latent heat may be significant, to an extent that the T-Q factor calculated using eqn. (4) may actually *exceed the limiting value of 1.* Rittel et al. (2006) investigated the mechanical response of pure iron subjected to high rate shear deformation. Iron is long known to undergo a phase transformation from α BCC to ε HCP phase under shock pressure (Barker and Hollenbach, 1974). However, it has been shown numerically that the transformation pressure (13GPa) can considerably be lowered in the presence of shear strains (Lew et al., 2006). The phase transformation is reversible (with some hysteresis) in this material, so that it cannot be probed after the test, but must be ascertained in real time, which is a highly challenging problem (dynamic tests must be conducted in a synchrotron environment).

To shed additional light on this phenomenon, Rittel et al. (2006) conducted their dynamic tests together with a simultaneous determination of the thermal evolution. A characteristic result is shown in Fig. 13. This result, which was obtained several times at the higher strain rates, clearly shows that the integral T-Q coefficient exceeds 1 over a given range of strains. This observation was suggested to be the result of the abovementioned exothermal phase transformation.

From the modeling point of view, phase transformations, that are well known to occur in 304 series stainless steel, were introduced into a fully coupled thermomechanical model of this material (Zaera et al., 2013). Using latent heat values taken from the literature, those authors reproduced this "anomaly" in the integral T-Q coefficient, as determined by eqn. (5) as shown in Fig. 13. Note that similar observations were also reported by Jovic et al. for austenitic steel (2006). As a final remark to this subject, the scientist is left to decide which definition to adopt concerning the T-Q coefficient, namely the more global (eqn. (4)) or the decoupled one (eqn. (5)). However, it seems at present that the accurate decoupling of the thermal contribution of the plastic work and that of a phase transformation may be delicate from an experimental point of view, as one cannot be sure that the same enthalpy is released for a given phase transformation induced by shock or by thermal means only.

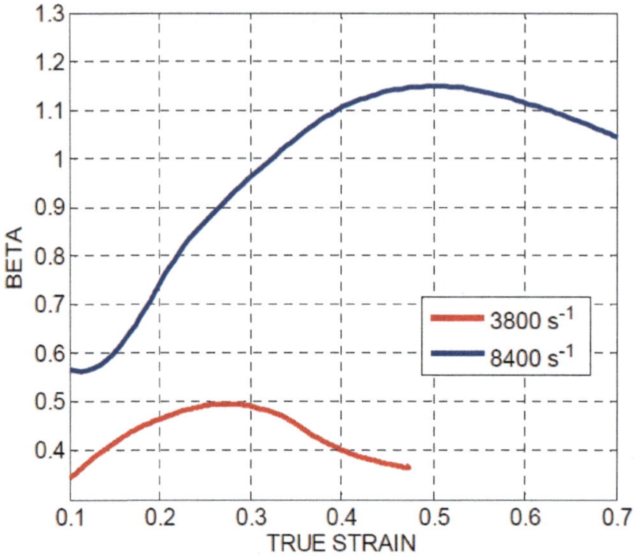

Fig. 13. Measurement of β_{int} in pure Fe at lower and higher strain-rates. Note that β_{int} exceeds 1, which is attributed to the latent heat released during the phase transition.

To conclude this chapter, let us look again at Taylor and Quinney's results on copper (1934) which led to the classical value of 0.9 for the T-Q factor. In a recent work, Rittel et al. (2012) set to investigate this factor as a function of the strain-rate for pure single copper crystals and also for polycrystalline OFHC. While the experimental methods were different, the measured quantities were the same and the goal of the experiment was identical, with one addition, namely the systematic investigation of single vs. polycrystalline material. Fig. 14 shows the dependence of the T-Q integral factor on the strain-rate for the two types of investigated material.

This figure reveals a linear dependence of the β_{int} on the strain rate for both materials. This dependence cannot be generalized to other materials in a straightforward manner, but it should nevertheless be further investigated as no data is available on the quantitative dependence in question, nor on its nature. Yet, this kind of information is required to improve the predictive capability of numerical models.

Another interesting observation is that irrespective of the strain-rate, a value of 0.9 is not reached and the *measured values are significantly lower.* The reasons for this result are not clear at present and cannot be further elaborated upon.

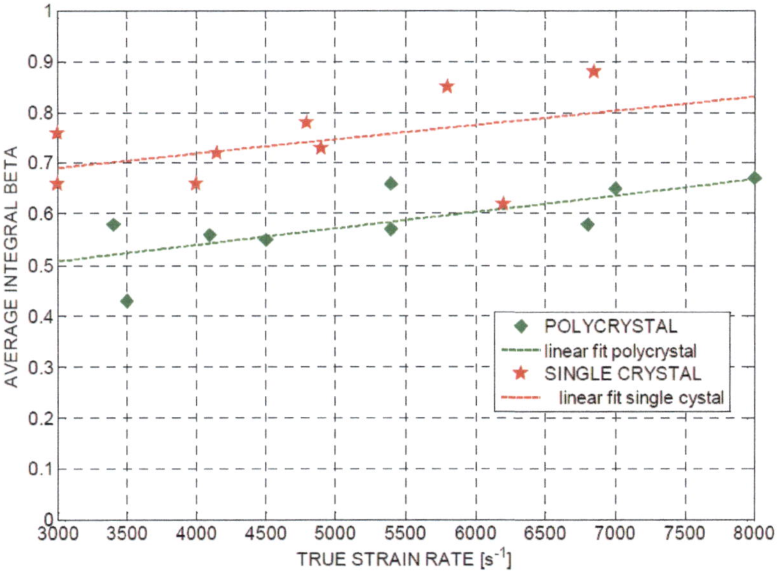

Fig. 14. Measurement of β_{int} as a function of the strain-rate for pure single and polycrystalline copper.

Finally, one can note that, while the two β_{int} lines are quite parallel, that of the single crystal is consistently superior to that of the polycrystal. This observation shows that the single crystalline copper stores less energy in its microstructure than its polycrystalline counterpart. Consequently, specimens of both types deformed in similar conditions were subjected to a transmission electron microscopy characterization in order to pinpoint differences in their dislocation substructure (and perhaps the presence of twins) that could be responsible for the different levels of energy storage.

The microstructural characterization showed absolutely no difference in the microstructures of the deformed single and polycrystalline materials, except of course for the presence of *grain boundaries* in the polycrystalline material. It was subsequently proposed that the stored energy in a polycrystal is shared between the grain boundaries and the bulk of the grains. Given that the baseline storage is provided by the results of the single crystals, one can suggest that the difference in energy levels is actually the grain boundary contribution to the overall energy storage. Since this is the only reported result of that kind which concerns only one material, one cannot generalize this observation, which could serve as a benchmark for crystal plasticity simulations. It is nevertheless an interesting observation that could be extended to grain-size considerations (large vs. nanograined materials), int order to maximize or minimize the amount of stored energy of cold work.

4 Concluding remarks

In this chapter, we have reviewed and discussed three selected topics that were presented in the course. Those topics and associated techniques were selected to address issues that complement the other topics mentioned during the course, such as the review of the Hopkinson bar, numerical approaches and constitutive models.

The first subject, dynamic fracture, has developed rapidly in the last decade (Cox et al., 2005), and has reached a state of experimental maturity (Jiang and Vecchio, 2009). What is missing today is a comprehensive database for various materials. Indeed, there are still very few reliable experiments carried out on various materials, so that a reliable database can be established and trends elucidated. We have shown here a very recent extension of the one-point impact technique to dynamic tensile (flexural) strength of brittle materials. This technique opens new directions to characterize a basic mechanical property of which little is really known. The technique is simple and allows for testing of a large sample size. One can thus hope for the future investigation of the dynamic failure statistical rules that will, among other things, establish whether weakest link assumptions really describe the dynamic strength of brittle solids, as there are second thoughts about that in the scientific community.

Pressure sensitivity is a central point of the mechanical properties of materials. Much work has been done, starting with Bridgman (1945), on the influence of the hydrostatic pressure on the static properties of materials. By contrast, much less is known about the dynamic properties, and we have described a couple of techniques which cover both quasi-static and dynamic testing in a seamless manner. The techniques have been discussed together with their limitations, including the fact that the level of triaxiality that can be reached in those experiments is still modest. Future work will have to investigate much higher triaxialities, in the dynamic regime, but the techniques still have to be developed.

Last but not least, we discussed the issue of thermomechanical couplings in solids. This subject has a tremendous importance for the establishment of reliable constitutive models and the assessment of the dynamic failure properties of materials, mostly when shear localization can develop. It has been shown that within certain assumptions, non-contact radiometric techniques can provide a valuable insight into the nature of the coupling in question. One of the main (recent) results is the fact that the Taylor-Quinney factor is not a constant, as it depends on both the strain and strain-rate. Moreover, the traditional value of 0.9 can be sometimes over-estimate the strength of the coupling, leading to much larger temperatures than those actually developing in the solids. A first step towards understanding the physics of energy storage in the dynamic regime was the identification of the role played by grain boundaries. Yet, this result is still the only one of its kind and much more work is needed to investigate this issue which will in turn relate to the grain size and associated mechanical properties. However, at this stage one may conclude that this technique is mature and should be employed without hesitation in thermomechanical high-rate studies.

References

Barker, L.M., Hollenbach, R.E., 1974. Shock-wave study of alpha reversible epsilon phase transition in iron. J. Appl. Phys. 45, 4872 -4881.

Belenky, A., Bar-On, I., Rittel, D., 2010. Static and dynamic fracture of transparent nanograined alumina. J. Mech. Phys. Solids 58, 484-501.

Belenky, A., Rittel, D., 2011. A simple methodology to measure the dynamic flexural strength of brittle materials. Exp. Mech. 31, 1325-1334.

Belenky, A., Rittel, D., 2012a. Static and dynamic flexural strength of 99.5% alumina: Relation to porosity. Mech. Matls. 48, 43-55.

Belenky, A., Rittel, D., 2012b. Static and dynamic flexural strength of 99.5% alumina: Relation to surface roughness Mech. Matls. 54, 91-99.

Bendersky, D., 1953. A special thermocouple for measuring transient temperatures. Mechanical Engng., 117-121.

Bever, M., Holt, D., Titchener, A., 1973. The Stored Energy of Cold Work. Pergamon Press, London.

Bloomquist, D.D., Sheffield, S.A., 1980. Thermocouple temperature measurements in shock compressed solids. J. Appl. Phys. 51, 5260-5266.

Boley, B.A., Weiner, J.H., 1960. Theory of Thermal Stresses. J. Wiley and Sons, New York

NY.Bridgman, P.W., 1945. Flow and Fracture. AIME Trans. 162, 569-583.

Brockenbrough, J.R., Suresh, S., Duffy, J., 1988. An analysis of dynamic fracture in microcraccking brittle solids. Phil. Mag.A 58, 619-634.

Chen, W., Ravichandran, G., 1997. Dynamic compressive failure of glass ceramic under lateral confinement. J. Mech. Phys. Solids 45, 1303-1328.

Cox, B.N., Gao, H., Gross, D., Rittel, D., 2005. Modern topics and challenges in dynamic fracture. J. Mech. Phys. Solids 53, 565-596.

Delvare, F., Hanus, J.L., Bailly, P., 2010. A non-equilibrium approach to processing Hopkinson Bar bending test data: Application to quasi-brittle materials. Int. J. Impact Engng. 37 Pages: , 1170-1179.

Dorogoy, A., Karp, B., Rittel, D., 2011. A shear compression disk specimen with controlled stress triaxiality under quasi-static loading. Exp. Mech., 1545-1557.

Duffy, J., Chi, Y., 1992. On the measurement of local strain and temperature during the formation of adiabatic shear bands. Matls. Sci. Eng. A 157, 195-210.

Freund, L.B., 1990. Dynamic Fracture Mechanics, Cambridge.Giovanola, J.H., 1986. Investigation and application of the one-point bend impact test", Fracture Mechanics, in: ASTM (Ed.), pp. 307-328.

Hanina, E., Rittel, D., Rosenberg, Z., 2007. Pressure sensitivity of adiabatic shear banding in metals. Applied Physics Letters 90, 021915.

Hartley, K., Duffy, J., Hawley, R., 1985. The torsional Kolsky (split-Hopkinson) bar, Mechanical Testing, 9 ed. ASM Metals Park, OH, pp. 218–230.

Hartley, K.A., Duffy, J., Hawley, R.H., 1987. Measurement of the temperature profile during shear band formation in mild steels deforming at high-strain rates. J. Mech. Phys. Solids 35, 283-301.

Hodowany, J., Ravichandran, G., Rosakis, A.J., Rosakis, P., 2000. Partition of plastic work into heat and stored energy in metals. Exp. Mech. 40, 113-123.

Jiang, F., Vecchio, K.S., 2009. Hopkinson Bar Loaded Fracture Experimental Technique: A Critical Review of Dynamic Fracture Toughness Tests. Applied Mech. Reviews 62, 060802-060841.

Jovic, C., Wagner, D., Herve, P., Lazzarotto, G.G.L., 2006. Mechanical behaviour and temperature measurement during dynamic deformationon split hopkinson bar of 304L stainless steel and 5754 aluminium alloy. J. Physique IV 134.

Kachanov, L.M., 1974. Foundations of the Theory of Plasticity. Mir, Moscow.

Kalthoff, J.F., Winkler, S., Beinert, J., 1977. The influence of dynamic effects in impact testing. Int. J. Fracture 13, 528-531.

Karp, B., Dorogoy, A., Rittel, D., 2013. A shear compression disk specimen with controlled stress triaxiality under dynamic loading. Exp. Mech. 53, 243-253.

Kolsky, H., 1949. An investigation of the mechanical properties of materials at very high rates of loading. Proc. Phys. Soc. London 62-B, 676-700.

Lew, A., Caspersen, K., Carter, E.A., Ortiz, M., 2006. Quantum mechanics based multiscale modeling of stress-induced phase transformations in iron. J. Mech. Phys. Solids 54, 1276-1303.

Maigre, H., Rittel, D., 1993. Mixed-mode quantification for dynamic fracture initiation: application to the compact compression specimen. International Journal of Solids and Structures 30, 3233-3244.

Marchand, A., Duffy, J., 1988. An experimental study of the formation process of adiabatic shear bands in a structural steel. J. Mech. Phys. Solids 36, 251-283.

Mason, J.J., Rosakis, A.J., Ravichandran, G., 1994. On the strain and strain rate dependence of the fraction of plastic work converted into heat: an experimental study using high speed infrared detectors and the Kolsky bar. Mechanics of Materials 17, 135-145.

Rabin, Y., Rittel, D., 1999. A model for the time response of solid-embedded thermocouples. Experimental Mechanics 39, 132-136.

Rabin, Y., Rittel, D., 2000. Infrared temperature sensing of mechanically loaded specimens: thermal analysis. Experimental Mechanics 40, 197-202.

Regev, A., Rittel, D., 2008. Simultaneous transient temperature sensing of impacted polymers using infrared detectors and thermocouples. Exp. Mech. 48, 675-682.

Rittel, D., 1998. Transient temperature measurement using embedded thermocouples. Experimental Mechanics 38, 73-79.

Rittel, D., 1999. The conversion of plastic work to heat during high strain rate deformation of glassy polymers. Mechanics of Materials 31, 131-139.

Rittel, D., Bhattacharyya, A., Poon, B., Zhao, J., Ravichandran, G., 2007. Thermomechanical characterization of pure polycrystalline tantalum. Matls. Sc. and Engng. A A447, 65-70.

Rittel, D., Brill, A., 2008. Dynamic flow and failure of confined polymethylmethacrylate. J. Mech. Phys. Solids in press.

Rittel, D., Dorogoy, A., 2008. A methodology to assess the rate and pressure sensitivity of polymers over a wide range of strain rates. J. Mech. Phys. Solids 56, 3191-3205.

Rittel, D., Hanina, E., Ravichandran, G., 2008. A note on the direct determination of the confining pressure of cylindrical specimens. Exp. Mech. in press.

Rittel, D., Kidane, A.A., Alkhader, M., Venkert, A., Landau, P., Ravichandran, G., 2012. On the dynamically stored energy of cold work in pure single and polycrystalline copper. Acta Mater. 60 3719-3728.

Rittel, D., Lee, S., Ravichandran, G., 2002. A shear compression specimen for large strain testing. Experimental Mechanics 42, 58-64.

Rittel, D., Ravichandran, G., Venkert, A., 2006. The mechanical response of pure iron at high strain rates under dominant shear. Matls. Sc. and Engng. A A432, 191-201.

Rittel, D., Silva, M.L., Poon, B., Ravichandran, G., 2009. Thermomechanical behavior of single crystalline tantalum in the static and dynamic regime. Mechanics of Materials 41, 1323-1329.

Rittel, D., Wang, Z.G., 2008. Thermo-mechanical aspects of adiabatic shear failure of AM50 and Ti6Al4V alloysMechanics of Materials 40, 629-635.

Rittel, D., Weisbrod, G., 2001. Dynamic fracture of tungsten base heavy alloys. Int. J. Fracture 212, 87-98.

Rokach, I.V., 1998. Modal approach for processing one- and three-point bend test data for DSIF-time diagram determination. Part I - Theory Fat. & Fract of Engng. matls. and Structures 21, 1007-1014

Rokach, I.V., Labedzki, P., 2009. Determination of the dynamic stress intensity factor for the four-point bend impact test Int. J. Fracture 160, 93-100.

Taylor, G.I., Quinney, H., 1934. The latent energy remaining in a metal after cold working. Proc. Royal Soc. London 143, 307-326.

Weisbrod, G., Rittel, D., 2000. A method for dynamic fracture toughness determination using short beams. Int. J. Fracture 104, 91-104.

Zaera, R., Rodriguez-Martinez, J.A., Rittel, D., 2013. On the Taylor-Quinney coefficient in dynamically phase transforming materials. Application to 304 stainless steel. Int. J. Plasticity 40, 185-201.

Zener, C., Hollomon, J.H., 1944. Effect of strain rate upon plastic flow of steel J. Applied Phys. 15, 22-32.

Dynamic Behavior of Materials. Constitutive Relations and Applications

A. Rusinek* and T. Jankowiak†

* Laboratory of mechanics, Biomechanics, Polymers, Structures, National engineering School of Metz, 1 route d'Ars Laquenexy, 57000 Metz, France

† Institute of Structural Engineering, Poznan University of Technology, Piotrowo 5, 60-965 Pozna, Poland

Abstract Abstract. In this chapter a particular attention has been directed on the dynamic behavior of materials and structures subjected to dynamic loading. Based on experimental observations, it is clear that the homogeneous material behavior of metals has several non linearities related for example to the strain rate and temperature sensitivity. Therefore, the material description and more precisely the constitutive relation used during numerical simulations for example must include all macroscopic observations. It has to be noticed that constitutive relations described in this chapter are defined in a macroscopic scale. Considering some examples, it is clear that the constitutive relation that was used is the key point to simulate a global problem allowing to avoid as frequently used some numerical tricks to obtain an agreement between experiments and numerical predictions. Following the concept described in this part, the reader will be available to propose new models to fit precisely their own materials for specific applications.

Introduction It is frequently observed that the structure is loaded dynamically by unique loading during special events as crashes in case of vehicles, impacts or perforations in case of military structures and blast in case of public and military structures. These kinds of loadings create in the structures large strains ε and strain rates $\dot{\varepsilon}$. It generates also the high increase of the temperature T. Many experiments show that behavior of materials is dependent on:

1. Strain sensitivity, ε
2. Strain rate sensitivity, $\dot{\varepsilon}$
3. Temperature sensitivity, T

In addition the loading process can be complex following several loading paths. For this reason, the material behavior $\sigma(\varepsilon, \dot{\varepsilon}, T)$ used to design a structure must be studied in a precise way to include all above effects.

However sometimes some additional effects must be considered to understand the macroscopic description of the material as for example the process of phase transformation, the pre-plastic deformation or the industrial process induced to the considered material. It has to be noticed that the kinetic of phase transformation is also depending on temperature, strain level, strain rate and hydrostatic pressure. Based on it, it is observed that the analytical description of the material is relatively complex and need a precise experimental study before modeling. As it is discussed later, experiments have to be done in a rigorous way to avoid artificial effects. Using the following case corresponding to a vehicle, Fig. 1-a. It is observed that several materials can be used and the strain rate applied will vary from quasi-static to dynamic loading $\dot{\varepsilon} \approx 10^3\ s^{-1}$. Moreover, on the impact side during a crash test large deformation will be observed, Fig. 1-b.

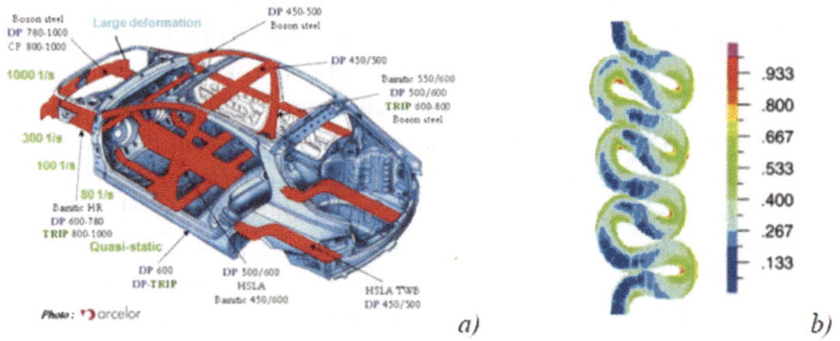

Fig. 1. Local observations during a crash test; a- Strain rate along the structure (Newletter Arcelor Auto, 2003); b- Plastic strain level induced to the crash box during buckling (Bathe et al., 1999)

In this case, the local temperature is relatively high and adiabatic condition can be assumed. However along the structure the strain rate is not constant and fare from the impact point, the strain rate is going to quasi-static loading, $\dot{\varepsilon} \rightarrow 10^{-3}\ s^{-1}$. In this case, as the process is slow the deformation is assumed as isothermal. Now regarding experiments for the range of strain rates considered previously, $10^{-3} < \dot{\varepsilon} < 10^3\ s^{-1}$, it is observed that the behavior is non linear in term of strain rate sensitivity, Fig. 2. In general for metals, the strain rate transition between isothermal and adiabatic condition is equal to $\dot{\varepsilon} \approx 10\ s^{-1}$(Rusinek and Klepaczko, 2001, Berbenni et al., 2004).

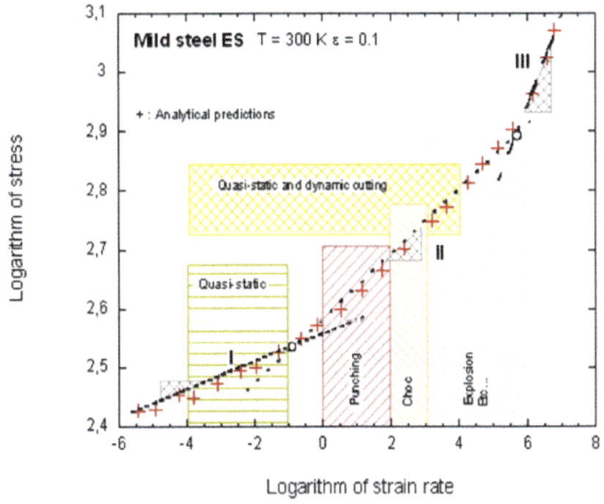

Fig. 2. Strain rate sensitivity definition and strain rate linked to loading process

For this reason to define properly the thermovsicoplastic behavior of materials, it is essential to include in the analytical description the non linear strain rate sensitivity. This observation according to the strain rate sensitivity is the same concerning the temperature sensitivity. Generally, the coupling between temperature and strain rate is defined using an Arrhenius equation. Using these basics results, it is clear that before simulating a complete process, it is necessary to take into account all materials constituting the structure and all phenomenon observed during experiments.

Constitutive relation description

Several approaches can be used to define the thermoviscoplastic behavior of materials, the first one is based on physical aspects and the second one on a macroscopic description trying to fit experimental measurements. It has to be noticed that the physical description is inducing an increase of the constants. A mix of these two approaches can be developed, as it will be discussed in this chapter. In a general way, the constitutive relations take into account microstructure description including for example the density of dislocations, grains size, density of twinning and others.

Therefore, before developing analytical description, it is necessary to know the needs and the microstructure description of the materials used to design the structure for example. The second parameter related to the microstructure is a key point since it will be linked to the analytical formulation of the constitutive relation describing the hardening during plastic deformation. In a schematic description to describe the strain rate sensitivity of the materials, for a FCC microstructure the material will be defined based on a multiplicative approach and for a BCC microstructure with an additive formulation. Thus, the material will behave as follow for a strain rate increase, Fig. 3.

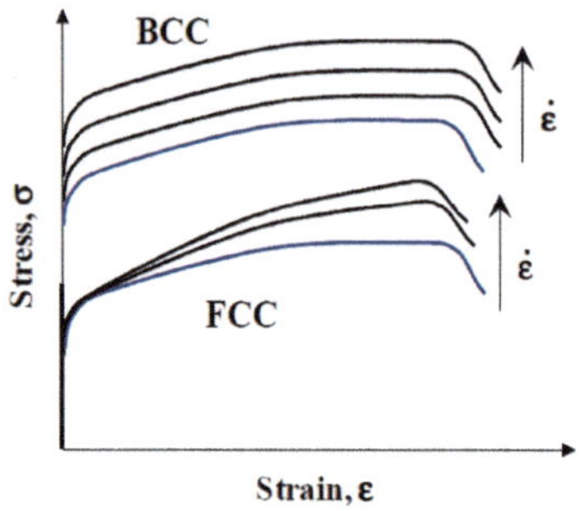

Fig. 3. Strain rate effect on the material behavior depending on the microstructure

It is frequently observed that the material behavior is changing with strain level, temperature and strain rate. Therefore, before choosing a constitutive relation $\sigma(\varepsilon, \dot{\varepsilon}, T)$or develoing new one, all the ranges of these parameters must be defined in term of :

1. Strain rate level, $\dot{\varepsilon}$
2. Strain level, ε
3. Temperature range, T

The two first ones stabilize the material behavior inducing a delay to trig instability and the last one is trigging quickly the loose of homogeneity due to thermal softening. Therefore if a material is used under dynamic loading a competition exists between hardening n, strain rate sensitivity m and temperature sensitivity ν. In a simple way taking into account all parameters described previously, the material behavior can be described using, Eq. 1. (Hollomon, 1945).

$$\sigma(\varepsilon, \dot{\varepsilon}, T) = K\varepsilon^n \dot{\varepsilon}^m T^{-\nu} \tag{1}$$

Where K is a constant of the material.

In the previous simple approach, all parameters characterizing the thermoviscoplastic material behavior are assumed as constants. They are defined as follows, Eq. 2.

$$m = \left.\frac{\partial \log \sigma}{\partial \log \dot{\varepsilon}}\right|_{\varepsilon,T} ; n = \left.\frac{\partial \log \sigma}{\partial \log \varepsilon}\right|_{\dot{\varepsilon},T} ; \nu = \left.\frac{\partial \log \sigma}{\partial \log T}\right|_{\varepsilon,\dot{\varepsilon}} \tag{2}$$

However based on experiments, it is observed that the hardening parameter due to thermal softening is depending on strain rate $\dot{\varepsilon}$ and temperature T, Fig. 4. Moreover, the strain rate sensitivity m is also depending on the temperature as reported, Fig. 5. It is observed an increase of the strain rate sensitivity at low temperatures. For these reasons, it is clear that the material behavior cannot be defined correctly for a large range of strain, strain rate and temperature if the parameters described previously, Eq. 2., are assumed as constants.

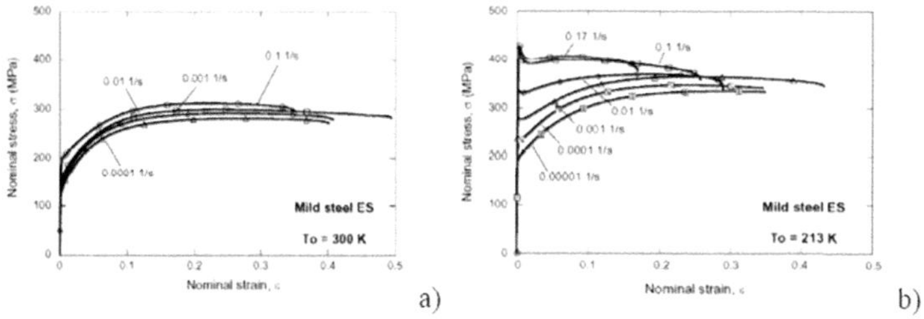

Fig. 4. Effect of the initial temperature on the strain rate sensitivity and thermal softening, a- Initial temperature T=300 K; b- Initial temperature T=213 K

To have a better definition, the strain rate sensitivity m must depend on the temperature T. Experimental results concerning the strain rate sensitivity as a function of the initial temperature are reported, Fig. 5. It is observed that a constant value frequently assumed equal to $m = 0.02$ for metals is correct at room or at high temperature. It is not the case when the material is subjected at high strain rates for low temperatures. Using these experiments, it is clear that the strain rate sensitivity must be defined as a function of the temperature $m(T)$. In conclusion to define precisely material behavior under dynamic loading a coupling between the strain rate $\dot{\varepsilon}$ and the temperature T is necessary.

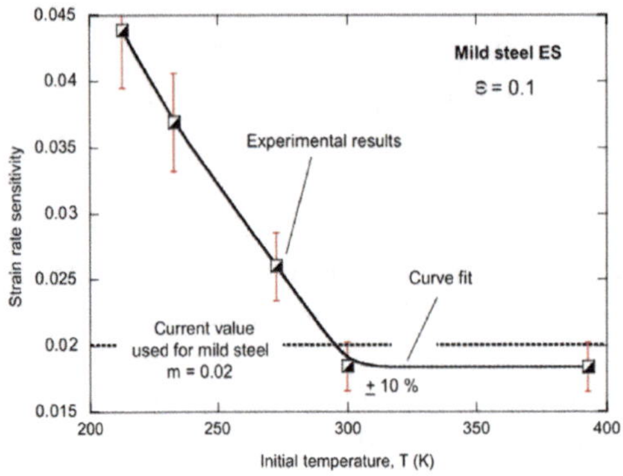

Fig. 5. Coupling between strain rate sensitivity and temperature, results defined for mild steel (Rusinek et al., 2007)

This phenomenon observed previously coupling strain rate $\dot{\varepsilon}$and temperature T is called the reciprocity. Thus, the definition of a material at room temperature and high strain rate is equivalent to the material behavior at low temperature and low strain rate in term of stress level σ if the material is without phase transformation which will induce an extra hardening increase (Rodriguez-Martinez et al., 2009). A schematic description in term of reciprocity is reported, Fig. 6.

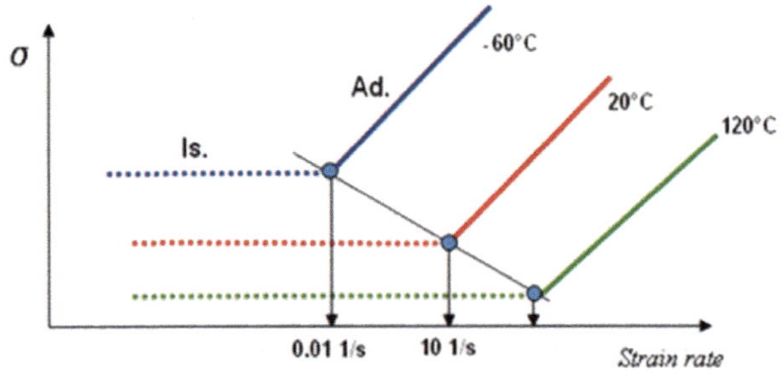

Fig. 6. Reciprocity between strain rate and temperature, definition of the transition strain rate between isothermal and adiabatic condition

Using the previous descriptions, Fig. 6., two ranges of strain rates sensitivity are observed. The first one is more or less constant or increasing slowly, $m \to 0$. The second one is characterized by a quick increase of the stress level with the strain rate. The strain rate characterizing, the change is defined as the strain rate transition $\dot{\varepsilon}_0$between isothermal and adiabatic conditions. Therefore, under dynamic loading the constitutive relation must be coupled with the heat equation assuming no conductivity ($k = 0$) to define the actual temperature T during plastic deformation, Eq. 3.

$$T = T_0 + \frac{\beta}{\rho C_p} \int_{\varepsilon_e}^{\varepsilon_{failure}} \sigma \, d\varepsilon \tag{3}$$

Where T_0 is the initial temperature, β is the Quinney-Taylor coefficient and C_p is the specific heat for a constant pressure. The Quinney-Taylor coefficient is generally ssumed equal to $\beta = 0.9$for metals at large deformation. However for small strain level the value is increasing as it was reported in (Rusinek and Klepaczko, 2009) to reach a plateau, a constant value. This effect, was also observed for different materials [Macdougall, 2000, Rittel, 2000., Arruda et al., 1995., Nasraoui et al., 2012].

As first conclusion and considering all previous experimental results, it is clear that the hardening n of the material is dependent on plastic strain ε, strain rate $\dot{\varepsilon}$ and temperature T. If the material is tested assuming a constant strain rate, the parameter n related to the hardening material can be defined as follow, Eq. 4. It is observed due to the negative strain rate sensitivity of materials that the hardening decreases with plastic deformation.

$$n_{adia} = n_{iso} - \frac{\nu}{\rho C_p} \sigma \tag{4}$$

Where n_{iso} is the hardening parameter under isothermal conditions

Assuming a simple approach, Eq. 1., to define the hardening of materials following a parabolic increase with plastic strain ε, the values used are generally close to, Tab. 1., Eq. 2.

n(-)	m(-)	ν(-)
0.2	0.02	-0.2

Tab. 1. Usual values used to define metals behavior based on Eq. 2

In addition to materials behavior, several fundamental works have been published mainly concerning instability to demonstrate the key point of the previous parameter (n, m, ν) as for example in (Fressengeas and Molinari, 1987). Thus as conclusion of this first part, it is demonstrated that several tests covering a large field of strain ε, strain rate $\dot{\varepsilon}$ and temperature T are necessary to estimate precisely the material behavior before modeling or when it is used for an analytical work. Without comparison with precise experiments, no analytical or numerical validation can be done.

Description of constitutive relations, from material science to phenomenological description

It is frequently reported in the international literature that different constitutive relations are used to define the material behavior at high strain rates and for a large field of temperatures. Some approaches as it was discussed previously takes into account some microstructures considerations, Fig. 3., and some of them, try to mimic experimental observations. Considering, the first approach, it allows to fix the bases of the physic and to have a better understanding of the mechanisms taking place during plastic deformation ε. Thus, in a general way, the thermoviscoplastic material behavior can be defined using this kind of constitutive relation, Eq. 5.

$$F\left(\sigma, \dot{\varepsilon}, p, T, S_j\right) = 0 \tag{5}$$

In this case of description, the microstructure is linked to some internal variables S_j related to the process of plastic deformation ε. The consequence is that the definition of the constants of the model are linked to some physical measurements. Generally, the microstructure evolution is defined using the density of total dislocations ρ. Based on the Taylor relation (1934), the description of the hardening law is defined by Eq. 6.

$$\sigma = \alpha \mu b \sqrt{\rho} \tag{6}$$

Where α is an interaction term, μ is the shear modulus and b is the Burger's vector.

Using the description of Gilman (1965) the density of dislocation is changing with plastic deformation ε using the following description, Eq. 7.

$$d\rho = M \, d\varepsilon \tag{7}$$

Where M is the coefficient of multiplication of dislocations.

Assuming that just a part of the total density of dislocation ρ can move during plastic deformation, the microstructure change can be described using the following relation, Eq. 8.

$$\rho_m = f_0 \left(\rho_0 + M\varepsilon\right) \exp\left(-AM\varepsilon\right) \tag{8}$$

Where A is corresponding to the process of dislocation annihilation (same Burger's vector), ρ_0 is the initial density of dislocation and f_0 is the initial fraction density of mobile dislocation in the material

Thus, the density of mobile dislocation ρ_m reach a maximum follow by an decrease since the free path of dislocation Λ for large plastic deformation is close to zero. Therefore, no dislocation can move for large plastic deformation.

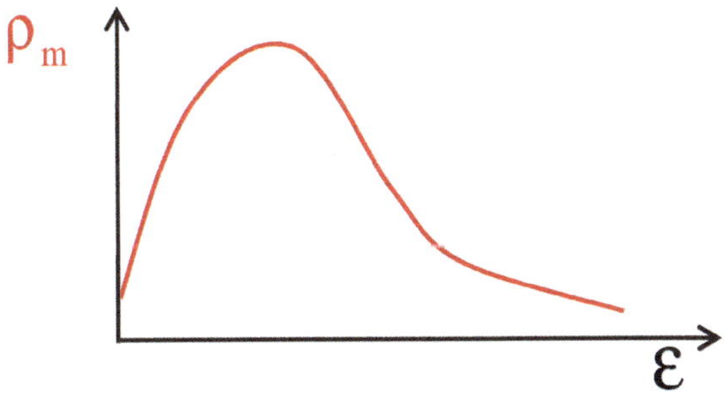

Fig. 7. Evolution of the density of mobile dislocation with plastic deformation, definition of the process of saturation for large plastic deformation

In general based on the previous description it is clear that the microstructure evolution of the material can be defined knowing two quantities, the total density of dislocation ρ and the density of mobile dislocation ρ_m. These two quantities are linked as follow, Eq. 9.

$$\rho(\varepsilon) = \rho_i(\varepsilon) + \rho_m(\varepsilon) \tag{9}$$

Where ρ_i is the density of immobile dislocation.

To have a complete description, the kinetic of the microstructure change is necessary, it means the velocity of dislocations. As Proposed by Conrad (1964,1970), the following relation can be used, Eq. 10.

$$V = V_0 \exp\left[-\frac{\Delta G(\tau^*, T)}{kT}\right] \tag{10}$$

Where ΔG is the free energy depending on the Peierls-Nabarro stress τ^* (Hartley and Duffy, 1984, Klepaczko, 1975), k is the Boltzman's constant and V_0 is called the pre-exponential term.

The free energy ΔG is defined based on the description of Ono (1968) or Kock et al. (1975), Eq.11.

$$\Delta G(\tau^*, T) = G(T)\left[1 - \left(\frac{\tau^*}{\tau_m^*(T)}\right)^p\right]^q \tag{11}$$

Where p and q are the obstacle shape that dislocations have to pass (short-range barrier), τ_m^* is the maximum stress level of the Peierls-Nabarro quantity (Klepaczko, 1975).

The free energy ΔG can be defined using the volume thermally activated which is also proportional to the strain rate sensitivity m. Concerning the strain rate $\dot{\gamma}$ induced, when a density of mobile dislocation is passing an obstacle, it is defined by Eq. 12.

$$\dot{\gamma} = \dot{\gamma}_0 \exp\left[-\frac{\Delta G\left(\tau^*, T\right)}{kT}\right] \tag{12}$$

Where $\dot{\gamma}_0$ is called the pre-exponential term linked to the average velocity of the mobile dislocations.

Using the previous equations, Eq. 11-12., the effective stress τ^* which defines the coupling between the strain rate and the temperature can be defined, Eq. 13. This stress is related to the strain rate sensitivity of the material during dynamic loading.

$$\tau^* = \tau_p^*\left(T\right) \cdot \left\{1 - \left[\frac{kT}{G\left(T\right)} \ln\left(\frac{\dot{\gamma}_0}{\dot{\gamma}}\right)\right]^{1/q}\right\}^{1/p} \tag{13}$$

Finally, the macroscopic stress level σ taking into account the microstructure changes is a contribution of two quantities, the internal stress σ_μ and the effective stress σ^*. The first one is defining the hardening due to the resistance of the dislocation motion, Eq. 6 and the second one the temperature and strain rate sensitivity, Eq. 13.

$$\tau = \tau_\mu + \tau^* \tag{14}$$

A description is reported on the following picture, Fig. 8. At low strain rate, the effective stress is close to zero, Fig. 8-b. In this specific case, the material behavior is defined using Eq. 6. When the strain rate increases, the stress level is larger, keeping or not the same hardening, Fig. 8-a. This observation is due to the positive strain rate sensitivity of the material. Moreover, when the strain rate increases, a non linear behavior is observed in term of strain rate sensitivity. This effect is defined by Eq. 13 and induce a stress increase, Fig. 8-b. In fact, this contribution is more important close to the strain rate transition or for a strain rate larger.

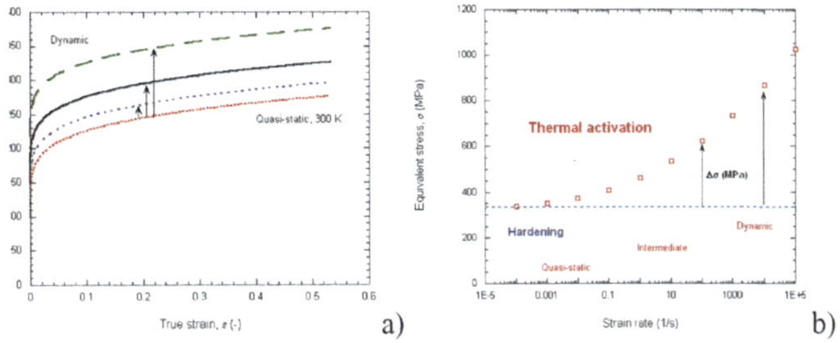

Fig. 8 a- Effect of the strain rate on the macroscopic stress level; b- Effect of the strain rate on the effective strain to define the non linear behavior

This description is mainly well defined for BCC microstructure as it was discussed previously with an additive decomposition of the macroscopic stress σ. It is observed based on the previous description that this kind of model can be useful to understand in a better way the changes of the microstructure including the strain level, the strain rate and temperature. However for computing, this kind of approach can be heavy since for each step of plastic deformation several differential equations must be solved. In this case the time computing is increased as for example with Eq. 15.

$$\begin{aligned} \frac{d\rho}{d\gamma} &= f_1\left(\dot{\gamma}, T, \rho, \rho_m\right) \\ \frac{d\rho_m}{d\gamma} &= f_2\left(\dot{\gamma}, T, \rho, \rho_m\right) \end{aligned} \tag{15}$$

Therefore, to solve real problem as for example a crash test, other constitutive relations are frequently used as it is discussed in the next part of this chapter. The main goal, is to reduce the time computing and the cost in term of experiments. In the second case, the number of constants is drastically reduced.

Thermoviscoplastic modeling, phenomenological approach

As discussed previously and as reported, it is possible to take into account the microstructure changes during dynamic or quasi-static loading. In this case, the changes are defined using some internal variables as the density of mobile dislocations and the total density ρ. However as it was concluded, the main problem is related to the time computing. For this reason new constitutive relations are frequently developed based on the previous physical considerations. This kind of approache is called semi-phenomenological. As before, the equivalent stress σ is defined as a summation of two quantities, the internal stress σ_μ and the effective stress σ^*, Eq. 16.

$$\sigma = \sigma_\mu + \sigma^* \tag{16}$$

The first component is describing the hardening of the material during plastic deformation and the second one the strain rate sensitivity. Generally σ_μ can be defined with a power law, Eq. 17.

$$\sigma_\mu = K\varepsilon^n \tag{17}$$

In this case no dependencies are defined with the strain rate and the temperature. This model can be used to describe for example the material behavior of metals under quasi-static loading corresponding to a strain rate varying from $10^{-4} \leq \dot{\varepsilon} \leq 10^{-3}\ s^{-1}$. However for dynamic loading this equation cannot be used to define properly experimental observations since the thermal softening inducing a hardening decrease is not included as the strain rate sensitivity. Indeed, Eq. 17., the hardening coefficient n is constant. As a first conclusion, it is observed that to define properly dynamic process the hardening coefficient n must be depending on the strain rate $\dot{\varepsilon}$. Moreover, the yield stress is also changing with the strain rate, Fig. 4. It is frequently reported in the literature that an increase of the strain rate induce an increase of the yield stress and vice versa. As previously the coefficient K must depend on the strain rate $\dot{\varepsilon}$. These comments concerning the strain rate are equivalent with the temperature since we have the phenomena of reciprocity discussed previously, Fig. 6. Therefore for dynamic loading, Eq. 17., must be defined as follow, Eq. 18.

$$\sigma_\mu = K(\dot{\varepsilon}, T)\,\varepsilon^{n(\dot{\varepsilon}, T)} \tag{18}$$

Concerning, the effective stress the rule will be to define the strain rate sensitivity at high strain rate. As reported previously at low strain rate, this component is not acting $\sigma^* = 0$, Fig. 8-b. In a general way and for metals, the existence of σ^* starts for a strain rate equal to $\dot{\varepsilon} \approx 10\ s^{-1}$. The previous strain rate defined the transition between isothermal and adiabatic conditions. As the effective stress σ^* is related to the strain rate, it must depend on the temperature. Thus, based on experimental observations, the equivalent stress must be defined as, Eq. 19.

$$\sigma = K(\dot{\varepsilon}, T)\,\varepsilon^{n(\dot{\varepsilon}, T)} + \sigma^*(\dot{\varepsilon}, T) \tag{19}$$

Based on these considerations, several approaches can be proposed. It is mainly depending on the application that the user want to simulate or compute. The most popular constitutive relation used in FE code is the model proposed by Johnson-Cook (1983,1985). The description is the following, Eq. 20.

$$\sigma\left(\bar{\varepsilon}^p, \dot{\bar{\varepsilon}}^p, T\right) = \left[A + B\left(\bar{\varepsilon}^p\right)^n\right]\left[1 + C\ln\left(\frac{\dot{\bar{\varepsilon}}^p}{\dot{\varepsilon}_0}\right)\right]\left[1 - \left(T^*\right)^{m^*}\right] \tag{20}$$

Where A is the yield stress, B and n are constants related to the hardening, C is the strain rate sensitivity, $\dot{\varepsilon}_0$ is the reference strain rate and m^* is the temperature sensitivity.

This model is well established for quasi-static loading or dynamic loading but it does not allow to define the complete field of strain rates between quasi-static and dynamic loading. The reason is mainly related to the formulation since the strain rate sensitivity is linear, Eq. 20. Therefore, the results are in disagreement with experimental observations, Fig. 2. In fact, the model can be used considering several ranges or domains. However, this model is popular since the number of constants is reduced and equal to five. The number of constants and the number of tests necessary to define the constants of a model is crucial for a final choose.

Considering phenomenological approach, a description was proposed in (Rodriguez-Martinez et al., 2009) to define all cases of hardening observed experimentally. The model is based on a hardening function H. However, the construction of the model is corresponding to Eq. 16 with respectively:

$$\sigma_\mu\left(\xi\right) = \sigma_y^{qs} + \frac{E\left(T\right)}{E_0}\int_{\varepsilon_e}^{\bar{\varepsilon}^p} H\left(\xi\right)d\xi \tag{21}$$

$$\sigma^*\left(\dot{\bar{\varepsilon}}^p, T\right) = \frac{E\left(T\right)}{E_0}\left\{\sigma_0^*\left[1 - D_1\left(\frac{T}{T_m}\right)\log\left(\frac{\dot{\varepsilon}_{\max}}{\dot{\bar{\varepsilon}}^p}\right)\right]^{m^*}\right\} \tag{22}$$

Where σ_y^{qs} is the yield stress under quasi-static loading, $E(T)/E_0$ is the Young's modulus ratio, Eq. 23., σ_0^* is the equivalent stress at $T = 0\ K$, D_1 is a constant of the material to define no strain rate sensitivity at low strain rate, m^* is the strain rate sensitivity, $\dot{\varepsilon}_{\max}$ is the upper limit of the model and T_m is the melting temperature.

The Young's modulus depending on temperature is defined as follow, Eq. 23. The description was proposed by (Klepaczko, 1975). A similar work was done by (Schreiber et al., 1973, Varshni,, 1970).

$$\frac{E\left(T\right)}{E_0} = \left\{1 - \frac{T}{T_m}\exp\left[\theta^*\left(1 - \frac{T_m}{T}\right)\right]\right\} \quad with \quad T > 0 \tag{23}$$

The dependency of the Young's modulus with temperature is also an important parameter for dynamic loading where temperature increase is relatively large. The fact of taking into account this coupling allows to describe precisely the process of elastic wave propagation inducing a decrease of the elastic speed wave C_0 when the local temperature is increasing, $C_0 = \sqrt{E/\rho}$. Several works were published in this field mainly to analyze the Critical Impact Velocity as reported in (Rusinek et al., 2005, Hu and Daehn, 1996).

Coming back to the material behavior definition, the hardening function H related to the internal stress σ_μ is defined using two ways, Eqs. 24.

$$\begin{aligned} \frac{\partial \sigma_\mu}{\partial \bar{\varepsilon}^p} &= H\left(\bar{\varepsilon}^p\right) = H_0^\varepsilon \exp\left[-\left(\frac{\bar{\varepsilon}^p}{\varepsilon_0}\right)^{p_\varepsilon}\right] \\ \frac{\partial \bar{\varepsilon}^p}{\partial \sigma_\mu} &= \left[H\left(\sigma_\mu\right)\right]^{-1} = \left\{H_0^{\sigma_\mu} \exp\left[-\left(\frac{\sigma_\mu}{\sigma_y^{static}}\right)^{p_{\sigma_\mu}}\right]\right\}^{-1} \end{aligned} \quad (24)$$

Using Eqs. 24, it is observed that all hardening shape can be defined including perfectly plastic behavior or linear hardening observed for example for FeMn (Allain et al., 2008). In addition to the flexibility of this analytical description to define all behaviors observed during experiments, it has to be noticed that the number of constants is reduced and equal to six.

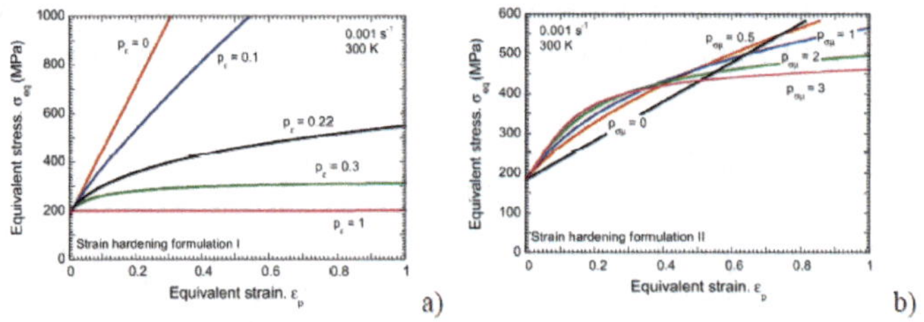

Fig. 9. Description of the hardening using Eq. 21-23 for two kinds of description, Eq. 24

Some cases are reported in (Rodriguez-Martinez et al., 2009) with all constants considered. As previously discussed, all equations Eq. 21-24 are coupled with Eq. 3. to define the thermal softening observed at high strain rate.

Even if this kind of approaches are frequently used, it exists some alternatives where the macroscopic behavior is defined taking into account micrsotructure considerations. In the next part of this chapter, a solution between physical approach and phenomenological description is discussed.

Thermoviscoplastic modeling, semi-physical approach

In the goal to describe dynamic behavior of material under dynamic loading mainly for numerical simulations, an alternative solution is to use semi-physical description. In this case, the computing time is reduced in comparison with previous physical description based on microstructure definition. For this main reason, several researchers were developing this kind of constitutive relations including strain rate and temperature sensitivities. One advantage is also the number of constants that is reduced and using a systematic identification of the constants (Klepaczko et al., 2009) the procedure is relatively easy in comparison with physical description where frequently constants are assumed and not measured systematically.

Several descriptions are based on the concept of thermal activation including in a hidden way the microstructure of the materials. Therefore it is frequent to have several formulations depending on the microstructure of the material. As for example, we can report the model proposed by Zerilli-Amstrong (1987), Eqs. 25-26.

$$\sigma\left(\bar{\varepsilon}^p, \dot{\bar{\varepsilon}}^p, T\right) = \Delta\sigma_G + B_0 \exp\left[\left(-\beta_0 + \beta_1 \ln \dot{\bar{\varepsilon}}^p\right) T\right] + K_0 \left(\bar{\varepsilon}^p\right)^n + k_\varepsilon d^{-1/2} \quad (25)$$

$$\sigma\left(\bar{\varepsilon}^p, \dot{\bar{\varepsilon}}^p, T\right) = \Delta\sigma_G + B_1 \left(\bar{\varepsilon}^p\right)^{1/2} \exp\left[\left(-\beta_0 + \beta_1 \ln \dot{\bar{\varepsilon}}^p\right) T\right] + k_\varepsilon d^{-1/2} \quad (26)$$

Where β_i, K_0, B_i, k_ε are the constants of the material and d is the average grains size of the microstructure.

In this previous description, the main parameter related to the microstructure is the grains size d and using two formulas it is possible to dissociate a BCC or FCC material. The last contribution of Eqs. 25-26. is the Hall-Petch effect. Based on Hall-petch considerations, it is possible to describe the work of Khan et al. (2006) where to model the viscoplastic response of a material including a bilinear Hall-Petch relation, the macroscopic relation is defined as, Eq. 27.

$$\sigma\left(\bar{\varepsilon}^p, \dot{\bar{\varepsilon}}^p, T\right) = \left[\sigma_0 + \frac{k}{\sqrt{d}} + B\left(1 - \frac{\ln \dot{\bar{\varepsilon}}^p}{\ln D_0^p}\right)^{n_1} \left(\bar{\varepsilon}^p\right)^{n_0}\right] \left(\frac{\dot{\bar{\varepsilon}}^p}{\dot{\varepsilon}^*}\right)^C \left(\frac{T_m - T}{T_m - T_r}\right)^m \quad (27)$$

Where T_r is the reference temperature, D_0^p is a upper limit defined by the user, $\dot{\varepsilon}^*$ is the reference strain rate, σ_0 is the friction stress needed to move individual dislocations and B, n_1, n_0, C, m are the constants of the materials.

Using the modified constitutive relation initially proposed in (Khan et al., 2004), material with nanostructure can be defined. The difference between Eq. 27. and the original description is based on the first term where

$\sigma_0 + k\, d^{1/2} = A$ where A is a constant of the material. A comparisons between this description and experiments are reported in (Khan et al., 2006).

One advantage is also the number of constants which is equal to seven since the parameter d is measured. Of course, several constitutive relations were proposed later. Several of them, are described in this part of the text. Thus, an alternative way to take into account the microstructure changes is to introduce an internal variable δ as in (Molinari and Ravichandran, 2005, Durrenberger et al, 2008) related to the density of dislocation ρ, size of dislocation or others. In addition the average grain size d of the structure is considered. The explicit form of the constitutive relation is the following, Eq. 28.

$$\sigma\left(\bar{\varepsilon}^p, \dot{\bar{\varepsilon}}^p, T\right) = \sigma_0(d)\left(\frac{1}{\delta}\right) + \sigma_{ref}\left(\frac{T_0}{T}\right)^{\nu^*}\left\langle a - b\ln\left(\frac{\dot{\varepsilon}_0}{\dot{\bar{\varepsilon}}^p}\right)\right\rangle^{1/m} \tag{28}$$

Where $\sigma_0\,(d)$characterizes the initial value of the resistance stress depending on the grain diameter$d$, σ_{ref} is a reference stress, a, b are the constants of the material, $m$is the strain rate sensitivity and ν^* is the temperature sensitivity

The internal variable δ is defined using a function of refinement δ_r and one for the process of saturation δ_S of the stress for large deformation. A complete description of this model is reported in (Molinari and Ravichandran, 2005) where the model was used to describe OFHC copper behavior at low and high strain rates. The parameter δ is defined as follow assuming a constant strain rate, Eq. 29. This description is similar to the one use to define the evolution of dislocation density (Mecking and Kocks, 1981).

$$\delta = \frac{\delta_S}{1 - (1 - \delta_S)\exp\left(-\delta_r \bar{\varepsilon}^p\right)} \tag{29}$$

with

$$\begin{aligned} \delta_r &= \delta_{r0}\left[1 + \alpha_r\left(\frac{\dot{\bar{\varepsilon}}^p}{\dot{\varepsilon}_{r0}}\right)^{\xi_r}\left(\frac{T}{T_0}\right)^{-\nu_r}\right] \\ \delta_s &= \delta_{s0}\left[1 - \alpha_s\left(\frac{\dot{\bar{\varepsilon}}^p}{\dot{\varepsilon}_{s0}}\right)^{\xi_s}\left(\frac{T}{T_0}\right)^{-\nu_s}\right] \end{aligned} \tag{30}$$

Where $\delta_{r0}, \delta_{s0}, \alpha_r, \alpha_s, \dot{\varepsilon}_{r0}, \dot{\varepsilon}_{s0}, \nu_S, \nu_R$ are the constants of the material allowing to define the microstructure change, ξ_i is the strain rate sensitivity and ν_i is the temperature sensitivity.

It is interesting to observe that the reciprocity between the strain rate and the temperature is also defined using the Arrhenius equation. Moreover as the microstructure change can be considered, the model may be used to take into a process of pre-plastic deformation as in (Durrenberger et al., 2008). The previous model, Eqs. 28-30., was used with success to model different kind of materials used in automotive industry as TRIP steel, DP steel and FeMn. Comparisons using the original description proposed by (Molinari and Ravichandran, 2005) for OFHC copper are reported in (Durrenberger et al., 2008).

Recently a model was proposed by (Voyiadjis and Almasri, 2008). The description is based on the theory of dislocations and the process of activation energy depending on temperature, strain rate and stress. All microstructure dependencies are included in the constants B_1 and B_2.

$$\sigma\left(\bar{\varepsilon}^p, \dot{\bar{\varepsilon}}^p, T\right) = \left[B\left(\bar{\varepsilon}^p\right)^n\right]\left[1 + B_1 T\left(\dot{\bar{\varepsilon}}^p\right)^{1/m} - B_2 T e^{A(1-T/T_t)}\right] + Y_a \tag{31}$$

Where T_t is the transition temperature beyond which the thermal activation energy maintains as constant, $B$and n are the hardening parameters and Y_a can be assumed as the yield stress.

A similar approach was used by (Voyiadjis and Abed, 2006) to take into account the FCC or BCC nature of the material considered. Comparisons between experiments and analytical descriptions are reported in (Voyiadjis and Abed, 2006) for different kinds of materials.

Considering the grain size as internal parameter, the following model can be considered, Eqs. 32-33. As previously, the material behavior definition is defined using a hardening function $\theta\left(\bar{\varepsilon}^p, \dot{\bar{\varepsilon}}^p, T\right)$. The constitutive relation is based on Eq. 16., using respectively for σ_μ and σ^*.

$$\sigma_\mu\left(\bar{\varepsilon}^p, \dot{\bar{\varepsilon}}^p, T\right) = \sigma_y + \frac{E(T)}{E_0}\int_0^{\bar{\varepsilon}^{ps}} \theta\left(\bar{\varepsilon}^p, \dot{\bar{\varepsilon}}^p, T\right) d\bar{\varepsilon}^p \tag{32}$$

$$\sigma^*\left(\dot{\bar{\varepsilon}}^p, T\right) = \frac{E(T)}{E_0}\left[\sigma^*\left(\log Z\right)^{1/\beta}\right] \tag{33}$$

Where Z is the Zener-Hollomon parameter (1944), σ_y is the yield stress and β is a constant.

The parameter Z allows to define the reciprocity between the strain rate and the temperature, Eq. 34., or a parameter where the temperature compensate the strain rate. The Zener-Hollomon is including the grain size of the metal (Li et al., 2009) and can be used to detect if the mechanism of plastic deformation is done by twinning (Li et al., 2009).

$$Z = \frac{\dot{\bar{\varepsilon}}^p}{\dot{\varepsilon}_0}\exp\left(\frac{T_0}{T}\right) \tag{34}$$

Thus as before, using a hardening function, it is possible to observe all kinds of hardening from stage II to IV (Durenberger et al., 2007). Two descriptions can be used depending if the process of saturation has to be included in the description behavior, Eqs. 35-36.

$$\theta\left(\bar{\varepsilon}^p, \dot{\bar{\varepsilon}}^p, T\right) = \theta_0 \left[1 - \frac{B}{\theta_0}\bar{\varepsilon}^p \left(\log Z\right)^m\right] \tag{35}$$

$$\theta\left(\bar{\varepsilon}^p, \dot{\bar{\varepsilon}}^p, T\right) = \theta_0 \exp\left[-\left(\log Z\right)^m \left(\frac{\bar{\varepsilon}^p}{\varepsilon^{p0}}\right)^2\right] \tag{36}$$

In the second case, a numerical method is necessary to define the material behavior in comparison with Eq. 35. where an analytical description can be defined. In comparison with the first formulation, the second approach is not linear and allows to consider all hardening stage (Durenberger et al., 2007).

Considering some of the descriptions and discussion about constitutive relation, it is observed that the number of it, is relatively large. For this reason, it is not easy to define the best one or the more appropriate to model the plastic flow of a material. For this reason, a solution is to develop a general description relatively flexible and to add some contributions depending on the material studied. In this goal, the following description was proposed in 2001 by (Rusinek and Klepaczko). The model is decoupled including a hardening description and the component of strain rate sensitivity with the reciprocity between strain rate $\dot{\bar{\varepsilon}}^p$and temperature T. It is observed that the Young's modulus dependency with temperature is considered, Eq. 37.

$$\sigma\left(\bar{\varepsilon}^p, \dot{\bar{\varepsilon}}^p, T\right) = \frac{E\left(T\right)}{E_0}\left[\begin{array}{l} B\left(\dot{\bar{\varepsilon}}^p, T\right)\left(\varepsilon_0 + \bar{\varepsilon}^p\right)^{n\left(\dot{\bar{\varepsilon}}^p, T\right)} \\ \qquad + \sigma_0^* \left[1 - D_1\left(\frac{T}{T_m}\right)\log\left(\frac{\dot{\varepsilon}^{\max}}{\dot{\bar{\varepsilon}}^p}\right)\right]^{1/m}\end{array}\right] \tag{37}$$

Where $B$is the tangent modulus, ε_0 is the strain level considering the yield stress, $n$is the hardening parameter, σ_0^* is the effective stress at $T = 0\ K$, $\dot{\varepsilon}^{\max}$ is the upper limit of the model fixed by the user and m is the strain rate sensitivity parameter.

As reported previously the hardening coefficient n is a function of the strain rate $\dot{\bar{\varepsilon}}^p$ and temperature T to define the process of thermal softening as the modulus of plasticity Bcorresponding to the yield stress change. The explicit descriptions are the following, Eqs. 38-39.

$$B\left(\dot{\bar{\varepsilon}}^p, T\right) = B_0\left[\left(\frac{T}{T_m}\right)\log\left(\frac{\dot{\varepsilon}_{\max}}{\dot{\bar{\varepsilon}}^p}\right)\right] \tag{38}$$

$$n\left(\dot{\varepsilon}^p, T\right) = n_0\left[1 - D_2\left(\frac{T}{T_m}\right)\log\left(\frac{\dot{\varepsilon}^p}{\dot{\varepsilon}^{\min}}\right)\right] \tag{39}$$

Where B_0 is a constant of the material, T_m is the melting temperature, n_0 is the hardening parameter at $T = 0\ K$, D_2 is a constant of the material and $\dot{\varepsilon}^{\min}$ is the lower limit in term of strain rate imposed by the user.

Comparing to other description it is observed that the yield stress and the hardening coefficient are not constant but depends on the strain rate and temperature. Therefore, the process of thermal softening observed for metal at $\varepsilon_{thermal} \geq 0.1$ will be described. As it will be discussed in the next part of this chapter considering different examples, it will be demonstrated that this hardening coefficient is the key to define without artifact, instabilities under dynamic loading. The model described, Eq. 37-39, was used to defined more than thirty materials including FCC, BCC and HC structures. The number of constants is relatively reduced and equal to eight. The main advantage of this model is that the set of constants for a material is unique and not depending on the user. The definition of the constants is not global but it is done step by step assuming physical assumption. A complete description to define the constants is reported in (Rusinek and Klepaczko, 2001).

The model discussed previously called RK, is later modified depending on the material considered and mainly to the observations done during experimental characterizations. In this model, it will be possible to include

1. Negative strain rate sensitivity
2. Phase transformation
3. Viscous drag effect at very high strain rate

Constitutive relation for high strain rate including phase transformation

Using the concept developed in the introduction considering that each physical contributions can be observed and modeled independently, no limits exist in term of constitutive relation development. Considering phase transformation, the following description can be used to develop a phenomenological description, Fig. 11.

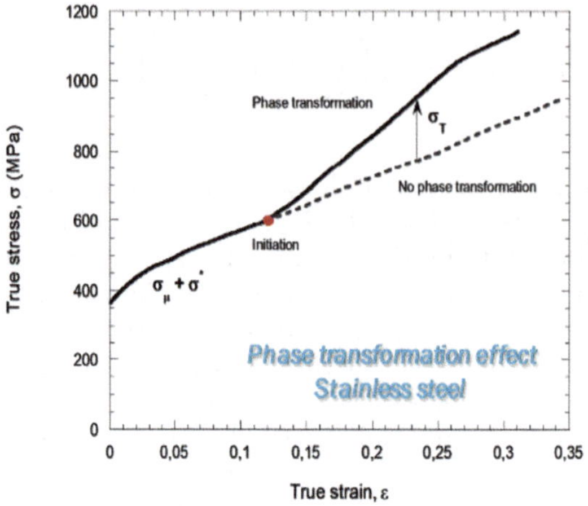

Fig. 11. Description of phase transformation effect on the macroscopic behavior of material

Indeed, the stress σ is decomposed as previously adding a new component σ_T related to the process of phase transformation. Consequently, the behavior is defined as follow, Eq. 40.

$$\sigma\left(\dot{\bar{\varepsilon}}^p, T\right) = \sigma_\mu\left(\bar{\varepsilon}^p, \dot{\bar{\varepsilon}}^p, T\right) + \sigma^*\left(\dot{\bar{\varepsilon}}^p, T\right) + \sigma_T\left(\dot{\bar{\varepsilon}}^p, T\right) \tag{40}$$

As the kinetic of phase transformation is related to the temperature and strain rate, the phase transformation component σ_T is defined as a function of these two contributions. The explicit form is the following, Eq. 41.

$$\sigma\left(\dot{\bar{\varepsilon}}^p, T\right) = \sigma_\mu\left(\bar{\varepsilon}^p, \dot{\bar{\varepsilon}}^p, T\right) + \sigma^*\left(\dot{\bar{\varepsilon}}^p, T\right) + \sigma_T\left(\bar{\varepsilon}^p, \dot{\bar{\varepsilon}}^p, T\right) \tag{41}$$

With

$$\sigma_T\left(\bar{\varepsilon}^p, \dot{\bar{\varepsilon}}^p, T\right) = \sigma_0^\alpha * f\left(\bar{\varepsilon}^p, \dot{\bar{\varepsilon}}^p\right) * g\left(T\right) \tag{42}$$

Where σ_0^α is the a fitting parameter depending on the material.

The process of phase transformation is governed by Eqs. 43. The phase transformation is depending on the plastic strain level and the strain rate, Eq. 43., as the temperature associated, Eq. 42

$$\begin{aligned} f\left(\bar{\varepsilon}^p, \dot{\bar{\varepsilon}}^p\right) &= \left[1 - \exp\left(-h\left(\dot{\bar{\varepsilon}}^p\right) * \bar{\varepsilon}^p\right)\right]^\xi \\ h\left(\dot{\bar{\varepsilon}}^p\right) &= \lambda_0 \exp\left(-\lambda \dot{\bar{\varepsilon}}^p\right) \end{aligned} \tag{43}$$

Where ξ is a material parameter and λ_0, λ are two shape fitting parameter.

Taking into account the temperature effect on the process of phase transformation, several solution can be used as reported in (Papatriantafillou et al., 2006). The first one is for example the following, Eq. 44.

$$\begin{array}{ll} g(T) = 1 - \Theta^{\eta} \quad if \quad T \leq M_S \rightarrow g_1(T) = 1 \\ \Theta = \frac{T - M_S}{M_D - M_S} \quad if \quad T \leq M_D \rightarrow g_1(T) = 0 \end{array} \tag{44}$$

Where M_S is the martensite-start temperature, is the temperature where no phase transformation is possible and η is the temperature sensitivity of the phase transformation.

The second one based on the work of (Koistinen and Marburger, 1959) is as follow, Eq. 45.

$$g(T) = \exp\left[-\left(\frac{T}{M_D - T_0}\right)^{\alpha}\right] \tag{45}$$

Where T is the current temperature, T_0 and α are material constants.

A complete parametric study about all coefficients is reported in (Rodriguez-Martinez et al., 2009)]. It is also possible to find some comparisons between experiments and modeling. A comparison is reported, Fig. 12., for an austenitic steel 301 Ln2B.

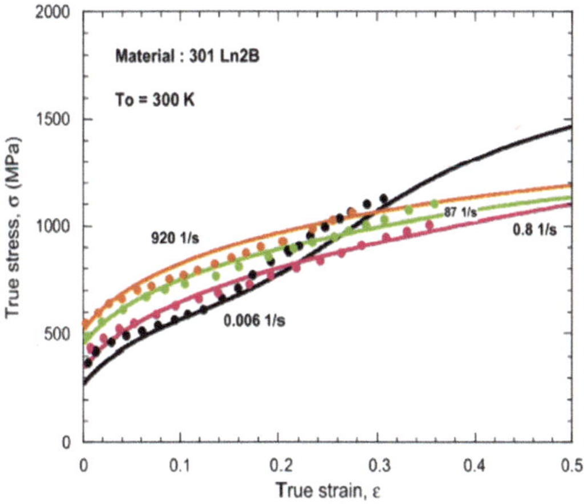

Fig. 12. Comparison between experiments and modeling for an austenitic steel with phase transformation

It is observed that using a phenomenological description, it is possible to model in an efficient way physical observations as phase transformation. Of course more complicated approaches can be used taking into account all the process of phase transformation. Some recent works can be considered as [Zaera et al., 2012, Han et al., 2004, Iwamoto and Tsuta, 2002, Kubler et al., 2011].

Viscous drag

In this part a constitutive relation is proposed to define the behavior of materials at high strain rates showing a viscous drag effect inducing a large strain rate sensitivity. For example, OFHC cooper is used since it was studied in several papers as for example (Nemat-Nasser and Li., 1998, Follansbee, 1985, Bhattacharyya et al., 2005).

$$\sigma\left(\dot{\bar{\varepsilon}}^p, T\right) = \frac{E\left(T\right)}{E_0}\left[\sigma_\mu + \sigma^*\left(\bar{\varepsilon}^p, \dot{\bar{\varepsilon}}^p, T\right)\right] + \sigma_{VS}\left(\dot{\bar{\varepsilon}}^p\right) \tag{46}$$

Where σ_μ is the yield stress of the material considered.

The viscous drag component is independent on temperature as discussed in (Kapoor and Nemat-Nasser, 1999, Guo and Nemat-Nasser, 2006). The main dependency is with strain rate. In addition the internal stress is assumed as constant as reported in (Voyiadjis and Almasri, 2008). The effective stress is defined based on the description, Eq. 37. The difference is due to the component σ_0^* which is not assumed as constant, Eq. 47.

$$\sigma^*\left(\dot{\bar{\varepsilon}}^p, T\right) = B\left(\dot{\bar{\varepsilon}}^p, T\right)\cdot\left(\bar{\varepsilon}^p\right)^{n\left(\dot{\bar{\varepsilon}}^p, T\right)}\left\langle 1 - \xi_1\left(\frac{T}{T_m}\right)\log\left(\frac{\dot{\varepsilon}_{\max}}{\dot{\bar{\varepsilon}}^p}\right)\right\rangle^{1/\xi_2} \tag{47}$$

Where ξ_1 and ξ_2 are material constants describing temperature and strain rate sensitivity of material.

The definition of $B\left(\dot{\bar{\varepsilon}}^p, T\right)$ and $n\left(\dot{\bar{\varepsilon}}^p, T\right)$ to include the strain rate sensitivity and the thermal softening are defined, Eq. 38-39. A systematic procedure to define the constants is reported in (Rusinek et al., 2010). In addition several comparisons are reported between Eq. 46-47 and the model proposed by (Nemat-Nasser and Li, 1998) and (Voyiadjis and Almasri, 2008) for OFHC Copper.

Experimental description

The crucial point for correct description of the materials for a wide range of strain rates and temperatures is to define properly the constants. For it, an optimization methods can be used as the least square method. The process of identification is coupled with experiments. Therefore, depending on the working field several tests must be performed. Concerning the quasi-static behavior for a range of $10^{-3} \le \dot{\varepsilon} \le 10^{-1}\ s^{-1}$, the tests are performed using a tensile screw machine. In dynamic, several methods can be used. The first one is to perform the tests with hydraulic machine but in this case the range of strain rates is varying from $10^{-3} \leq \dot{\varepsilon} \leq 500\ s^{-1}$. The problem with this technique even if the velocity of the jack is high is the eigen frequency of the cell of force close to 1 kHz. Therefore, the behavior measure at high strain rate is not really related to the material but a combination between the material, the set-up to fix the specimen and the structure of the machine. On the following picture, Fig. 13., an example is reported to define the material behavior of a sheet steel at high strain rate. Even if this problem is well known, this kind of results are reported in the literature and an average function is used to define the material behavior.

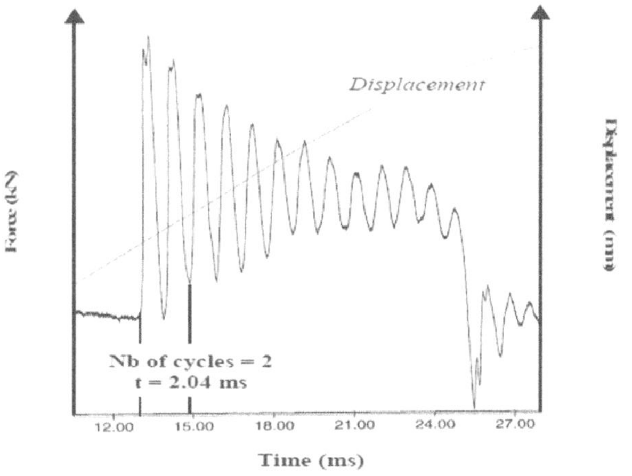

Fig. 13. Example of experimental measurements obtained at high strain rate using a fast hydraulic machine (Penzes, 1998)

To avoid the problem of high oscillations disturbing the measurement of the intrinsic behavior of materials, the Split Hopkinson Pressure Bars (SHPB) technique can be used. The set-up was initially proposed by Hopkinson (1914). The mechanical behavior of the material is defined knowing the force and the displacement imposed to the specimen. The previous quantities are obtained thank to the elastic waves as described for example in (Jankowiak et al., 2011). It consists to sandwich between two long elastic bars a short specimen and it is reported schematically Fig. 14., to measure the incident ε_i, reflected $\varepsilon_r$and transmitted ε_t waves.

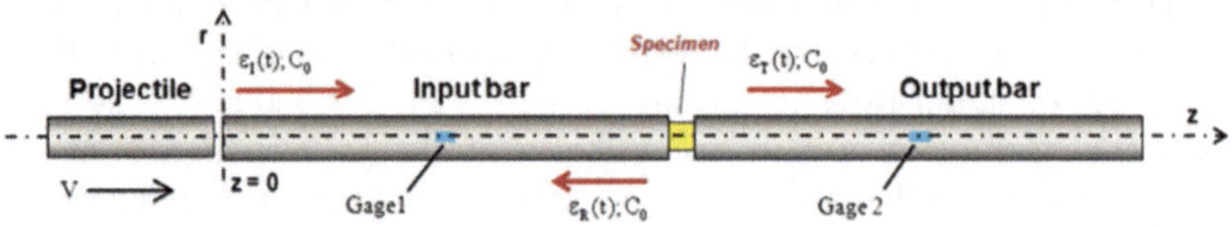

Fig. 14. Schematic description of the Hopkinson Pressure Bar set-up

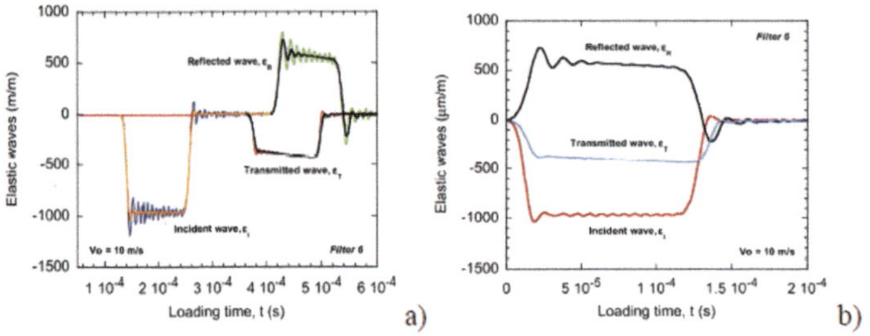

Fig. 15. a-Elastic waves traveling along SHPB; b- Elastic waves after initiation to zero for elastic waves analyze

Thanks to the elastic waves, the forces and the displacements imposed to the specimen can be defined, Eq. 48-49.

$$\begin{aligned} F_{input}(t) &= A_0 E\left[\varepsilon_i(t) + \varepsilon_r(t)\right] \\ F_{output}(t) &= A_0 E \varepsilon_t(t) \end{aligned} \tag{48}$$

Where A_0 is the area section of the Hopkinson bar, E is the Yong modulus of the bar.

$$\begin{aligned} U_{input}(t) &= C_0\left[\varepsilon_i(t) + \varepsilon_r(t)\right] \\ U_{output}(t) &= C_0 \varepsilon_t(t) \end{aligned} \tag{49}$$

Where C_0 is the elastic wave speed proportional to the Young modulus and the density of the bar material.

Assuming an equilibrium of the specimen ($\sum_{i=1,2} F_i = 0$) during the loading time, the stress σ, strain ε and strain rate $\dot{\varepsilon}$ can be defined, Eq. 50

$$\begin{aligned} \sigma(t) &= E\left(\frac{A_0}{A_S}\right)\varepsilon_t(t) \\ \varepsilon(t) &= -\frac{2C_0}{L_0}\int_0^{t_{\max}} \varepsilon_r(\zeta)d\zeta \\ \dot{\varepsilon}(t) &= -\frac{2C_0}{L_0}\varepsilon_r(\zeta) \end{aligned} \tag{50}$$

Where L_0 is the initial length of the specimen.

All these quantities are defined as nominal. Before using results for modeling all values must be changed for true.

However several recommendations can be found in the international literature explaining how the tests must be analyzed to avoid all artifact superposition. As discussed previously, the stress-strain curve is also showing a oscillating behavior. This effect is related to the Pochammer-Chree phenomena discussed in details in (Jankowiak et al., 2011). The dispersion is connected with different wave propagation velocity which has different frequencies. These oscillations are related to a geometric parameter, the bar diameter. The problem was studied in details by Skalak (1957). The problem is defined by Eq. 51-53.

$$\varepsilon_z(z,t) = \frac{\partial U_z(z,t)}{\partial z} = \frac{V_0}{C_0}\left[\frac{1}{6} + \int_0^{\alpha'} (A_i)(\alpha)\, d\alpha + \frac{1}{6} + \int_0^{\alpha'} (A_i)(\alpha)\, d\alpha\right] \tag{51}$$

Where is the Airy's function defined by Eq. 52.

$$(A_i)(\alpha) = \frac{1}{\pi}\int_0^{\infty} \cos\left(\alpha\eta + \frac{1}{3}\eta^3\right) d\eta \tag{52}$$

Where the upper limits of the integrals are defined as follows, Eq. 53.

$$\alpha' = \frac{z - C_0 t}{\sqrt[3]{3\phi_b t}} \quad and \quad \alpha'' = \frac{-z - C_0 t}{\sqrt[3]{3\phi_b t}} \tag{53}$$

Where z is the distance from the impact side and ϕ_b is the Hopkinson bar diameter.

Solving previous equations, the results are reported in the following picture, Fig. 16., for different diameter ϕ_b.

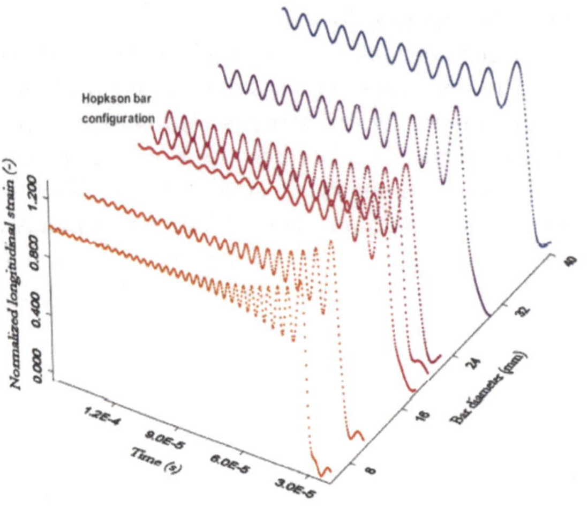

Fig. 16. Effect of the bar diameter on the dispersion effect, Eq. 51-53

The problem in term of oscillation is not a real problem for ductile material since quickly the stress will oscillate around an average value, Fig. 16. Moreover the slope corresponding to the Young modulus can be corrected knowing the theoretical value. The problem is more complicated when the material study is behaving as brittle. In this case, it is not so easy to define the maximum stress level of the failure stress level. Another problem related to SHPB is the friction effect. In fact it is not possible to obtain a condition where the friction coefficient is equal to $\mu = 0$. For this reason, an overstress state is observed. Using experimental measurements, the stress level is including the material behavior and an additional component related to the friction, Eq. 54. Therefore before defining the constants of the material based on a constitutive relation, the friction effect must be corrected.

$$\sigma = \sigma_{material} + \sigma_{friction} \tag{54}$$

Recently based on the work of Klepaczko-Malinowski (1977) a proposition has been done to correct friction effect on the flow curve σ. In a general way, it is inducing a decrease of the stress level. The correction is showing a linear behavior follows by a plateau. As discussed in (Klepaczko and Malinowski, 1977) the overstress value is depending on the geometry of the specimen used during the SHPB tests defined by ϕ_0 and L_0, Eq. 55.

$$\sigma_{material} = \sigma_{measured}\left[1 - \frac{\mu}{3}\left(\frac{\phi_0}{L_0}\right)\right] \tag{55}$$

Where ϕ_0 is the diameter of the specimen.

A complete analyze of the previous effects is reported in (Jankowiak et al., 2011). In addition the problem of punching is also considered as the shape of the contact zone between the projectile and the incident bar. A complete description of the punching effect is studied in (Jankowiak et al, 2011).

On the following curve is reported the dynamic behavior of OFHC copper under dynamic loading taking into account friction or not, Fig. 17. It is observed as discussed previously and considering Eq. 55., that friction effect induces an overstress state.

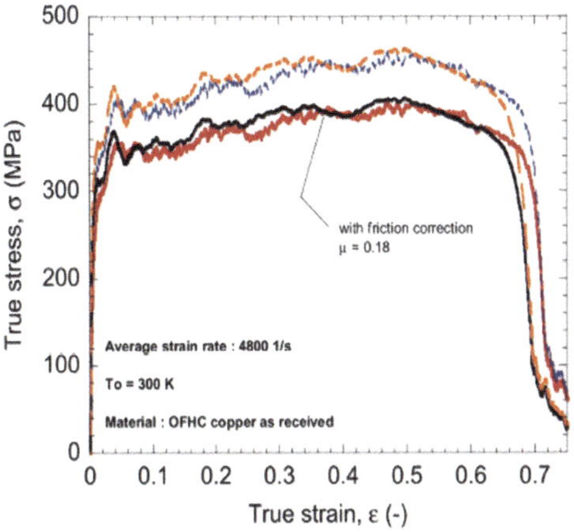

Fig. 17. Dynamic behavior of OFHC copper including or not friction correction (Jankowiak et al., 2011)

Therefore if the constants of the constitutive relation are defined without correction, the friction effect will be included in the model.

Considering the previous experimental descriptions, it is observed that the strain rate in an average way is varying from $10^{-4} \leq \dot{\varepsilon} \leq 5 \cdot 10^3\, s^{-1}$. This range is enough for dynamic loading as impact or crash. In the case of high strain rate observed during perforation problem of explosion, the maximum strain rate is higher than $5 \cdot 10^3\, s^{-1}$. Therefore, to analyze the mechanical behavior of material at 10^4-10^5 s^{-1} and the mechanism associated, it is necessary to use other techniques, Fig. 18.

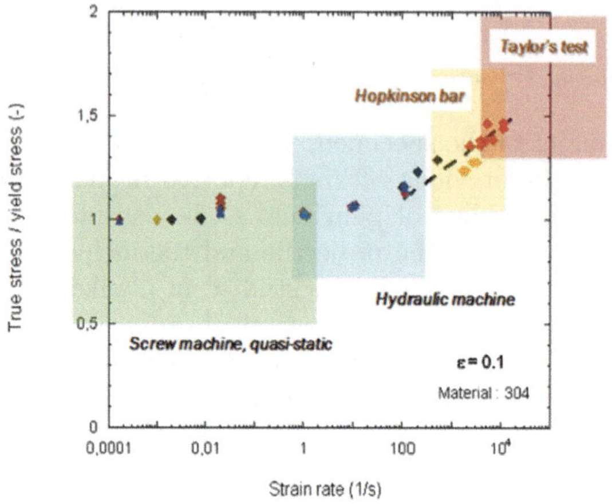

Fig. 18. Description of the rage of strain rates depending on the set-up used

Thus, an alternative technique to reach high strain rate up to 10^4-10^5 s^{-1} is to used Taylor's test (1948). It consists launching a cylindrical specimen against a rigid wall to avoid deformation of it. When the projectile impacts the rigid wall, the specimen deforms. The mushroom shape of the specimen allows defining the stress level imposed to the specimen, the strain rate and the strain level. Therefore, the analyze is post mortem, Fig. 19.

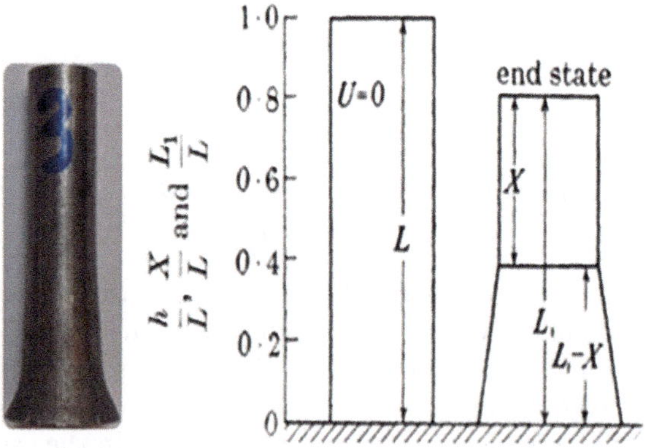

Fig. 19. Schematic description of the Taylor's specimen after impact test, description of the mushroom shape and description of the geometric quantities

Several papers were published on Taylor's test proposing a technique to estimate thanks to the final length of the specimen L_1 and the diameter ϕ_1. However as it is reported in (Taylor, 1948) the formulas proposed do not allow to define properly the average stress level. Using this test, the stress, the strain and the strain rate are defined as follow, Eq. 56-58.

$$\sigma_y = \frac{1}{2}\rho V^2 \frac{(L-X)(L-L_1)}{\ln\left(\frac{L}{X}\right)} \tag{56}$$

$$\varepsilon = \ln\left(\frac{L_1 - X}{L - X}\right) \tag{57}$$

$$\dot{\varepsilon} = -\left(\frac{V}{2(L-L_1)}\right)\ln\left(\frac{L_1 - X}{L - X}\right) \tag{58}$$

Where L is the initial length of the specimen, X is the final elastic length after impact, L_1 is the final length of the specimen after impact, V is the impact velocity and ρ is the density of the material tested.

To validate if the relations fit properly the material behavior description based on this experiment, some tests have been performed using brass. This material was used since it is not strain rate sensitive. In this specific case, it is possible to rebuilt completely the hardening curve $\sigma - \varepsilon$. Based on this analyze, it was observed that the analytical description does not allow to define properly the material behavior. In the original description, Eq. 56-58, no hardening is included as strain rate sensitivity. For this reason, the test as to be linked to a precise description as reported in (Jones et al., 1998). The correct approximation of the flow stress is defined using Eq. 59.

$$\sigma(\varepsilon) = (1+\varepsilon)\left\{\sigma_0 + \frac{[1-\beta(\varepsilon)]^2}{\varepsilon}\rho V^2\right\} \tag{59}$$

$$\beta(\varepsilon) = 1 + m \cdot \varepsilon \tag{60}$$

$$\sigma_0 = \frac{\sigma}{1+\varepsilon} \tag{61}$$

Where m is defined by the slope of the curve $L_1/X - L_1/L_f$, ρ is the density of the material and V is the impact velocity.

In the previous description it is observed that the hardening effect is included thanks to $\beta(\varepsilon)$. A comparison is reported on the following picture using the original approximation, Eqs. 56-58. and Eq. 59-61. In addition the formulation of (Wilkins and Guinan, 1973) is also reported, Eq. 62.

$$\sigma(\varepsilon) = -\frac{\rho V^2/2}{\ln\left(\frac{L_1/L - 0.12}{0.88}\right)} \tag{62}$$

It is observed based on this analyze that using Eqs. 59-61, the dynamic behavior of materials at low and high strain rate can be defined allowing to model more precisely the behavior of materials.

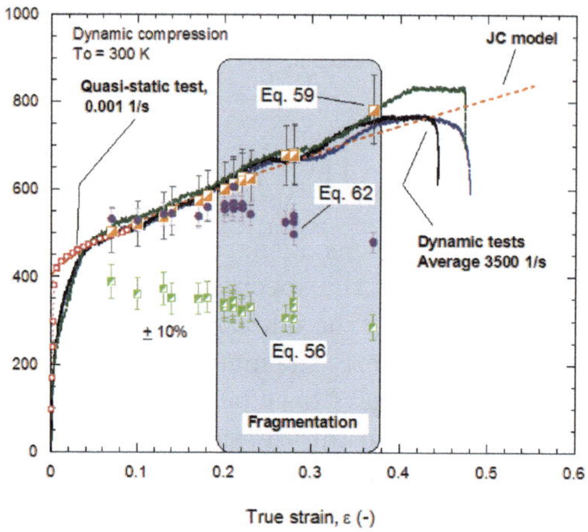

Fig. 20. Comparison between analytical descriptions based on Taylor's test (Julien et al., 2013)

In addition, with the Taylor's test it is not only possible to define the material behavior (homogenous flow stress) but also the process of dynamic fracture depending on the initial impact velocity V_0, Fig. 21.

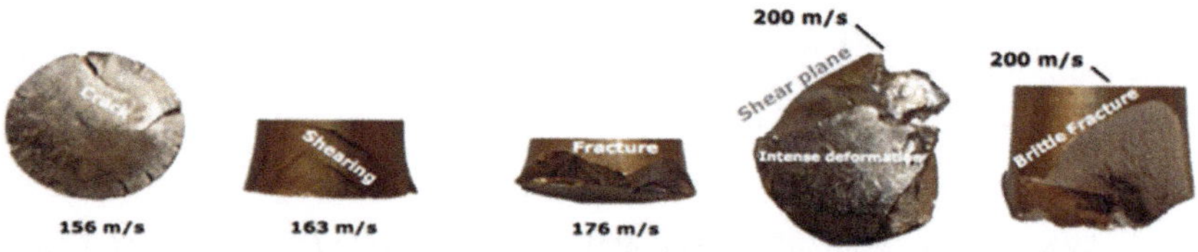

Fig. 21. Description of the process of failure depending on the initial impact velocity for brass (Julien et al., 2013)

In conclusion coupling screw machine, hydraulic machine, SHPB technique and Taylor's test the rage of strain rates is varying from 0.001 s^{-1} to

$10^4\ s^{-1}$. In all cases discussed previously the tests are performed at room temperature. To reach low and high temperature the tests can be easily done using screw machine and hydraulic machine. It has to be noticed that the temperature effect is included in the material behavior by thermal softening effect. It takes place for metals when the strain level is higher than 0.1. Therefore explicitly performing tests at high strain rates allow to define the strain rate sensitivity and temperature effect.

Algorithm for constitutive implementation in FE codes

The final step when the constitutive relation is defined and when the constants are defined based on experiments, it is frequently necessary to implement them in FE codes. For it several approach can be used. The famous one, is the theory proposed by (Perzyna, 1963). In this case, the strain rate must be defined explicitly, Eq. 63.

$$\dot{\bar{\varepsilon}}^p = \frac{1}{\eta} \langle \Phi \left[\bar{\sigma} - \sigma_Y \left(\bar{\varepsilon}^p, T, ... \right) \right] \rangle \tag{63}$$

Where η is the viscosity parameter assumed as constant.

The main problem with this description if the overstress state induced. Due to this effect during unloading the plastic deformation increase. Moreover, it is necessary to define explicitly the strain rate. It is relatively easy with the Johnson-Cook model or with a power law but more complicated with other models as Eq. 37-39. To avoid this problem, several authors have proposed some descriptions as (Zaera and Fernandez-Saez, 2006, Wang et al., 1997, Winnicki et al., 2001). In a general way a consistent approach is formulated and for each step of plastic deformation, the following equation is solved, Eq. 64.

$$f_{n+1} = f \left(\bar{\sigma}_{n+1}, \bar{\varepsilon}^p_{n+1}, \dot{\bar{\varepsilon}}^p_{n+1}, T_{n+1} \right) = 0 \tag{64}$$

It has to be noticed that the strain rate is included as the hardening and temperature sensitivity. To solve the problem a return mapping algorithm is used (Zaera and Fernandez-Saez, 2006). In some cases and assuming some equalities, the model proposed by (Perzyna, 1963) is equivalent to the consistency approach as discussed in (Heeres et al., 2002). A complete description of Eq. 64. coupled to RK model, Eqs. 37-39. is reported and described in details in (Zaera and Saez, 2006, Rusinek et al., 2007). A comparison is reported in this chapter studying an industrial process as High Speed Machining. The complete analyze is published in (Lodygowski et al., 2012).

Instabilities under dynamic loading

Crash test of structure, strain rate sensitivity effect

When a constitutive relation is used to simulate some processes it is mainly to define the behavior of complex structures. In general, the structure is defined using different materials. Therefore to obtain correct agreement between the structure behavior and numerical simulations, the materials must be described in a precise way. The description is mainly due to the analytical model and the constants identified from experiments. Thus, we have a strong coupling between material, experiments performed and the method used to define the constants of the constitutive relation. If one of these steps is not well done, the simulations will be in disagreement with experimental observations. Moreover, it will be complicated to catch instabilities and as final stage the dynamic failure if initially the homogeneous behavior is not well defined. In several cases reported in the literature it is observed that a bad description of the behavior does not allow quantifying local observations. It just allows to mimic the trends in term of force, displacement for example. Coming back to the problem of crash test discussed in introduction of this chapter the constitutive relation used to simulate the process is the key point. As reported in (Rusinek et al., 2008) it is demonstrated that during a crash test a strong competition is existing between the collapse by plasticity and elastic wave propagation, Fig. 22. Therefore, if the non linear strain rate sensitivity is not included in the material description the collapse is due to the process of elastic wave propagation inducing on the embedded side an increase of the stress level. In this precise case the collapse is on the opposite side of the crash-box. Now if the strain rate is correctly defined and assuming that the stress level is bigger than two times the stress level induced by elastic wave propagation, the collapse site will be on the impact site.

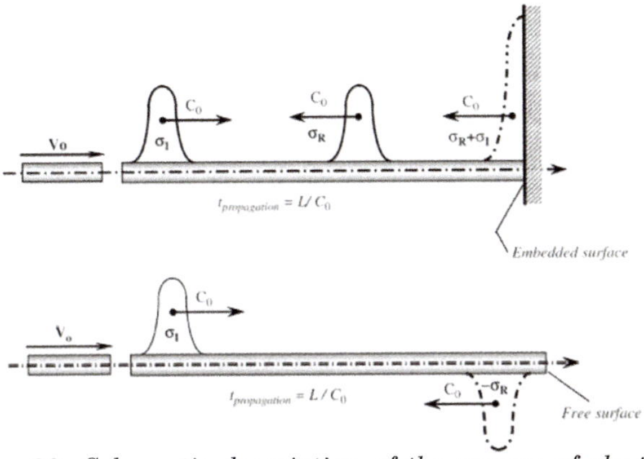

Fig. 22. Schematic description of the process of elastic wave propagation depending on the boundary conditions

Based on this precise example corresponding to design structures, it is precisely reported that the constitutive relation is governing the process of collapse. If the constitutive relation is not well defined, numerical artifacts are necessary as for example a mesh defect. It has to be noticed that frequently this option of mesh defect is used during numerical simulations to fit or to have agreement between numerical predictions and experiments. On the following picture, Fig. 23., is reported the strain rate sensitivity of material for different value of C, Eq. 20.

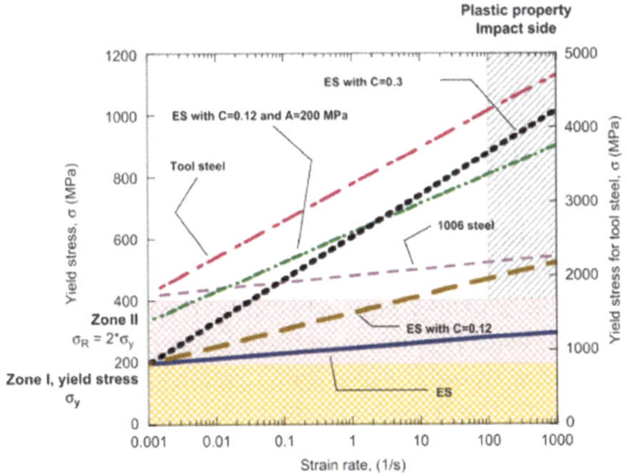

Fig. 23. Effect of C parameter based on the model of Johnson-Cook, Eq. 20

It is observed for $C = 0.3$a large value of the dynamic stress level corresponding to $\dot{\varepsilon} = 10^3\ s^{-1}$. Therefore, the stress value is larger than the intensity due to the wave reflexion on the opposite side of impact. For this reason the collapse of the crash-box is on the impact zone. On the contrary, if the strain rate is low as for steel close to $C = 0.022$, the collapse is located on the embedded side since the stress level induced by reflexion is bigger than the stress level induced at $\dot{\varepsilon} = 10^3\ s^{-1}$. These two examples are reported on the following pictures, Fig. 24.

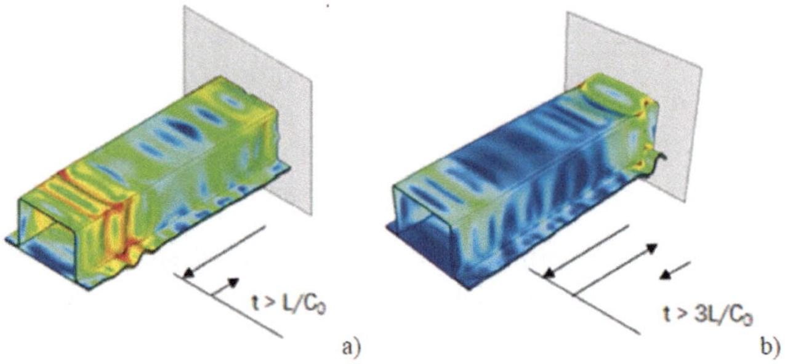

Fig. 24. Constitutive relation effect on the collapse description and effect of the strain rate sensitivity definition; a- C=0.022;b- C=0.3

Thus, if a linear strain rate sensitivity is used as in the case of Johnson-Cook model, it is clear that the strain rate sensitivity definition is crucial. To avoid the problem discussed previously is to use two set of constants to define the non linearity, Fig. 2. Another solution is to use a constitutive relation including the non linear behavior in term of strain rate and temperature sensitivity as for example (Rusinek and Klepaczko, 2001).

Ring extension, strain rate and thermal softening effect

As second example, dynamic instability taking place under ring expansion is discussed. This problem was frequently studied using an experimental, analytical or numerical way. It consists in including a fast radial velocity to a specific specimen, a ring to induce dynamic tension. Depending on the radial velocity V_0 imposed to the internal part of the ring, the failure mode is caused by an unique necking as under quasi-static loading but increasing the radial velocity a process of fragmentation talks place increasing the number of fragments with the velocity. A complete description is reported in (Rusinek and Zaera, 2007).

In this example considering ring expansion under dynamic loading, the constitutive relation effect is demonstrated. Therefore several models were used to predict instabilities with radial velocity imposed to the ring. This test is also used to define the material behavior of material at high strain rate and large deformation (Rusinek and Zaera, 2007). The radial velocity experimentally can be applied using a magnetic field or an explosive. In the first case, the radial velocity V_0 can be assumed as constant, Fig. 25. In this work a mild steel ring is considered. The area of the ring is assumed equal to 1x1 mm^2. The equivalent strain rate applied to the specimen is defined as follow, Eq. 65.

$$\dot{\varepsilon} = \frac{V_0}{R_0} \tag{65}$$

Where R_0 is the initial radius of the ring

The mesh used to analyze the process is without defect assuming an homogeneous section of it. The mild steel is described using different constitutive relation.

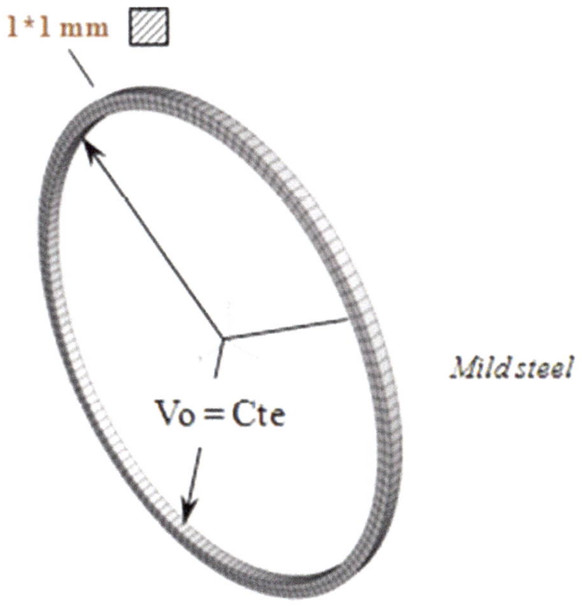

Fig. 25. Geometry description and boundary condition definition

The first model used is the Johnson-Cook model. However, two specific cases are described. The first one is assuming a thermoviscoplastic behavior

Eq. 20. and the second one considering no strain rate sensitivity, $C = 0$. For comparison, RK model was used Eqs. 37-39 and an analytical description proposed by (Mercier and Molinari, 2003). In the last case, the constitutive relation used to describe the viscoplastic of the mild steel is defined as, Eq. 66.

$$\sigma\left(\bar{\varepsilon}^p, \dot{\bar{\varepsilon}}^p\right) = K \cdot \left(\bar{\varepsilon}^p\right)^n \cdot \left(\dot{\bar{\varepsilon}}^p\right)^m \tag{66}$$

Where n is the hardening coefficient, m is the strain rate sensitivity and K is a constant of the material

It has to be noticed that the temperature sensitivity is not included in the analytical description. Therefore, three cases are considered

1. Thermoviscoplastic ($\bar{\varepsilon}^p$,$\dot{\bar{\varepsilon}}^p$,$T$)

2. Viscoplastic ($\bar{\varepsilon}^p$,$\dot{\bar{\varepsilon}}^p$)

3. Thermoplastic ($\bar{\varepsilon}^p$,T)

All constitutive relations were linked to a failure criterion to define fragmentation as in (Pandolfi et al., 1999). The criterion was defined thanks to a critical failure strain level $\bar{\varepsilon}^p_{failure}$. It is observed when the velocity is increased, the number of fragments N along the ring is larger. However, for low applied velocity close to quasi-static loading, the number of necks is equal to $N = 1$. For comparison, it is comparable with a tensile test at low strain rate since no fragmentation takes place. In general the process takes place for an applied velocity larger than $V_0 > 10\,m/s$. Considering, the previous constitutive relations, the following results have been obtained, Fig. 27. It is observed that each models are predicting an increase of N with V_0. However for large values of V_0, a process of saturation is observed when $V_0 > 100\ m/s$ assuming the Johnson-cook model or the power law. Considering a complete description Eq. 37-39, the number of fragments continue to increase with the speed. The numerical predictions are in agreement with experiments obtained for steel (Diep et al., 2004).

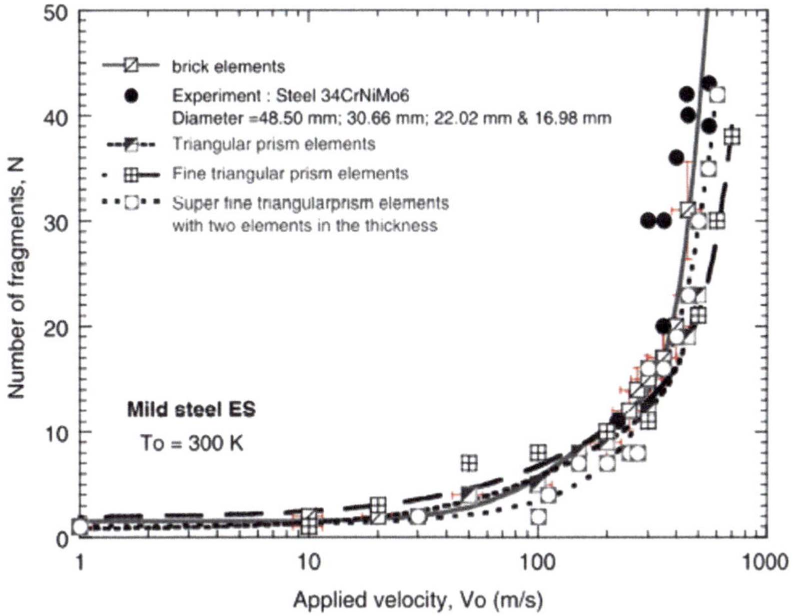

Fig. 27. Evolution of the number of fragments with the radial applied velocity depending on the constitutive relation used, comparison with experiments (Diep et et., 2004)

The reason and to explain the differences between constitutive relation is the following. For all cases considered, the hardening coefficient is assumed as constant; Only the model Eqs. 37-39 is considering a hardening coefficient depending on strain rate and temperature. As reported in several papers, to trig instability a completion exists between the strain rate sensitivity, the temperature sensitivity and the hardening. Some experiments reported in (Altinova et al., 1996) demonstrate that lower hardening is inducing more fragments. A parametric study changing in Eq. 39., is reported, Fig. 28.

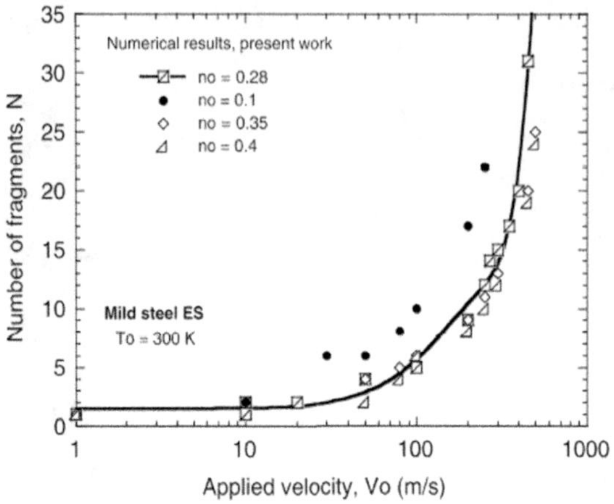

Fig. 28. Hardening effect on the number of fragments with applied velocity

Based on the previous experiments, a local analyze is done using numerical simulations. Thus plotting the local temperature T along the ring with V_0, it was observed for a radial speed equal to $V_0 > 100\ m/s$an increase of the temperature.

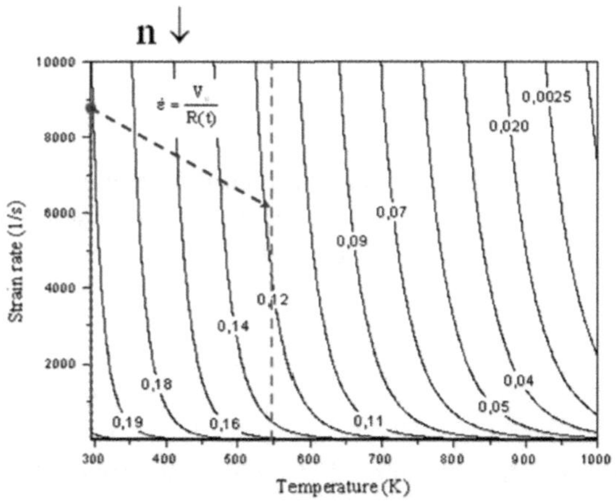

Fig. 29. Effect of strain rate and temperature on hardening definition, Eq. 39

This transition is corresponding to the rate increase of N, Fig. 27. Thus, the reason is defined. For smaller applied velocity, the temperature is more or less constant and the hardening parameter stay more or less the same. For high applied velocity as the temperature increases the value of n decrease during the loading process, Fig. 29. For this reason the number of fragment N is larger in agreement with experiment, Fig. 27.

Based on this example it is observed that the coupling strain rate, temperature sensitivity is an important parameter and mainly the definition of the hardening coefficient. Indeed if the hardening n was kept as constant, the process of fragmentation would not be defined correctly at high applied velocity. It is clear that this parameter is a key point to study instability and as it was demonstrated by experiment this coefficient cannot be assumed as constant for dynamic loading. Using numerical results it is observed that all necks will not induce a complete failure. Therefore, the process of complete failure and arrested neck can be defined in agreement with experiments of (Grady and Benson, 1983).

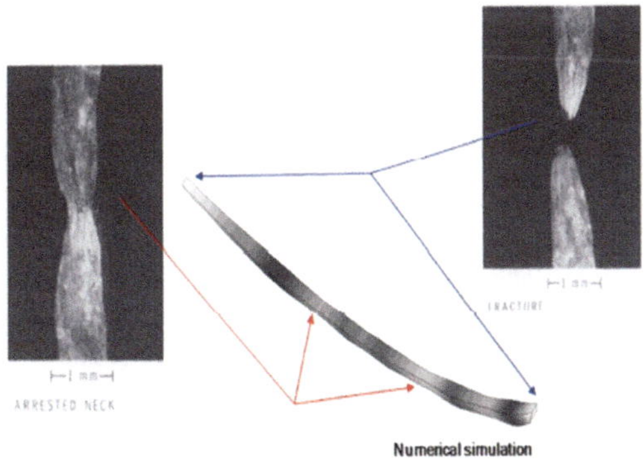

Fig. 30. Description of a numerical fragment comparing with experiments (Grady and Benson, 1983)

In the next part a discussion is reported concerning an industrial process, the High Speed Machining frequently called HSM. This problem was studied by several authors using experiments (Moufki et al., 2004), analytical modeling (Moufki et al., 1998) or numerical simulations (Molinari et al., 2011).

High speed machining, strain rate and temperature sensitivity

The main purpose of the High Speed Machining (HSM) modeling and computing is to estimate properly the cutting forces depending on other parameters involved during machining as reported by Molinari and others (2002). The configuration considered is shown in Fig. 31 together with the cutting tool geometry, see Fig. 31-b. The model is three dimensional and it consists only one layer of finite elements out of plane thickness as presented by odygowski and others (2012). The material used during the cutting process is modeled using the JC and the RK model to describe it thermoviscoplastic behavior. The main part at the beginning before simulation is to identify the material parameters as discussed in introduction. The constants related to this problem are in odygowski and others (2012).

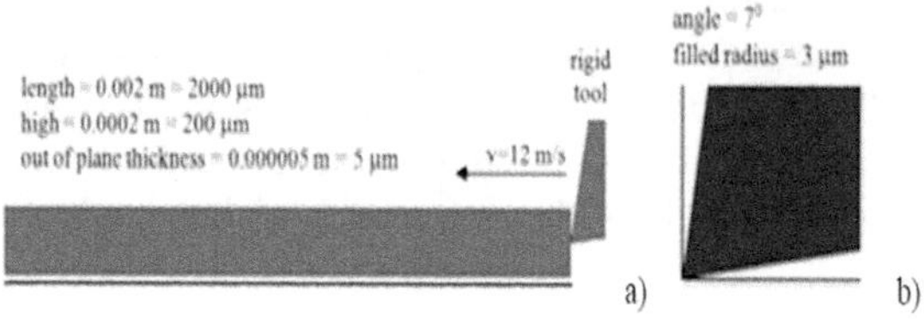

Fig. 31. a) Configuration of the High Speed Machining; b) Geometry of the rigid tool

Both discussed material models, Eq. 20 and Eqs. 37-39 have been used with Abaqus/Explicit finite element code to demonstrate how the constitutive relation $\sigma(\varepsilon, \dot{\varepsilon}, T)$ can change the numerical predictions. Many aspects have been analyzed by odygowski and others (2012) for example: the yield stress effect, the strain hardening effect, the strain rate hardening, the temperature sensitivity and friction effect. In term of numerical aspects, it was discussed the finite element size and the type of elements effect. In this chapter only one aspect is discussed, the cutting depth considering the JC and the RK model. Two depths are assumed, 50 μm and 100 μm. The results in term of ship shape, strain distribution and forces are presented, Fig. 31. Concerning the forces, a difference depending on the constitutive relation applied is observed. The average force difference between JC and RK for the two depths of cut considered is close to 12%. The main reason to explain the force difference is the strain rate sensitivity definition. Using RK model for an imposed local strain rate, the stress level is larger in contrary with the JC model where the strain rate sensitivity is linear. Therefore, using a non linear strain rate sensitivity and due to the stress increase with strain rate, the cutting force is higher, Fig. 31-c.

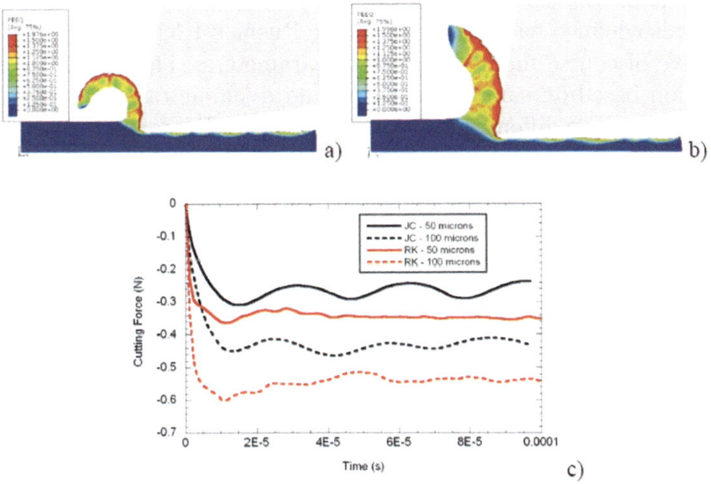

Fig. 32. The maps of equivalent strains for JC model and for different depth of cut: 50 μm, b) 100 μm, c) cutting force comparison depending on the constitutive relation used for FE simulations

Using of these constitutive relations (JC or RK) in numerical simulations depends significantly on the possibility of proper identification of material parameters. It is a reason that in engineering application very often the simple material models with smaller number of parameters are used.

Conclusions

The different problems connected with the dynamic behavior of structure and its dynamic failure were presented in the chapter. We focused on the metal applications but of course extension of the topic to other materials is possible. The most important and favor aspect of the book was presentation that the material properties are dependent on strain (deformations), strain rates (velocity of deformations) and on increasing of the temperature during dynamic events. It is crucial if we want to predict the real behavior of the structures. We presented some constitutive relations which can be used for simulation of the dynamic structure behavior. We followed with readers from material science into the phenomenological description. At the beginning the models based on the density of dislocation are presented and all the evolutionary rules are presented in general. The other possibility is phenomenological approach because of using simplicity these kinds of material models in full scale simulations. The following models among other things were discussed: Power law, Johnson-Cook, Zerilli-Armstrong,

Hall-Petch and others. The effects as phase transformation or as viscous drag were also defined with extension of the Rusinek-Klepaczko model. The models have of course many numbers of parameters. The important is its identification based on same quasi-static and dynamic tests. The different tests are accepted to different range of strain rates. We focused on the split Hopkinson pressure bars technique and Taylor test which near quasi-static test can cover full range of strain rates. We especially presented the effect of friction during the compression test and also the rising time was discussed. In the Taylor test mainly the different formulation for description of the strains, strain rates, and stresses were discussed. We present which formulations of stresses give the best accuracy. Coupling the experiments and constitutive models together with identification of material's parameters leads to possibility of the simulation of the read processes. Only one additional step is necessary - to implement the constitutive material model into finite elements code and later using of some numerical method to calculations for example by finite element method, finite difference method or for example by smoothed particles hydrodynamics. Of course in computer cods same material models are ready to use and pre-implemented other needs user implementation. We presented some industrial application which can finally limit the number of real experimental test and may limit the costs. First was simulation of the dynamic behavior of the crash box responsible for the safety of the all cars. The second application concerned the expansion of the ring. Based on this example the process of dynamic failure is discussed together with material fragmentation. The last example applies to the machining process as high speed machining or orthogonal cutting. We presented the influence of the material model on the results (cutting force and chip shape) dependent on the deep of cutting.

Finally the chapter presented in details some important aspects for metal applications, dynamic analyses and we hope it is the important point for somebody (engineer or scientist) who wants to start the adventure in material science engineering.

Acknowledgements: Author thanks all Phd students who were contributing in this topic of constitutive relation and dynamic behavior of materials: J.A. Rodriguez-Martinez, L. Durrenberger, R. Cheriguene, P. Chwalik, T. Fras and colleagues, specially Prof. J.R. Klepaczko, Prof. R. Zaera, Prof. A. Molinari and Dr. A. Arias.

References

Allain, S., Cugy, P., Scott, C., Chateau, J.P., Rusinek, A., Deschamps, A. (2008). *The influence of plastic instabilities on the mechanical properties of*

a high-manganese austenitic FeMnC steel. Int. J. Mater. Res. 99, 7
Altinova M, Hu X, Dahen GS. Increased ductility in high velocity electromagnetic ring expansion. Metall Trans A 1996;27: 1837–44.
Arruda EM, Boyce MC, Jayachandran R. (1995). Effects of strain rate, temperature and thermomechanical coupling on the finite strain deformation of glassy polymers. Mech Mater. 19:193–212.
Bardenheier, R., and Rogers, G. (2006). *Dynamic impact testing with servohy-draulic testing machines.* Journal of Physic IV France. 134:693–699.
Bathe, K. J., Walczak, J., Guillermin, O., Bouzinov, P.A., and Chen, H. Y. (1999). *Advances in crush analysis.* Computers and Structures, 72:31–47.
Berbenni, S., Favier, V., Lemoine, X., and Berveiller, M. 2004. *Micromechanical modeling of the elastic-viscoplastic behavior of polycrystalline steels having different microstructures,* Materials Science and Engineering .A 372:128–136
A. Bhattacharyya, D. Rittel, G. Ravichandran. (2005). Effect of strain rate on deformation texture in OFHC copper, Scr. Mater. 52 (2005) 657–661
Cheng, J., Nemat-Nasser, S., and Guo, W. (2001). *A Unified constitutive model for strain-rate and temperature dependent behavior of molybdenum.* Mechanics of Materials. 33:603-616.
Conrad, H. (1970). *The athermal component of the flow stress in crystalline solids.* Materials Science and Engineering. 6:260-264.
Conrad, H., (1964). *Thermally activated deformation of metals.* J. Metals 16, 582–588.
Diep, Q. B., Moxnes, J. F., and Nevstad, G. (2004). *Fragmentation of projectiles and steel rings using numerical 3D simulations.* In: The 21th International Symposium of Ballistics. Adelaide. Australia.
Durrenberge, L., Molinari, A., and Rusinek, A. (2008). *Internal variable modeling of the high strain-rate behavior of metals with applications to multiphase steels.* Materials Science and Engineering: A. 478.1–2.15:297-304.
Durrenberger, L., Klepaczko., J.R., Rusinek, A. (2007), *Constitutive Modeling of Metals Based on the Evolution of the Strain-Hardening Rate.* Transactions of the ASME. 129:550-558
Follansbee PS. (1985). High-strain-rate deformation of FCC metals and alloys. In: Metallurgical applications of shock-wave and high-strain-rate phenomena. p.451–79
Fressengeas, C., and Molinari, A. (1987). *Instability and Localization of plastic flow in shear at high strain rates.* Journal of the Mechanics and Physics of Solids. 33:185–211.
Gilman J.J., (1965). *Microdynamics of Plastic Flow at Constant. Stress.* J. Appl. Phys. 36: 2772-2777

Gloger, S., and Wanke, T. (2005). *Structural requirements and material selection in body development.* In: von Hagen, I., Wieland, H. J., editors. Steels in cars and trucks. Wiesbaden/Germany. Dusseldorf/Germany. Verlag Stahleisen. 32-35.

Grady, D. E., and Benson, D. A. (1983). *Fragmentation of metal rings by electromagnetic loading.* Experimental Mechanics. 12:393-400.

Guo, W. G., and Nemat-Nasser, S. (2006). *Flow stress of Nitronic-50 stainless steel over a wide range of strain rates and temperatures.* Mechanics of Materials. 38. 1090-1103.

Han, J. B., and Tvergaard, V. (1995). *Effect of inertia on the necking behaviour of ring specimens under rapid axial expansion.* European Journal of Mechanics - A/Solids. 14:287–307.

Han, H.N., Lee, C.G., Oh, C., Lee, T., Kim, S., (2004). A model for deformation behavior and mechanically induced martensitic transformation of metastable austenitic steel. Acta Materialia 52, 5203–5214.

Hartley, K.A., Duffy, J. (1984). Strain rate and temperature history ffects during deformation of fcc and bcc metals, International Physical Conf. Ser. 21-30

Heeres, O. M., Suiker, A. S. J., and de Borst, R. (2002). *A comparison between the Perzyna viscoplastic model and the consistency viscoplastic model.* European Journal of Mechanics - A/Solids. 21:1–12.

Hollomon JH. (1945). *Tensile Deformation*, Trans AIME 162:268-290

Hopkinson, B. (1914). A method of measuring the pressure produced in the detonation of explosives or by the impact of bullet. Phil. Trans. A. 213:437

Hu X, Daehn GS. (1996). *Effect of velocity on low localization in tension.*Acta Mater. 44:1021–33.

Iwamoto, T., Tsuta, T., 2002. Computational simulation on deformation behavior of CT specimens of TRIP steel under mode i loading for evaluation of fracture toughness. International Journal of Plasticity 11, 1583–1606.

Jankowiak, T., Rusinek, A., and odygowski, T. (2011). *Validation of the Klepaczko-Malinowski model for friction correction and recommendations on Split Hopkinson Pressure Bar.* Finite Elements in Analysis and Design. 47.10: 1191-1208.

Johnson, G.R., Cook, W.H. (1983), *A constitutive model and data for metals subjected to large strains, high strain rates and high temperatures.* In: Proceedings of 7th International Symposium on Ballistics. 541–547.

Johnson, G.R., and Cook, W.H. (1985). *Fracture characteristics of three metalssubjected to various strains, strain rates, temperatures and pressures.* Eng Fract Mech. 21:31–48.

S.E. Jones, J.A. Drinkard, W.K. Rule, L.L. Wilson, (1998). An elementary theory for the Taylor impact test, Int. J. Impact Engng. **21**:1 -13

Kapoor R, Nemat-Nasser S. (1999). Comparison between high strain-rate and low strain-rate deformation of tantalum. Metall Mater Trans. 31A: 815–823.

Khan, A.S., Suh, Y.S., Kazmi, R., (2004). *Quasi-static and dynamic loading responses and constitutive modeling of titanium alloys.* International Journal of Plasticity. 20: 2233–2248.

Khan, A.S., Suh, Y.S., Chen, X., Takacs, L., and Zhang, H. (2006). *Nanocrystalline aluminum and iron: Mechanical behavior at quasi-static and high strain rates, and constitutive modeling*, International Journal of Plasticity 22:195–209

Klepaczko, J. R., Rusinek, A., Rodrguez-Martnez, J. A., Pcherski R. B., and Arias, A. (2009). *Modeling of thermo-viscoplastic behavior of DH-36 and Weldox 460-E structural steels at wide ranges of strain rates and temperatures, comparison of constitutive relations for impact problems.* Mechanics of Materials. 41.5:599-621.

Klepaczko, J.R. (1975). *Thermally activated flow and strain rate history effects for some pollycristalline FCC metals*, Mat. Sci. and Engng, 18:121

Klepaczko, J.R., Malinowski, J.Z. (1977). Dynamic frictional effects as measured from the split Hopkinson pressure bar, in: High: Velocity Deformation of Solids, IUTAM Symposium, Tokyo, Japan. Springer-Verlag, Berlin, 1977, p. 403Kocks, U.F., Argon, A.S., and Ashby, M.F. (1975). *Thermodynamics and kinetics of slip*, Pergamon Press Oxford

Koistinen DP, Marburger RE. (1959). *A general equation prescribing the extent of the austenite–martensite transformation in pure iron–carbon alloys and plain carbon steels.* Acta Mater. 7:59–60.

Kubler, R.F., Berveiller, M., Buessler, P., (2011). Semi phenomenological modelling of the behavior of TRIP steels. International Journal of Plasticity 27, 299–327.

Larson, M., Needlemen, A., Tvergaard, V., and Storakers, B. (1982). *Instability and failure of internally pressurized ductile metal cylinders.* Journal of the Mechanics and Physics of Solids. 30.3:121–154.

Li, H. Y., Wang, X. F., Duan, J. Y., and Liu, J. J. (2013). *A modified Johnson Cook model for elevated temperature flow, behavior of T24 steel*, Materials Science and Engineering: A. 577:138–146.

Li, Y. S., Zhang, Y., Tao, N. R., and Lu, K. (2009). Effect of the Zener–Hollomon parameter on the microstructures and mechanical properties of Cu subjected to plastic deformation. Acta Materialia. 57:761–772.

odygowski, T., Rusinek, A., Jankowiak, T., and Sumelka, W. (2012). *Selected Topics of High Speed Machining Analysis.* Engineering Transactions. 60. *1:69-96.*

Macdougall D. (2000). Determination of the plastic work converted to heat

using radiometry. Exp Mech. 40:298–306
Molinari, A., Musquar C., and Sutter G. (2002). *Adiabatic shear banding in high speed machining of Ti-6Al-4V: experiments and modeling.* International Journal of Plasticity. 18:443-459.
Molinari, A., Ravichandran, G. (2005). *Constitutive modeling of high-strain-rate deformation in metals based on the evolution of an effective microstructural length*, Mech. Mater. 37: 737–752.
Molinari, A., Cheriguene, R., and Miguelez, H. (2011). Numerical and analytical modeling of orthogonal cutting: The link between local variables and global contact characteristics, International Journal of Mechanical Sciences. 53:183–206
Moufki, A., Molinari, A., and Dudzinki, D., (1998). *Modelling of orthogonal cutting with a temperature dependent friction law*, J. Mech. Phys. Solids. 46:2103-2138
Moufki, A., Devillez, A., Dudzinki, D., and Molinari, A. (2004). *Thermomechanical modelling of oblique cutting and experimental validation*, International Journal of Machine Tools & Manufacture. 44:971–989
Mecking, H., and Kocks U. F. (1981). Kinetics *of flow and strain-hardening.* Acta Metallurgica. 29.11:1865–1875.
Mercier, S., and Molinari, A. (2003). *Predictions of bifurcation and instabilities during dynamic extension.* International Journal of Solids and Structures. 40:1995–2016.
Mercier, S., and Molinari, A. (2004). *Analysis of multiple necking in rings under rapid radial expansion.* International Journal of Impact Engineering. 30:403-19.
Nasraoui, M., Forquin, P., Siad, L. and Rusienk, A. 2012. Influence of strain rate, temperature and adiabatic heating on the mechanical behaviour of poly-methyl-methacrylate: Experimental and modelling analyses. Materials and Design. 37:500–509
Nemat-Nasser, S., Guo, W. G., and Kihl, D. P. (2001). *Thermomechanical response of AL-6XN stainless steel over a wide range of strain rates and temperatures.* Journal of the Mechanics and Physics of Solids. 49:1823–1846.
Nemat-Nasser, S., Guo, W. G., Nesterenko, V. F., Indrakanti, S. S., and Gu, Y. B. (2001). *Dynamic response of conventional and hot isostatically pressed Ti-6Al-4V alloys: experiments and modeling.* Mechanics of Materials. 33:425-439.
Nemat-Nasser, S., and Guo, W. G. (2003). *Thermomechanical response of DH-36 structural steel over a wide range of strain rates and temperatures.* Mechanics of Materials. 35:1023–1047.
Nemat-Nasser, S., and Guo, W. G. (2005). *Thermomechanical response of*

HSLA-65 steel plates: experiments and modeling. Mechanics of Materials 37.379–405.

Nemat-Nasser S, Li Y. (1998). Flow stress of FCC polycrystals with application to OFHC. Copper. Acta Mater. 46:565–77.

Newslette arcelor Auto (2003), Arcelor group, n°3

Ono, K. (1968). *Temperature dependence of dispersed barrier hardening.* Journal of Applied Physics. 39:1803-1968

Ortiz, M., and Pandolfi, A. (1999). *A class of cohesive elements for the simulation of three dimensional crack propagation.* International Journal for Numerical Methods in Engineering. 44:1267-1282.

Pandolfi, A., Krysl, P., and Ortiz, M. (1999). Finite element simulation of ring expansion and fragmentation. International Journal of Fracture. 95:279-97.

Papatriantafillou I, Agoras M, Aravas N, Haidemenopoulos G. (2006). *Constitutive modeling and finite element methods for TRIP steels.* Comput Methods Appl. Mech Eng. 195:5094–114

Peng, X., Guo, H., Shi, Z., Qin, C., and Zhao, Z. (2013). *Constitutive equations for high temperature flow stress of TC4-DT alloy incorporating strain, strain rate and temperature.* Materials and Design. 50:198–206.

Penzes, G. (1998). Dynamic constitutive relation under dynamic tension, Master report ENIM-Sollac, Appendix 1

Perzyna, P. (1963). *The constitutive equations for rate-sensitive plastic materials.* The Quarterly of Applied Mathematics. 20:321–32.

Rittel D. (2000). An investigation of the heat generated during cyclic loading of two glassy polymers. Part I: Experimental Mech Mater. 32:131–47

Rodrguez-Martnez, J. A., Rusinek, A., and Klepaczko, J.R. (2009). *Constitutive relation for steels approximating quasi-static and intermediate strain rates at large deformation.* Mechanics Research Communications. 36.4:419-427.

Rodrguez-Martnez, J. A., Rusinek, A., Klepaczko, J. R., and Pcherski, R. B. (2009). *Extension of R-K constitutive relation to phase transformation phenomena.* Materials and Design. 30.7:2513-2520.

Rusinek A., and Klepaczko, J. R. (2001). *Shear testing of a sheet steel at wide range of strain rates and a constitutive relation with strain-rate and temperature dependence of the flow stress.* International Journal of Plasticity. 17.1: 87-115.

Rusinek, A., Zaera, R., Klepaczko, J. R., and Cheriguene R. (2005). *Analysis of inertia and scale effects on dynamic neck formation during tension of sheet steel*, Acta Materialia, 53.20:5387-5400.

Rusinek, A., and Zaera, R. (2007). *Finite element simulation of steel ring fragmentation under radial expansion.* International Journal of Impact Engineering. 34.4:799-822.

Rusinek, A., Zaera, R., Klepaczko, J. R. (2007). *Constitutive relations in 3-D for a wide range of strain rates and temperatures – Application to mild steels.* International Journal of Solids and Structures. 44.17:5611-5634.

Rusinek, A., Zaera, R., Forquin, P., and Klepaczko, J. R. (2008). *Effect of plastic deformation and boundary conditions combined with elastic wave propagation on the collapse site of a crash box.* Thin-Walled Structures. 46.10:1143-1163.

Rusinek, A., and Klepaczko J. R. (2009). *Experiments on heat generated during plastic deformation and stored energy for TRIP steels.* Materials and Design. 30:1:35-48.

Rusinek, A., Rodrguez-Martnez, J. A., Klepaczko, J. R., and Pcherski, R. B. (2009). *Analysis of thermo-visco-plastic behavior of six high strength steels.* Materials and Design. 30.5:1748-1761.

Rusinek, A., and Rodrguez-Martnez, J. A. (2009). *Thermo-viscoplastic constitutive relation for aluminum alloys. modeling of negative strain rate sensitivity and viscous drag effects.* Materials and Design, 30.10:4377-4390.

Rusinek, A., Rodrguez-Martnez, J.A., and Arias, A. (2010). A *thermo-viscoplastic constitutive model for FCC metals with application to OFHC copper.* International Journal of Mechanical Sciences. 52.2:120-135.

Schreiber, E., Anderson, O.L., and Soga, M. (1973). Elastic constants and their measurement, Mc Graw-Hill, New York

Simo JC, Hughes TJR. Computational inelasticity. New York: Springer; 1998 [chapter 8].

Simulia (2012). *Abaqus Analysis User's Manual.* HTML version 6.12.

R. Skalak, Longitudal impact of a semi-infinite circular elastic bar, J. Appl. Mech. 24 (1957) 59–64.

Sorensen, N. J., and Freund, L. B. (2000). *Unstable neck formation in a ductile ring subjected to impulsive radial loading.* International Journal of Solids and Structures. 37:2265–83.

Taylor, G. (1934). *The mechanism of plastic deformation of crystals.* Proceedings of the Royal Society of London A, 145:362–387

Taylor, G., (1948). *The use of flat-ended projectiles for determining dynamic yield stress I. Theoretical considerations*, Proceedings of the Royal Society of London. Series A, Mathematical and Physical Sciences, **194:**289-299

Varshni, Y. P., 1970. Temperature Dependence of the Elastic Constants, Phys. Rev. B, 2, pp. 3952–3958.

Voyiadjis, G. Z., and Abed, F. H. (2005). *Microstructural based models for BCC and FCC metals with temperature and strain rate dependency.* Mechanics of Materials. 37:355-378

Voyiadjis, G. Z., and Abed, F. H. (2006). *A coupled temperature and strain rate dependent yield function for dynamic deformations of BCC metals.* International Journal of Plasticity. 22:1398-1431.
Voyiadjis, G. Z., and Almasri, A. H. (2008). *A physically based constitutive model for FCC metals with applications to dynamic hardness.* Mechanics of Materials. 40:549–563.
Wang WM, Sluys LJ, de Borst R. (1997). *Viscoplasticity for instabilities due to strain softening and strain-rate softening.* Int J Numer Methods Eng. 40:3839 3864
Wilkins, M.L., Guinan, M.N. (1973). *Impact of cylinders on a rigid boundary*, J. Appl. Phys., 44:1200-1207
Winnicki A, Pearce CJ, Bicanic N. (2001). *Viscoplastic Hoffman consistency model for concrete.* Comput Struct. 79:7–19
Zaera, R., Rodriguez-Martinez, J.A., Fernandez-Saez, J., Rusinek, A., Pesci. R. (2012). *Constitutive model for analyzing martensite formation in austenitic steels deforming at high strain rates.* International Journal of Plasticity. 29:77–101
Zaera, R., and Fernandez-Saez, J. (2006). *An implicit consistent algorithm for the integration of thermoviscoplastic constitutive equations in adiabatic conditions and finite deformations.* International Journal of Solids and Structures. 43:1594-612
Zener, C., and Hollomon, J. H., (1944). *Effect of Strain Rate Upon the Plastic Flow of Steel.* J. Appl. Phys., 15: 22–32
Zerilli, F.J., and Armstrong, R.W. (1987). *Dislocation-mechanics-based constitutive relations for material dynamics calculations.* J. Appl. Phys. 61: 1816–1825.

Analysis of high-speed impact problems in the aircraft industry

Ángel Arias, Jorge López-Puente, José Antonio Loya, David Varas and Ramón Zaera

Department of Continuum Mechanics and Structural Analysis, University Carlos III of Madrid, Spain

Abstract

1 Introduction

The high cost of the energy needed to propel aircraft and ground vehicles has meant that reducing the weight in these systems is vital in order to reduce operational costs. This factor has a significant influence on the design of structures in the aeronautical industry and more recently in others such as high-speed rail networks and road haulage. This is a particularly sensitive issue for the civil aviation industry, given that the cost of fuel is one of the main expenses incurred by passenger airlines. Bearing in mind that fuel represents up to 40% of the total weight of an aircraft, a reduction of its weight results in a concurrent reduction in the amount of fuel needed as well as a significant reduction of the gross weight taken into account.

Due to their excellent specific mechanical properties, carbon fiber composites (CFRPs) are being used more and more as primary structures in components for wings and fuselage panels (Varas et al., 2012). The Boeing 787 and the Airbus A350 are typical examples of this, in which the composite materials can amount to up to 50% of the total weight. When designing aeronautical structures, it is particularly important to bear in mind the kinds of loads they will be subjected to during their service life in order to optimise structural response with minimum weight. Thus, one of the critical aspects to consider when using CFRPs is their vulnerability to out-of-plane impulsive loading, as a result of the reduced thickness of the laminate configurations and the brittle behaviour of the carbon fibers. If impulsive loading is due to high-speed impacts, the study of this problem takes on particular importance in the aeronautical field (Mines et al., 2007; Johnson et al., 2009), where situations in which aircraft are subjected to loads of this kind are not uncommon. During takeoff and landing, certain structural components of the aircraft can be hit by stones, tyre fragments, or any other kind of solid found on the runway; similarly, the fuselage, wings, stabilisers, radomes and nacelles can be hit during flight by birds, hail, ice, or even small fragments of material from the turbines. Of all the different kinds of impact, those involving hail and birds are the most serious due to their high probability of occurrence as well as their potential consequences (Pérnas-Sánchez et al., 2012). Furthermore, if the impact of the object is felt on wing structures, where the fuel deposit is found on many aircraft, this can lead to the phenomenon known as Hydrodynamic Ram, which is extremely important because of its effect on the vulnerability of aircraft (Artero-Guerrero et al., 2013). Vulnerability to impacts has thus become a relevant issue both from a regulatory perspective, as well as from the point of view of research into aeronautical safety. The regulatory certification requirements for European Aircraft (Joint Aviation Requirements, JAR) and for the American Federal Aviation Regulations (FAR) include specific requirements for preventing severe failures caused by impact and for guaranteeing a certain level of functioning in the event of an incident, so that aircraft can continue to fly until landing safely (Grimaldi et al., 2013). Similarly, from the current EU Framework Programme for Research (FP7), as well as the new European funding framework for scientific projects, Horizon-2020 (from 2014 to 2020), the promotion of research into aeronautical safety is considered a priority. As a result of this, and bearing in mind its particular importance, this chapter will analyse the most relevant aspects related to behaviour resulting from impact of CFRP structures, the Hydrodynamic Ram phenomenon, bird impact, ice impact and tyre impact.

The analysis of the behaviour of carbon fiber composites subjected to impact began in the 1980s (Caprino et al., 1984; Cantwell and Morton, 1989) in studies that demonstrate that the response of CFRP panels varies notably depending on the impact velocity. Behaviour at low speeds has been widely studied (Cantwell and Morton, 1989; de Moura and Marques, 2002; Caprino et al., 2003), mainly through impact experiments conducted on drop weight towers or pendulums. By contrast, the number of studies on the behaviour of CFRP panels subjected to high-speed impact (hundreds of meters per second, typically carried out through a single-stage gas gun) is relatively low and, given its relevance in relation to the vulnerability of aircraft, will be developed further in the following section. The research discussed here demonstrates the influence of the type of configuration (tape or woven), the temperature, the impact velocity, and the obliquity of the projectile (López-Puente et al., 2003; del Río et al., 2005; López-Puente et al., 2008).

The phenomenon of Hydrodynamic Ram (HRAM) must be taken into account, particularly in aircraft designed to reduce weight and whose fuel tanks tend to be integrated in the structure of the airplane. These tanks take up a large part of the wings, which are, in turn, the surface area most at risk of being involved in an impact. Thus, an object that impacts and penetrates the inner part of a fluid tank will transfer part of its kinetic energy to the fluid and, as a result of this, to the walls of the tank in such a way that the interaction between the fluid and the structure may cause serious structural damage that could lead to a catastrophic failure of the tank and the adjoining components. Over recent years, studies into the phenomenon of HRAM have increased (Townsend et al., 2003; Poehlmann-Martins et al., 2005; Disimile et al., 2009; Varas et al., 2009a) and the analysis of this complex problem through appropriate modelling has proved particularly interesting. Given its particular relevance in terms of attempting to describe the process, this chapter discusses the analysis of the numerical model proposed by Varas et al. (2009b, 2012), whose predictions help to advance knowledge in the phenomenon of HRAM.

Bird strike accounts for around 90% of all incidents related to structural damage due to impact on aircraft (Meguid et al., 2008). Because of this, the aviation authorities demand that all aeronautical components facing the direction of flight (radomes, wing leading edges, fuselage, tail wing, engines and window frames) have a certain resistance to bird impact which can be demonstrated through certification. This kind of certification testing, carried out with real birds in flight conditions, is extremely expensive and also increases the time taken to develop aircraft components. Thus, many aeronautic companies use artificial birds or substitute material in pre-certification experimental tests (Budgey, 2000). As another alternative, numerical methods have also been developed since the 80s in order to optimise the design of structures against bird impact and to guarantee that the first prototypes used in certification tests enable satisfactory results to be obtained (Heimbs, 2011a). The complexity of this problem will be analysed in detail below.

The threat of ice impact has become a subject of regulation for the aeronautical authorities (JAR-E 970), especially in aircraft with open-rotor engines belonging to the new generation of aircraft used for medium-range routes (Pérnas-Sánchez et al., 2012). In the last decade, interest in research into this problem has been demonstrated by the publication of different studies related to ice impact on carbon fiber panels (Kim et al., 2003; Johnson et al., 2006; Park and Kim, 2010) and on thin sheets of aluminium (Chuzel, 2009; Combescure et al., 2011). The studies confirm the influence of strain rate sensitivity of the ice in the process as well as the need to develop numerical models in order to approach the problem more exhaustively.

To that end, the authors present in this chapter the constitutive model of Pérnas-Sánchez et al. (2012) which is valid for high strain rates.

Studies on the impact of tyre fragments and their influence on the design of aeronautic structures took on particular importance after the repercussions of the accident involving the supersonic Concorde plane in Paris in the year 2000. With this kind of impact, which occurs at the operational speeds used for takeoff and landing (in the range of 10^2 m/s), the kinetic energy transferred to the structure impacted strongly depends on the behaviour of the projectile during impact, given the significant deformations that this can undergo (Mines et al., 2007; Karagiozova and Mines, 2007). As a result of this, research into this phenomenon conducted over the last decade and discussed in this chapter, has focused on the experimental characterisation of the type of rubber used in tyres (Mines et al., 2007; Guégan et al., 2010; Neves et al., 2010) as well as the proposal and validation of constitutive models of rubber and reinforcements (Peyraut, 2004; Mines et al., 2007) to enable adequate predictions about their behaviour to be obtained.

Throughout the chapter, the reader will be able to compare the complementary experimental and modelling approaches with those used in recent advances in relation to the kinds of problems considered. The physical phenomena associated with impact damage and its potential progression to the point of causing a failure is extremely complex (non-linear constitutive equations, large deformations, shock waves, inertial effects, fluid-structure interaction) and therefore appropriate modelling is necessary as a complementary tool of analysis. At the same time, and to ensure that the model is accurate, it is essential that quantitative experimental results are used which will also provide the information needed to understand the impact event.

2 Impact on CFRPs

With recent developments in the civil aviation industry, about 50% weight of the structure in modern aircraft designs, such as the Boeing 787 and Airbus A350, consists of laminated composite materials, but there is still sufficient margin to increase this percentage and further decrease the weight of the aircraft. The composite materials most used in the aviation industry are carbon fiber/epoxy matrix laminates, which possess desirable mechanical properties. With a density of 1500 kg/m^3, their tensile strength, for example, is higher than that of most metal alloys. However, although they exhibit excellent mechanical properties in the plane of the laminate, these materials have low impact resistance when the impact is perpendicular to the plane of the material. Such stresses may cause delamination that would increase substantially if the laminate is subjected to compression or shear in-plane loads. Such delamination causes a significant decrease in the stiffness and strength of the material, and could lead to catastrophic failure of the structure.

Impacts on composite laminates are usually classified on a scale ranging from low to high velocity. Low velocity impact corresponds to masses of an order of magnitude of kilograms and velocities up to tens of meters per seconds. An example of this kind of event is a dropped tool during maintenance operations. High velocity impacts occur at velocities of hundreds of meters per second. An example of this type of impact is uncontained engine rotor failure. In the following sections, these phenomena are described, giving more attention to the latter group since high velocity impacts are, in general, more detrimental to the structural integrity. We give a short review of the behavior of carbon fiber composite laminates under impact from two points of view, experimental and numerical.

2.1 Experimental results

Experimental tests of low velocity impacts on composite laminates are usually carried out using a drop weight tower; high velocity impacts are performed using a single stage gas gun. As noted above, composite laminates exhibit internal failure after impact in the through-thickness direction (under both low and high velocity impact). The amount of damage is usually measured using non-destructive techniques; these include ultrasound (C-scan, B-scan), X-rays, or thermography. These methods enable the area of the internal damage to be precisely quantified as a representative variable. Another quantitative parameter evaluated by many authors is residual strength, measured by subjecting impacted specimens to various types of loads such as uniaxial tension or compression, bending, or fatigue.

Impacts on CFRPs at low velocities

The first studies of low velocity impact on carbon fiber composite laminates appeared in the 1980s. Caprino et al. (1984) studied the response of various composite laminates, including carbon/polyester. They showed that the governing parameter is the total energy applied during the impact, rather than the speed or mass of the impact body. Cantwell et al. (1986) analyzed the extension of the damaged area caused by low velocity impacts in carbon/epoxy laminates, and impacted specimens were tested under tension to quantify the residual strength. Some years later, the same authors published two relevant studies in which comparisons between low and high velocity impacts were made (Cantwell and Morton, 1989, 1990). They concluded that the damage and energy absorbed by low velocity impact was determined by the size and shape of the target, whereas in high velocity impacts the damage has a localized shape, and is not governed by the specimen size. Another interesting conclusion was that high velocity impacts are in general more detrimental to the integrity of a composite structure than low velocity impact loading for the same energy range.

One of the main characteristics of low velocity impacts is the general lack of penetration by the impactor. The subsequent decade (1990s) saw several studies which determined the failure mechanism under low impact conditions, and established its relative importance. Wang and Vukhanh (1994) analyzed how mechanical properties affect the extension of the damaged area of transverse cracks and delamination. According to their observations, delamination results from a mode-II unstable fracture. Pavier and Clarke (1995) analyzed the complex damage patterns of matrix cracks and delamination that appear under low velocity impacts and how their distribution affects residual strength. The effect of the architecture under this loading condition was studied by Kim et al. (1996); they concluded that one of the major advantages offered by woven-fabric composites over non-woven composites made from unidirectional tapes is a greater resistance to interlaminar fracture.

In subsequent years, an increasing number of scientific articles on low velocity impact on composite laminates were published. Sala (1997) analyzed how the type of matrix (in a carbon reinforced composite laminate) affects the extension of the damaged area in this kind of impact. Siow and Shim (1998), and later de Moura and Marques (2002) studied damage and residual strength of laminates with different ply sequences; uni-directional, cross-ply and quasi-isotropic. Wang and Vukhanh (1998) analyzed what material parameters control the extension of the damaged area during low velocity impacts, in particular residual stresses during manufacturing and the effect of adjacent plies. Caprino et al. (1999) performed low-velocity impact tests on carbon/epoxy woven laminates of different thickness. The contact force was measured by placing a strain gauge in the impactor. This technique enabled a better understanding of the impact process, and the authors concluded that delamination was produced mainly by shear interlaminar stresses. Zhou et al. (2001) studied the effect of four geometric factors on the behavior of carbon/epoxy laminates under low velocity impacts: impactor size and shape, laminate size, and boundary conditions.

Indentation and perforation could also occur with low velocity impact if the energy is high enough; this phenomenon has been studied by Caprino et al. (2003). Structures made with composite laminates are often subjected to uniaxial or biaxial stress prior to impact; this subject has been investigated by Whittingham et al. (2004), who demonstrated that energy absorbed during impact is independent of pre-stress level at low impact energies, but significant as impact energy increases. Hybrid composites, including various reinforcements (e.g., glass and carbon fibers), are also of great importance for aeronautic structures; the effect of hybridization has been analyzed by Hosur et al. (2005), Sayer et al. (2010b), Sayer et al. (2010a) and Badie et al. (2011).

Since carbon/epoxy composite laminates are widely used in aeronautic and aerospace industries, it is necessary to know what their behavior is under impact at low temperatures. This has been examined by del Río et al. (2005). Carbon/epoxy laminates with different stacking sequences and architectures (unidirectional, cross-ply, quasi-isotropic and woven laminates) were impacted using a drop weight tower at temperatures ranging from 20° to −150°C. After the impact tests, the extension of the damaged area was measured by C-scan ultrasonic inspection, and the failure mechanisms were studied by optical microscopy. Figures 1 and 2 shows the C-scan images of cross-ply and quasi-isotropic laminates respectively at the different temperatures. It is evident that the lower the temperature, the greater the extension of the damaged area in laminates with tape architecture. Due to the different orientations of plies, low temperatures produce residual thermal interlaminar stresses because the thermal expansion coefficient is much smaller in the direction of the fibers than in the transverse direction. These stresses are high enough to contribute substantially to the failure mechanism upon the impact process, and hence lead to greater areas of delamination.

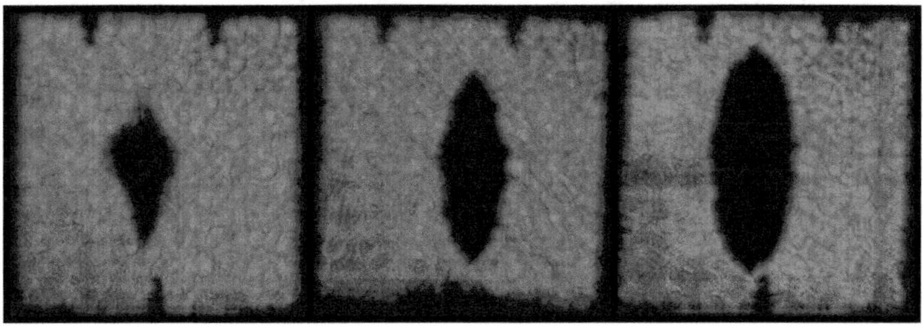

Figure 1. C-scan image showing damage extension in cross-ply laminate impacted at 4 J. 20° (left), −60° (center) and −150° C (right) (del Río et al., 2005).

Figure 3 shows C-scan images for a woven laminate impacted at 4 J at three different temperatures. In this case the differences in the size of the

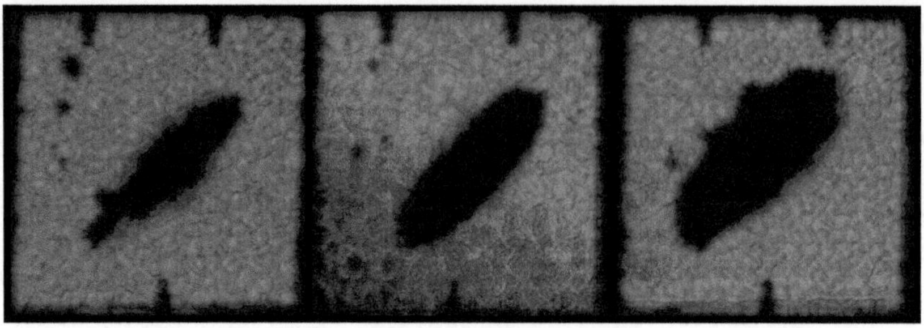

Figure 2. C-scan image showing damage extension in quasi-isotropic laminate impacted at 4 J. 20° (left), −60° (center) and −150° C (right) (del Río et al., 2005).

damaged area are smaller because the orientation of the fibers is parallel in every ply, and consequently thermal variation did not induce interlaminar stresses.

Figure 3. C-scan image showing damage extension in the woven laminate impacted at 4 J. 20° (left), −60° (center) and −150° C (right) (del Río et al., 2005).

Impacts on CFRPs at high velocities

The behavior of carbon fiber reinforced composites under high velocity impacts were first examined by military researchers around 1970. They studied the superior strength properties of these materials compared to metals, and investigated their behavior under these loading conditions. The Advisory Group for Aerospace Research and Development (AGARD) conducted preliminary analyses, describing the results of impacts with cube-shaped projectiles at speeds of up to 500 m/s. In particular, they analyzed the extent and types of damage that occur in laminates and the degradation of material properties.

In the late 1980s some scientific articles were published on high velocity impacts on composite laminates. Cantwell (Cantwell, 1988a,b) published two papers in which ultrasound was used to examine the damage in carbon/epoxy laminates under a wide range of impact velocities from 10 to 500 m/s. The influence of specimen thickness was analyzed. In these articles Cantwell highlighted the fact that the damage under high velocity impact is extremely localized around the impact point and hence the specimen size has a negligible effect. Sun and Potti (1996) performed several ballistic tests with different laminate thicknesses to measure energy absorbed during impact and ballistic limit (minimum impact velocity to perforate the laminate). The results were used to develop an analytical model to predict residual velocity in the event of perforation. Larsson (1997) analyzed the influence of stitching on the behavior of carbon/epoxy tape laminates under high velocity impact. He concluded that stitching provides up to 50% greater damage tolerance. Fujii et al. (2002) conducted tests on CFRP laminates at high speed (500 to 1230 m/s) using a projectile of mass 0.2 g. The study concluded that damage to the laminate decreases with increasing impact velocity. Tanabe et al. (2003) investigated the effect of the interfacial strength of carbon/epoxy laminates on their behavior under high velocity impact. Interfacial treatments were found to be effective for absorbing the kinetic energy of a projectile in a certain range of impact velocities; however, the range varies depending on both thickness and velocity.

As in low velocity impact, studying the behavior of carbon/epoxy laminates under high velocity impact at low temperatures is also important. López-Puente et al. (2002) analyzed the effect of impact velocity and low temperature (between 25° and −150°C) on the extension of the damage (measured by C-scan) produced by high velocity impacts on quasi-isotropic and woven carbon/epoxy laminates. Figure 4 shows the C-scan damage contours for woven laminates at different impact velocities and temperatures. Figure 5 shows the results under the same range of variables for quasi-isotropic laminates. Comparing Figures 4 and 5, it is easy to see that the woven laminates exhibit a smaller damage area than the quasi-isotropic laminates. Delamination occurs mainly between plies with different fiber orientation and hence different bending stiffness. To analyze the influence of the impact velocity it is necessary to distinguish between cases below and above the ballistic limit. When the impact velocity is not enough for the impactor to perforate the laminate, the size of the damaged area increases with impact velocity. For velocities above the ballistic limit, the effect is the opposite; the damaged area decreases as the impact velocity increases. At higher velocities the response in the laminate is more localized because impact induces shear plugging with no global bending. Regarding the effect of temperature, it is clear that for quasi-isotropic laminates lower temperature causes a larger damaged area under the same impact velocity. For woven laminates, temperature does not significantly affect the size of the damaged area because all plies have the same orientation and hence no thermal-residual stresses are induced. This behavior is the same as that found in low velocity impacts.

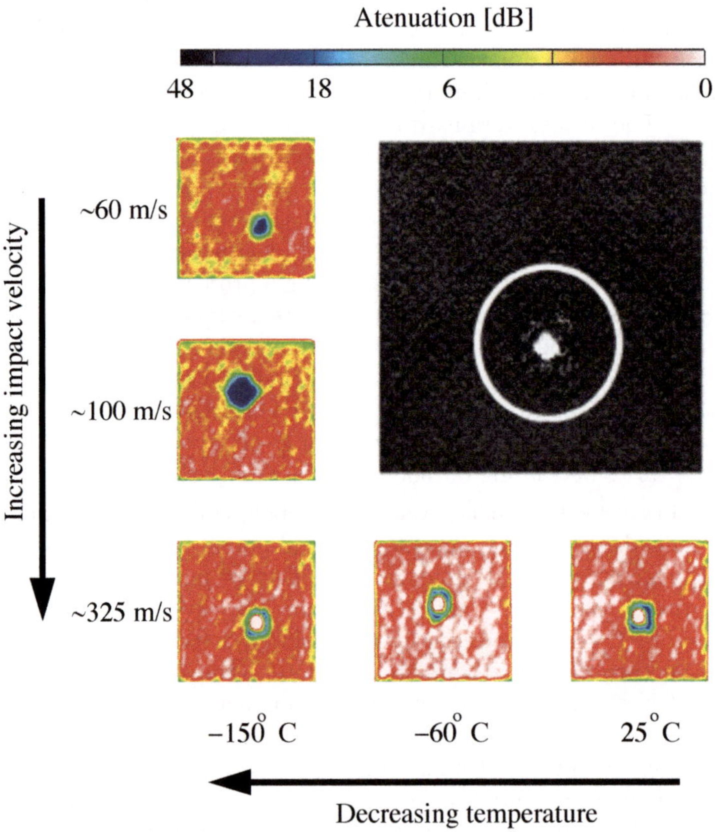

Figure 4. C-scan damage contours in woven laminates as a function of impact velocity and temperature. External appearance of the damage with perforation (López-Puente et al., 2002).

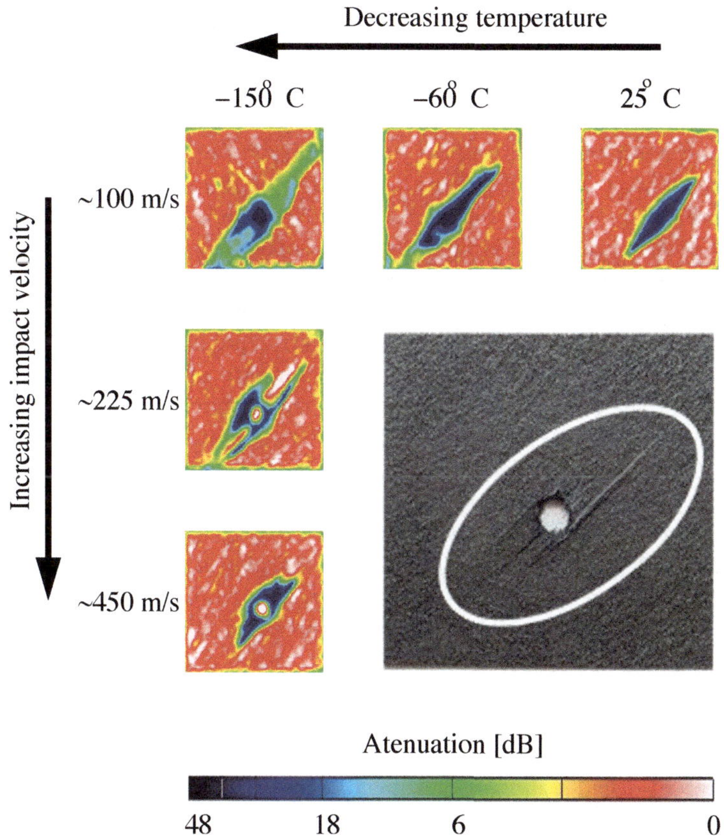

Figure 5. C-scan damage contours in quasi-isotropic laminates as a function of impact velocity and temperature. External appearance of the damage with perforation (López-Puente et al., 2002).

High velocity impacts on in-service structural composite components can occur at any angle, and hence it is important to study the effect of obliquity. López-Puente et al. (2008) analyzed the influence of impact velocity and obliquity on woven carbon/epoxy laminates. In this study, high velocity impact tests were carried out at a wide range of velocities (from 70 to 500 m/s) and comparing two impact angles (0° and 45°). After impact, specimens were inspected using C-scan. Results for impacts below the ballistic limit are summarized in Figure 6. At this range of velocities, the kinetic energy of the projectile is completely absorbed by the laminate through failure. The damaged area is proportional to the energy of the projectile, i.e. to the square of the impact velocity of the projectile. The extent of the damage is greater for normal impacts than oblique impacts; the component of the projectile velocity parallel to the laminate induces almost no damage. Above the ballistic limit, the size of the damaged area decreases with velocity for both normal and oblique impacts (Figure 7). At high velocities, the main mechanism to absorb the impact energy is transfer of momentum from the projectile to the ejected plug, and little delamination is observed. As the impact velocity decreases, penetration takes more time, and local bending (and hence delamination) starts to play a role. In this range of velocity, the damaged area is slightly larger for oblique impacts than for normal impacts because the former produces an elliptical hole, while the latter produces a circular hole.

2.2 Modeling of impacts on CFRPs

In recent decades simulation tools have become a key factor in the development of composite structures. Models using various materials have been proposed for this purpose. Because of the brittleness of the carbon fibers, those models do not take into account plastic strains, and elastic behavior until failure is usually considered.

The first models used only one scalar variable that defined the status of the material (intact or completely failed) as a function of current stress and strength properties. Some examples include the Tsai-Hill (Azzi and Tsai, 1965) and the Tsai-Wu (Tsai and Wu, 1971) criteria. In the subsequent decade, Chang and Chang (1987) proposed a more elaborate criterion, in which three different failure mechanism were distinguished; fiber failure, matrix failure under tension and matrix failure under compression. All three are in-plane failure mechanisms and hence only valid for 2-D approximations. Delamination is an important failure mechanism that can only be accounted for using a 3-D approximation. Hou et al. (1997) developed a 3-D material model for composite laminates. Based on the Chang and Chang

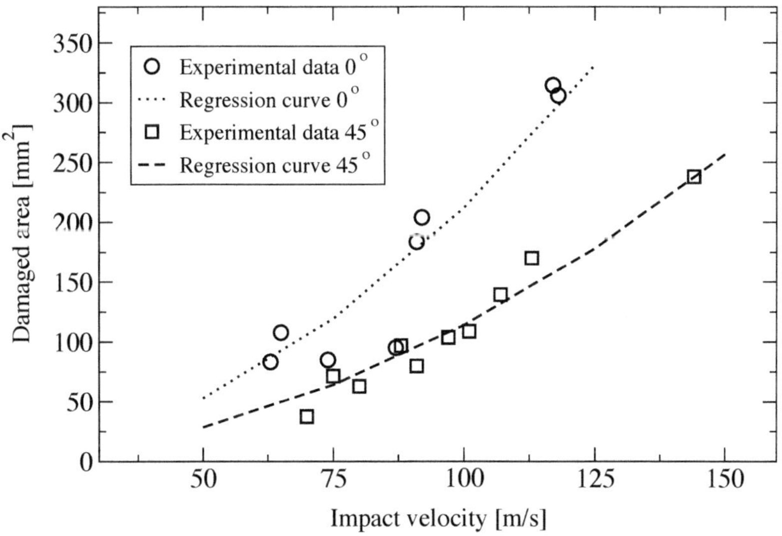

Figure 6. Damaged area vs impact velocity for woven CFRPs below the balistic limit (López-Puente et al., 2008).

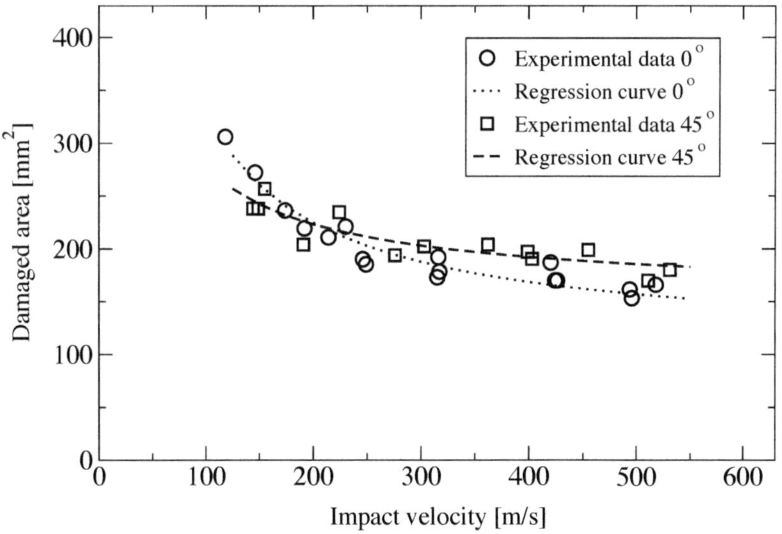

Figure 7. Damaged area vs impact velocity for woven CFRPs above the balistic limit (López-Puente et al., 2008).

(1987) model, they added a delamination criterion developed by Brew and Lagace (1988).

The above models were developed for tape laminates. The first model based on stresses for woven laminates was developed by López-Puente et al. (2003). In a more recent article (López-Puente et al., 2008), these authors presented a validation of this model for a wide range of velocities and under two different impact angles, comparing residual velocity and damaged area. Figure 8 shows the correlation between experimental tests and numerical results for residual velocity. At speeds close to the ballistic limit, small differences could be observed, but consistently lower than 5%.

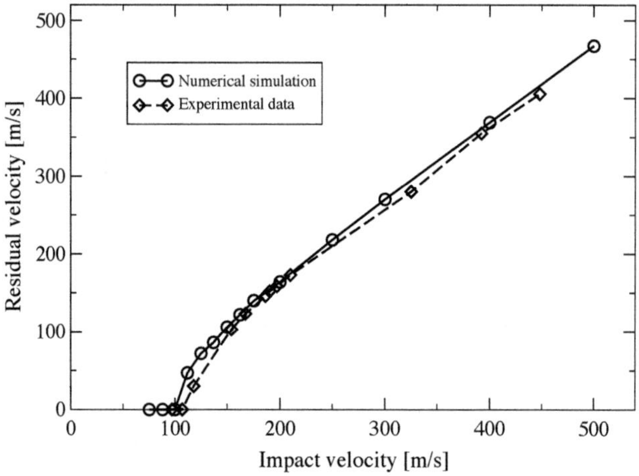

Figure 8. Residual velocity for different normal impact velocities. Experimental and numerical results for CFRP woven laminates (López-Puente et al., 2008).

Figures 9 and 10 show damaged area versus impact velocity for impacts at 0° and at 45°. For impact velocities below the ballistic limit, damage increased with impact velocity, as observed experimentally. Since the component of velocity normal to the laminate decreases as obliquity increases, and these types of materials are particularly sensitive to out-of-plane impacts, a reduction of the extent of damage is also observed as the impact angle increases. At high velocities, when penetration takes place, higher impact angles are associated with a larger damaged area due to the longer trajectory of the projectile through the laminate. In all cases the damaged surface decreases as impact velocity increases.

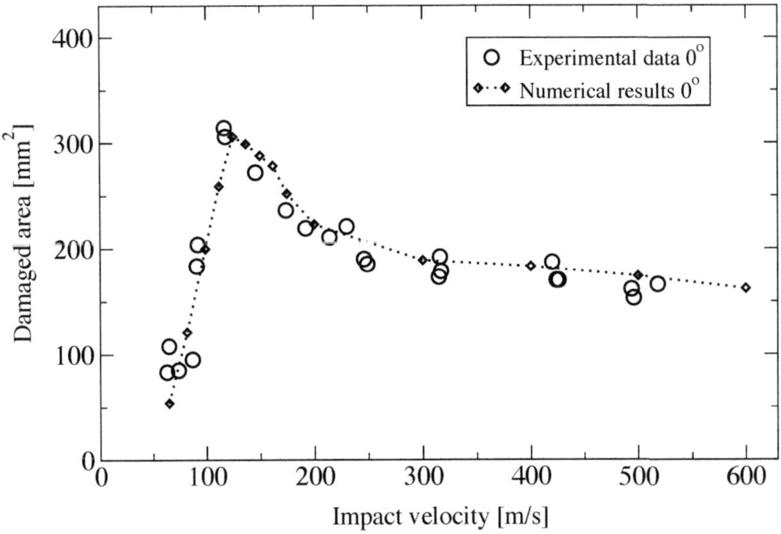

Figure 9. Damaged area vs impact velocity for different impact angles. Experimental and numerical results for CFRP woven laminates (López-Puente et al., 2008).

3 Hydrodynamic RAM in fuel tanks

Hydrodynamic ram (HRAM) is a phenomenon that occurs when a high-energy object penetrates a fluid-filled container. The projectile transfers its momentum and kinetic energy through the fluid to the surrounding structure, increasing the risk of catastrophic failure and extensive structural damage. Vulnerability to high-velocity impact loads is a critical issue for the design of aircraft structures. Of all the vulnerable components of an aircraft exposed to the threat of impact, the component with largest area is the wing. HRAM effects caused by impact on a fuel tank in the wings are therefore among the most important factors in the vulnerability of an aircraft. The significance of this phenomenon was amply demonstrated in the Vietnam War, in which many low-flying aircraft were downed by small arms and automatic fire. More recently, in Desert Storm, 75% of all aircraft losses were related to the fuel system (Addessio et al., 1997). Although vulnerability of aircraft to the HRAM phenomenon has usually been related to military aircraft, commercial airplanes are not immune from this risk. In

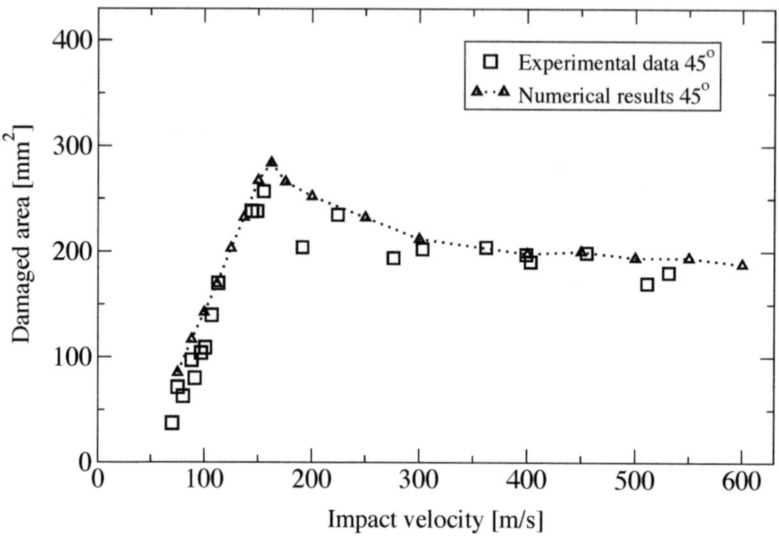

Figure 10. Damage area vs impact velocity for different impact angles. Experimental and numerical results for CFRP woven laminates (López-Puente et al., 2008).

1990, the Federal Aviation Administration (FAA) established the Aircraft Catastrophic Failure Prevention Research Program. One of the research areas of the program was the analysis of the effects of penetration of aircraft fuel tanks by uncontained turbine engine fragments (Santini et al., 1998). An example of the importance of this phenomenon is the Air France Concorde crash in 2000. The final investigation report revealed that HRAM had played a significant role in the failure of the aircraft. HRAM is especially dangerous for aircraft designed according to extreme lightweight principles. These aircraft commonly use wet wings (integral fuel tanks) and their structural strength cannot be improved by strengthening the airframe, since this would counteract the requirements of a lightweight design.

In recent years, interest in the study of HRAM has increased (Townsend et al., 2003; Poehlmann-Martins et al., 2005; Disimile et al., 2009; Varas et al., 2009a,b, 2011); however, modeling the HRAM phenomenon continues to be a challenge. The following section introduces the basic concepts of the HRAM phenomenon and its importance for the aircraft industry. In addition, before describing how the HRAM phenomenon can be simulated by means of a specific model with two different approaches, some experimental results are shown which give an idea of the difficulty of simulating such a complex problem. Finally, we show how the results obtained in the numerical model can help to understand the HRAM phenomenon.

3.1 Experimental analysis

A hydrodynamic ram event consists of four principal stages: shock, drag, cavitation and exit (Figure 11). Each stage contributes to structural damage through a different mechanism and to a different extent. When the projectile penetrates the wall of the fluid-filled structure, the energy of the impact is transferred to the fluid and generates a high pressure hemispherical shock wave. This leads to damage primarily in the vicinity of the impact location. During the drag phase, the projectile travels through the fluid, and its kinetic energy is partially transformed into fluid motion as the projectile is slowed by viscous drag. Displacement of the fluid from the projectile path generates a radial pressure field. In contrast to the pressure field developed during the shock phase, the fluid is accelerated gradually rather than suddenly. This causes less intense peak pressures, but they are of greater temporal extent. Displacement of fluid during the drag stage forms a cavity behind the projectile. The subsequent expansion and collapse (oscillations) of the cavity are known as the cavitation stage. These oscillations of the cavity can cause significant pressure pulses. The final stage of the hydrodynamic ram event occurs when the projectile exits the container. In contrast to the perforation of the front wall, the projectile exits through a pre-stressed wall. The pre-stress is caused by the initial shock stage and subsequent loading by the fluid.

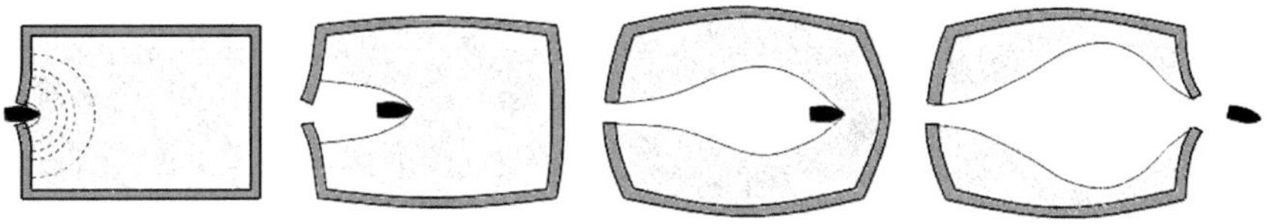

Figure 11. The phases of hydrodynamic ram.

In order to achieve a better understanding of the HRAM problem and associated fluid-structure interaction, several groups in the US Defence Department mounted a considerable research effort in the 1970s. The Naval Weapons Center (NWC, China Lake, California), conducted an interesting hydrodynamic ram project, in which a series of ballistic tests were performed to obtain fluid pressure measurements at several locations for a variety of projectiles (Lundstrom and Stull, 1973). At the same time, the Naval Postgraduate School (NPS) conducted an analytical and experimental hydrodynamic ram program in conjunction with the NWC project. Fluid pressures and entry-wall strains were measured for various projectile sizes and energy levels. The tests and their results are reported in different documents (Bates Jr, 1973; Holm, 1973; Soper, 1973; Fuhs and Ball, 1974; Mueller, 1974; Kappel, 1974; Holm, 1974; Page, 1975; Patterson, 1975). The University of Dayton Research Institute also performed several impact studies on fuel tanks during the 1970s. Bless et al. (1977) conducted experiments using spherical projectiles with diameters of 11.1 and 14.3 mm at impact velocities ranging from 1.5 to 2.4 km/s, and obtained data on entrance and side panel displacement, fluid pressure and the projectile trajectory. Lundstrom and Anderson (1989) conducted an experiment in which 23 and 30 mm high-explosive incendiary (HEI) rounds were shot into an open rectangular thick steel tank filled with water. Fluid pressure data from several points were reported. Sparks et al. (2005) carried out an experiment to examine the HRAM loads generated by a 12.7 mm projectile entering a water-filled container. The container was similar to that used in previous studies. Data pressure at different points was obtained and the trajectory of the projectile recorded with two high-speed digital cameras. All these tests employed open containers with a steel structure, and their behavior could be completely different from that of a closed container, as in the case of a fuel tank. Varas et al. (2009a) performed experiments on closed water-filled aluminum square tubes. These containers were subjected to impact by steel spherical projectiles (12.5 mm diameter) at impact velocities in the ballistic range. In addition, the aluminum tanks were filled at different volumes to study the way in which an air layer inside the tank influences the behavior under impact. The test boxes were instrumented with strain gauges and two pressure transducers at different fluid depths. The process of cavity formation was recorded with a high-speed camera. A diagram of the experimental device used for impact tests is shown in Figure 12. The test containers consisted of 6063-T5 square aluminum tubes (Figure 13). The tubes were closed with two PMMA windows, fixed to the tubes with four steel bars; these transparent panels allowed the impact process to be recorded. The contact points between the PMMA windows and tubes were sealed with silicone to avoid fluid leakage.

Figure 14 shows the sequence of an impact in a partially filled tank as recorded by the high speed camera. The projectile penetrates the fluid from left to right, creating a cavity behind it as it travels through the fluid. Since penetration of the projectile into the fluid is subsonic, spherical wave fronts propagating in the direction of the projectile as well as their rarefactions with the free surface can be seen in the photos. In partially filled containers,

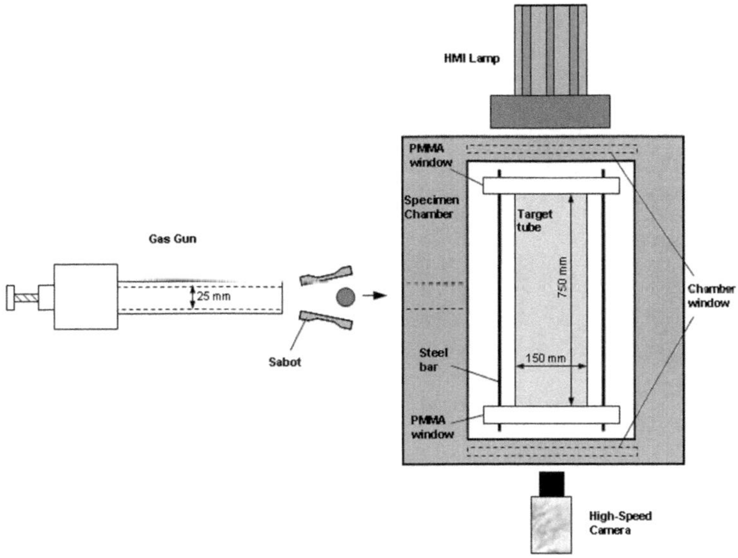

Figure 12. Diagram of the experimental device used for impact tests (Varas et al., 2009a).

the cavity raises the free surface of the fluid and a layer of fluid accelerates upwards. This layer will impact the upper wall of the tank at a high velocity.

The time history of the HRAM pressure wave inside the fluid was measured by means of two pressure gauges at different locations, near (PTn) and far (PTf) from the impact point. Figure 15 shows the typical curves found for pressure measurements at PTn and PTf at different velocities and fill volume percentages. As shown by these curves, the pressure time history changes as a function of the location of the pressure gauge and the velocity of the projectile. The pressure level is highly sensitive to the projectile velocity, the peak pressure at 900 m/s being nearly double the pressure reached at 600 m/s. Consequently, velocity will strongly affect the level of permanent deformation of the tube. Since the energy of the spherical pressure wave decreases with distance from the source, the position of the sensor also influences the pressure it records. The main difference observed as a function of filling percentage is the duration of the pressure pulse, which decreases as the fluid level falls. This is due to the effect of the rarefaction wave formed at the free surface, as explained in Varas et al. (2009a).

Varas et al. (2009a) also observed that strains farther away from the impact area are more influenced by filling percentage than by velocity. That

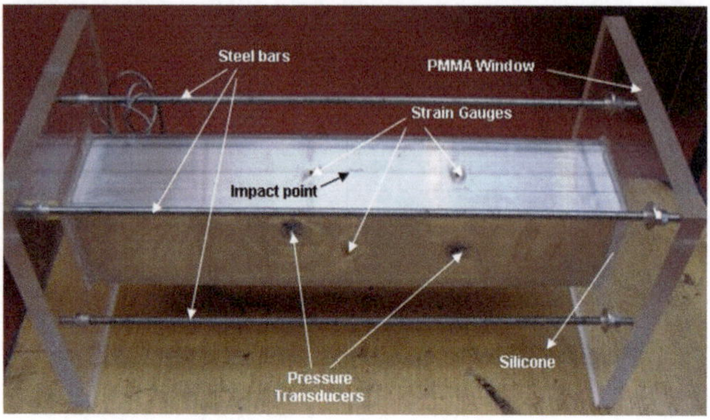

Figure 13. Closed test box (Varas et al., 2009a).

Figure 14. Sequence of projectile penetration into 75% full tube. $V = 600$ m/s. Images taken at 18 μs, 73.4 μs, 128.8 μs, 267.3 μ s, 461.2 μs, 599.7 μs. $t = 0$ corresponds to initial contact.

is, higher strains were found in a completely filled tank than in a partially filled tank, even though the latter was impacted at a higher velocity than the former. At points close to the trajectory of the projectile, this was not the case. Strains at these points are influenced more by velocity than by filling percentage. Therefore it can be concluded that the effect of velocity on the tubes is to generate a localized bulge, while the filling level produces a distribution of the deformation on the tubes. Figure 16 shows the effects of filling percentage schematically.

Although most studies of HRAM are conducted in metallic tanks, some

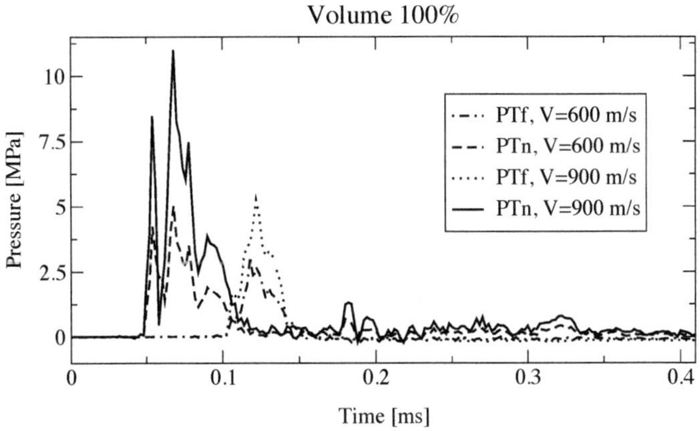

Figure 15. Pressure time history in a 100% filled tube impacted at 900 and 600 m/s (Varas et al., 2009a).

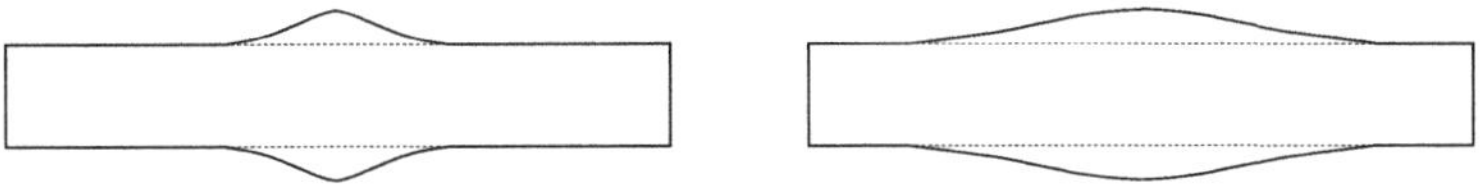

Figure 16. Left: region affected by the variation in impact velocity. Right: region affected by the variation in filling percentage (Varas et al., 2009a).

experimental studies using composite containers can be found. One of the first published studies using composite fuel tanks was from the Advisory Group for Aerospace Research and Development. Impact tests were conducted in 1988 to study the influence of projectile shape, fluids, and impact angles on aluminum and carbon epoxy tanks (Avery, 1981). Recently, Varas et al. (2011) studied high velocity impacts in CFRP tanks, which clearly revealed the importance of the fluid layer generated by the cavity in partially filled CFRP tanks.

3.2 Modeling HRAM

HRAM events have been simulated in various ways for over 30 years. The first methods were based on the use of the *Piston Theory* (Ball, 1974, 1972) and the *Variable Image Method* (Lundstrom, 1977) for the fluid-structure interaction. In general, none of these methodologies provided a realistic coupling between the fluid and the structure. Other codes such as HRSR (Hydraulic Ram Structural Response), ERAM and EHRSR were developed

(Herlin and Avery, 1981; Freitas et al., 1996), but all of them showed limitations and lacked accuracy for predicting the consequences of an HRAM event, since none of them fully coupled the mechanisms of the fluid-structure interaction, nor did they allow for a complex engineered structure.

One of the first documented examples of an HRAM simulation using the full set of continuum equations (i.e., conservation of mass, momentum, energy and material description) was developed by Kimsey (1980). Kimsey applied a Lagrangian finite-element method to simulate the penetration of a steel rod into a cylindrical tank. Distortions were a problem in Kimsey's simulation; nevertheless, qualitatively good results were obtained. The complicated physics and mechanics of HRAM phenomena were not satisfactory solved until higher-order numerical algorithms were incorporated into the codes in the late 1980s.

Coupled Euler-Lagrange methods have been under development since the early-to-mid 1990s, combining the desirable characteristics of Lagrangian and Eulerian formulations. These methods are used in various industries for a wide variety of analyses involving fluids interacting with structures or when high distortions may appear (Santini et al., 1998; Seddon et al., 2004; Anghileri et al., 2005d; Jarzab et al., 1988; Souli et al., 2002; Fasanella et al., 2006; Børvik et al., 2009; Lu and Wang, 2006; Kim and Shin, 2008), including airbag and tire-water dynamics in the automobile field (Marklund and Nilsson, 2002; Koishi et al., 2006), the impact of bird strikes on aircraft (Hanssen et al., 2006; Jenq et al., 2007), and the effects of sloshing on ships (Zhang and Suzuki, 2007). The Arbitrary Lagrangian Eulerian (ALE) technique is widely used in problems of this kind.

Another technique developed for calculating fluid flow and large deformation in structures is Smoothed Particle Hydrodynamics (SPH) (Lucy, 1977; Gingold and Monaghan, 1977). In the SPH method, the conservation equations are applied to discrete particles in the computational domain. The particles are not associated with a traditional grid system; therefore, they are not fixed by connectivity constraints as is the case in finite elements. This enables them to freely move and deform in any manner, making SPH particularly suitable for simulating processes in which large deformations appear. The SPH technique combined with a Lagrangian method provides an alternative to the ALE approach in which small distortions are computed using the Lagrangian method while large distortions are handled with the SPH technique. Examples of the use of this approach to simulate fluid-structure interaction or HRAM problems can be found in the literature (Anghileri et al., 2005d; Sparks et al., 2005; Vignjevic et al., 2002; Pentecôte and Kohlgrueber, 2004). Although the application of SPH to fluid-structure problems is relatively recent, this method has been applied for some time to problems in which large deformations occur, such as hyper-velocity impacts (Chen et al., 1997; Chen and Medina, 1998; Medina and Chen, 2000; Knight et al., 2000; Shintate and Sekine, 2004).

The capability to develop an accurate numerical model is of great importance since analytical solutions can provide only a limited understanding of the nature of the behavior. However, the modeling of coupled fluid-structure interaction problems such as HRAM has proven to be a complicated task and is still quite challenging. Recently Varas et al. (2009b) performed, through the development and validation of different numerical models, a thorough analysis of the suitability and predictive capabilities of the above techniques (ALE and SPH) to HRAM problems, that will be summarized next.

Comparison of ALE and SPH techniques for the HRAM problem

Varas et al. (2009b) simulated the impact of a spherical projectile into an aluminum container filled with fluid using several modeling techniques (ALE and SPH), and comparing their ability to represent this complex phenomenon. Some aspects of the methodology that Varas et al. (2009b) followed to develop the numerical models are described below. The commercial finite element code LS-DYNA v.971 (LSTC, 2007) was used in the development of the models.

Structural FE model. the models consider only a quarter of the whole tank because there are two planes of symmetry (Figure 17, left). Any reduction in the size of the model is very desirable since the nature of this simulation demands a high mesh density. The container is divided into three parts; the walls impacted by the projectile (entry and exit walls), the lateral wall, and the PMMA window that encloses the whole tube.

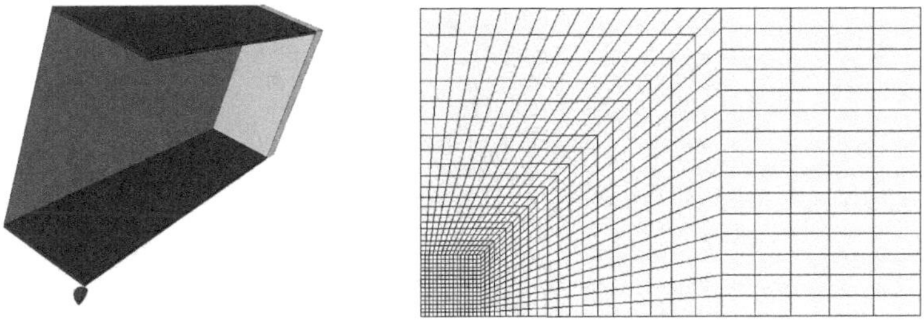

Figure 17. Left: box model geometry used for the analysis. Right: detail of the entry wall mesh (Varas et al., 2009b).

The impacted walls and the PMMA window are discretized using eight-node solid hexahedron Lagrangian elements with reduced integration. There is a refined mesh around the impact zone and a mesh that grows progressively coarser as the distance to the hit point increases, as can be observed in Figure 17, right. Four-noded Belytschko-Tsay shell elements were used to discretize the lateral wall in order to reduce the number of elements.

The Johnson-Cook hardening relation (Johnson and Cook, 1983), which takes into account the equivalent plastic strain, equivalent plastic strain rate and temperature effects, was selected to model the aluminum of the box. An elastic material model was used for the PMMA window (Vesenjak et al., 2005).

Fluid FE approaches. The deformations that the fluid inside the tank will undergo are too large to consider a pure Lagrangian description as an appropriate option. For this reason a multi-material ALE formulation or, alternatively, a mesh-free approach (SPH) were chosen for the treatment of the fluid.

- ALE approach. The fluid inside the box is discretized by means of eight-node solid hexahedron elements. In this case, the air surrounding the box should be considered, being modeled using the same elements as the water. Modeling this air region is essential to allow the water to flow into it, deforming the walls of the structure (Figure 18). It is important to take into account that element size can affect modeling of the fluid/solid contact. Refined meshes can lead to instability and leakage problems at the fluid/solid interfaces. Those problems can be avoided by choosing the same element sizes as the Lagrangian mesh at the interfaces.

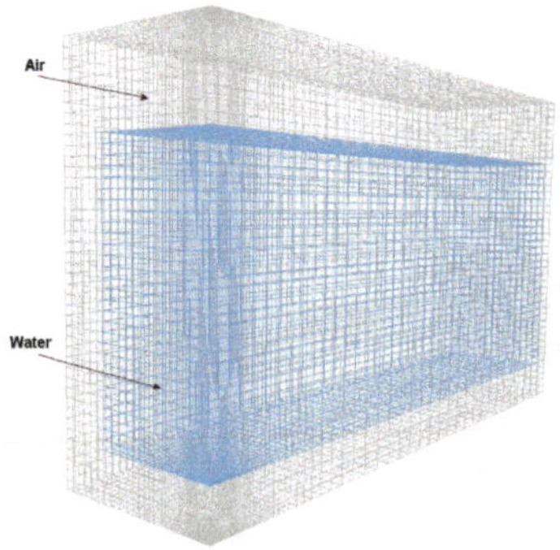

Figure 18. Mesh of the fluids in the ALE approach (Varas et al., 2009b).

The water was modeled using the equation for viscous fluids

$$\sigma_{ij} = 2\upsilon_d \dot{\varepsilon}'_{ij} - PI \tag{1}$$

in which υ_d is the dynamic viscosity, $\dot{\varepsilon}'_{ij}$ the deviatoric strain rate tensor, and I the identity tensor. The pressure P is calculated as a function of the compression and of the internal energy per unit volume, using the Mie-Gruneisen equation of state based on a cubic shock velocity-particle velocity. The air was modeled using the same constitutive equation (1), with the properties of the air, and a linear polynomial equation of state (LSTC, 2007) that models the air as an ideal gas.

The fluid-structure interaction is achieved by means of a penalty-based ALE-Lagrangian coupling algorithm implemented within LS-DYNA. This allows the fluid material to flow around the structure but prevents its penetration into the structure mesh, with penalty forces being applied to the fluid and the structure.

- SPH approach. The SPH method requires a large number of uniformly distributed particles to provide reasonably accurate results (Figure 19). In the SPH method, it is not necessary to model the surrounding air since the particles can flow freely in any direction, deforming the walls of the structure. The constitutive law and the equation of state used to model the water are the same as in the ALE approach.

Figure 19. Left: mesh of the water in the SPH approach. Right: detail of the SPH mesh (Varas et al., 2009b).

Here, the fluid-structure coupling algorithm is different from that applied in the ALE approach. In all the contact algorithms, the SPH particles are considered to be nodes. As other authors suggest (Schwer, 2004), the contact interface should be chosen with special care in order to model the interaction of several SPH particles with each face of the elements of the Lagrange contacting body. This fact, along with the need for a homogeneous mesh to obtain accurate results, means that a higher number of particles are used to discretize the water domain than in the ALE approach. The interaction of the SPH particles and the walls of the structure is modeled by means of a constraint interface that ties the SPH particles to the solid elements of the surrounding structure. This has been proven to be the best way to transmit the fluid movement to the structure and avoid contact instability.

Varas et al. (2009b) observed that the pressure level and the beginning of the pressure pulse predicted by both numerical models correlated quite well with the experimental results, as shown in Figure 20. The evolution of the cavity predicted by the various approaches matches well with images from the experiments, as can be seen in Figure 21. This is of great importance since the cavity is the main cause of the deformation of the walls. The way in which the walls are deformed is interesting because of its influence on the behavior of contiguous cells that can be part of a whole fuel tank system. The deformation at points all along the entry and exit walls of the tanks was therefore also compared experimentally and numerically. It was observed that the shape and maximum deformation of the walls near the impact point were well predicted by the simulations, but the deformation at points farther away from the impact point were over-estimated in the simulations (see Figure 22).

In the light of the results of the validation, Varas et al. (2009b) concluded that both approaches, ALE and SPH, were capable of reproducing cavity evolution, pressure and wall deformation of the hydrodynamic ram phenomenon. Finally, it is worth noting some considerations about the CPU time required for the approaches used. The finer meshing of the SPH approach, necessary for effective simulation, requires very long CPU times; however the results are not more accurate than the ALE results. Thus it seems that the ALE approach is more suitable for simulating a problem of this kind where a large physical domain needs to be modeled.

Numerical analysis of the HRAM phenomenon

The broad range of quantitative results provided by the numerical codes can be analyzed to obtain better knowledge about the fluid/structure interaction phenomenon that takes place during HRAM events. In this kind of problem, the numerical analysis enables identification of the beginning and end of each stage of the HRAM and their influence on the permanent dis-

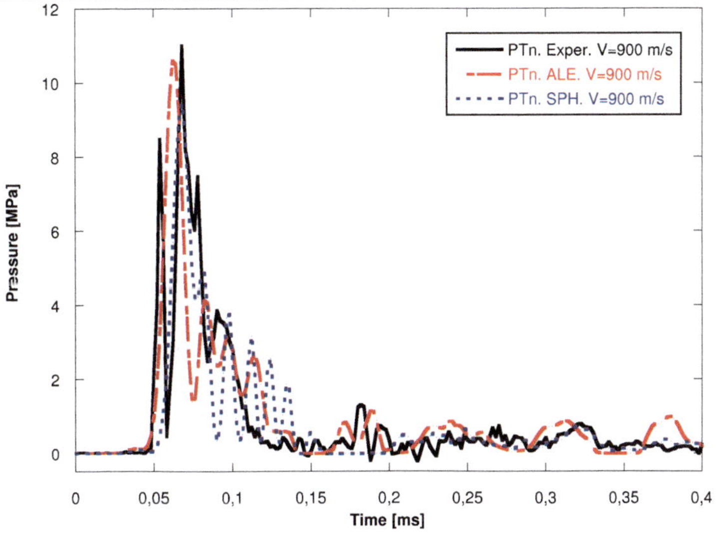

Figure 20. Pressure time history of a pressure gauge close to impact point (PTn) in a 100% filled tube impacted at 900 m/s. (Varas et al., 2009b)

placement of the structure, resulting in a complete and detailed view of the HRAM phenomenon. In the work of Varas et al. (2012), the link between projectile advance and deformation of the tank is established by means of a study of the interactions between these two solids and the fluid inside the tank, paying attention to the transfer of energy and momentum between projectile, fluid and tank walls. Some of the results are presented here.

Figures 23 and 24 depict the kinetic and internal energies of the various components of the problem; projectile, water, and the walls of the tank. A first stage can be seen that consists of the passage of the projectile through the tube. In this stage, the kinetic energy of the projectile is transferred to the fluid and, through it, to the tank walls. Once the projectile exits the tank, the work done over the tank-fluid group ends and there is now a transfer between the different energy terms of the structure and fluid.

When the projectile exits the tank, most of the energy has been transferred to the fluid. This is mainly due to the kinetic energy that the projectile has communicated to the water, rather than to the increase in water pressure. After the exit of the projectile, however, the kinetic energy in the fluid decreases but the internal energy due to pressure keeps a constant and higher value for a much longer time than that taken by the projectile to pass through the tank. The kinetic energy in the water begins to decrease

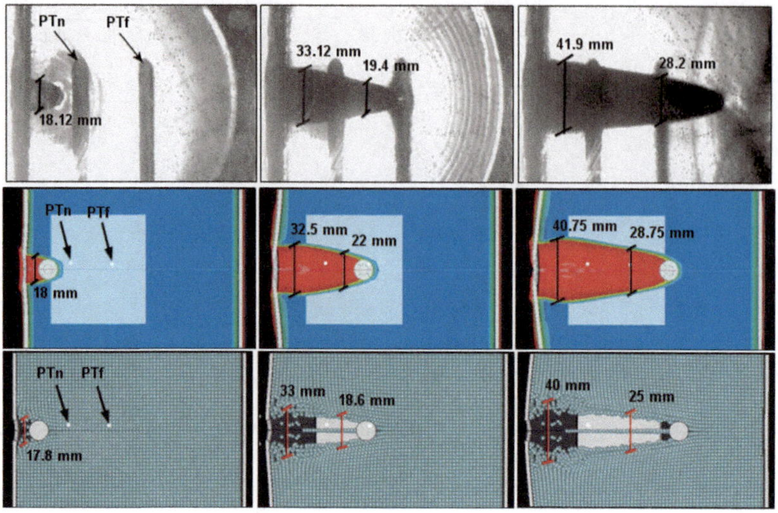

Figure 21. Cavity evolution at times 0.028, 0.084 and 0.140 ms obtained experimentally (top), ALE simulation (centre) and SPH simulation (bottom) (Varas et al., 2009b).

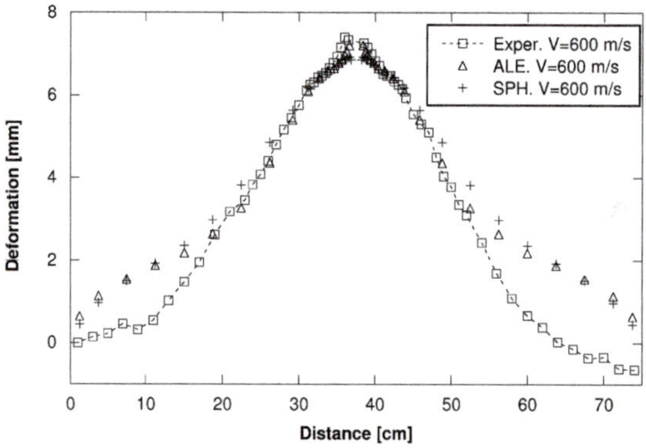

Figure 22. Deformation of the entry wall of a completely filled tank impacted at 600 m/s (Varas et al., 2009b).

before the projectile's exit, which means that the fluid has already started to transmit part of its energy to the surrounding walls.

When the internal and kinetic components of energy in the various walls were analyzed in detail, Varas et al. (2012) observed that the walls had a greater kinetic energy value than the internal energy during the stage of

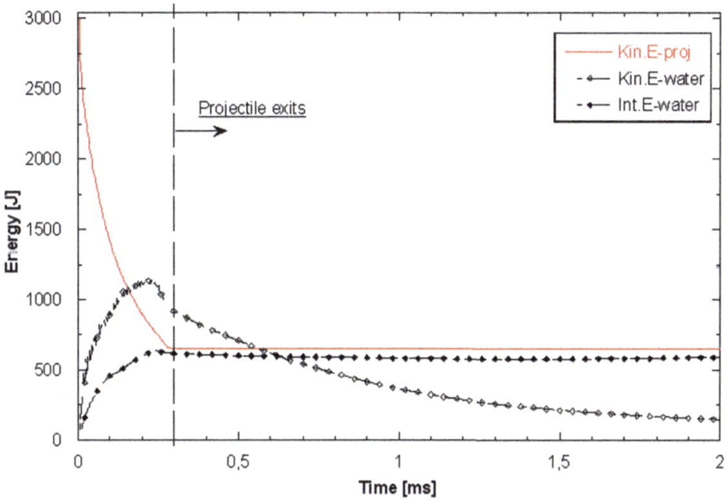

Figure 23. Time history of energy in projectile and fluid. Fully filled tank and V=900 m/s (Varas et al., 2012).

projectile penetration into the tank (Figure 24). This fact confirmed the essentially dynamic nature of the first stage of HRAM, with the prevalence of inertial forces, as compared to a behavior characterized by a more uniform pressure value in the fluid and slower wall deformation once the projectile has exited the tank.

For a better understanding of the behavior of the tank walls, Varas et al. (2012) also studied the pressure contours generated in the fluid during the progress of the projectile (Figure 25). The pressure contours clearly show the existence of two zones of overpressure. The first advances at the speed of sound in the fluid and is bounded between a spherical front centered on the impact point and a rarefaction front that is created as a result of the interaction with boundary conditions on the entry wall. The second zone advances in front of the projectile, at the same speed, and its value and extension decrease as the projectile decelerates. The first wave front, due to its larger extension, affects both the exit and the lateral wall. The second wave front, with a smaller area, mainly affects the exit wall when the projectile approaches it, and to a smaller extent the lateral walls. These two overpressure fronts explain the evolution in kinetic energy of the tank walls. It can be seen, for example, that the times at which the first wave

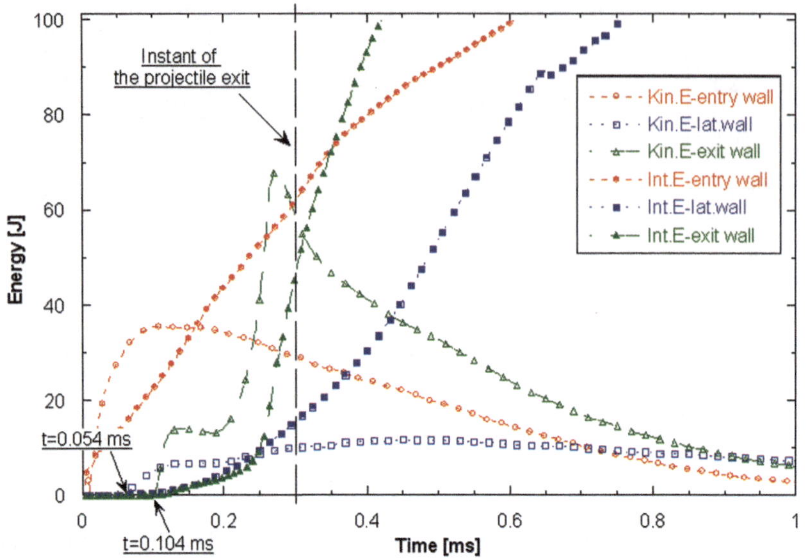

Figure 24. Detail of time history of energy in aluminum walls. Fully filled tank and V=900 m/s (Varas et al., 2012).

front reaches the lateral and the exit wall match the times at which the kinetic energy of these walls begins to rise. When the exit wall is no longer affected by the first overpressure front, a plateau is observed in the kinetic energy of the exit wall. When the second overpressure front begins to act on the exit wall, the kinetic energy rises much more than before. Thus the second overpressure front seems to have more influence on exit wall deformation than the initial wave. The maximum kinetic energy value in the exit wall occurs at the instant at which the projectile makes contact with it. At that moment it begins to decrease as the kinetic energy progressively turns into internal energy. It is worth noting that the gap between the two overpressure fronts becomes more noticeable at lower impact velocities because the first front always moves at the same speed, while the second one travels at the velocity of the projectile. The lateral walls are affected by both overpressure fronts because of their proximity to the area of influence of the second front. On fuel tanks whose lateral walls are further away from the projectile trajectory, the second overpressure front should have no influence on lateral wall deformation.

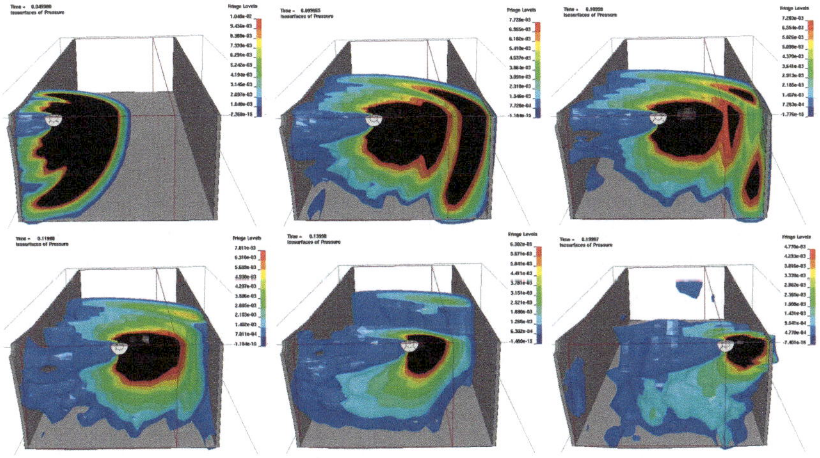

Figure 25. Pressure contours in the fluid at 0.05, 0.10, 0.11, 0.12, 0.14 y 0.20 ms. Fully filled tank and V=900 m/s (Varas et al., 2012).

After the exit of the projectile, the deformation of the walls continues due to the overpressure caused by the formation of the cavity. This overpressure is less intense than that generated by the two previously mentioned fronts (as can be seen from the pressure values obtained both numerically and experimentally in the previous works of Varas et al. (2009a,b)), but of greater temporal extent due to the fact that this overpressure is not related to either the propagation of the sound waves propagation or to the forward movement of the projectile. Therefore its effect on the deformation of the tank walls is important.

In addition to the pressure contours, Varas et al. (2012) analyzed the momentum time history, both in the fluid and in the structure, to obtain information for a better understanding of the HRAM phenomenon. The momentum of the different parts of the problem was clearly related to the two overpressure fronts. It was also shown that the stages of increasing momentum in the exit wall matched the stages of decreasing momentum in the fluid, confirming the transfer of momentum between the water and the tank walls.

As a conclusion of the above discussion, we note that the availability of an accurate validated numerical tool not only enables the HRAM phenomenon to be understood, but is also helpful for the design of future structures that can attenuate HRAM effects.

4 Bird strike

Collisions between an aircraft and one or more birds are known as bird strike. One of the first records of this phenomenon was reported by the Wright brothers in 1905 (MacKinnon, 2004), but the first fatal aircraft accident resulting from a bird strike was in 1912 (Lewis, 1995). As of 2004, bird strikes had caused the loss of 90 civilian aircraft and cost 242 lives worldwide (Stoll and Brockman, 1997). Some studies (ICAO, 2001) estimate more than 30000 bird strikes a year on civil aircraft, but the majority cause little or no damage.

In recent decades, bird populations have been increasing due to greater environmental awareness efforts and the birds' ability to accommodate to human environments. By some estimates, the population of Canada geese in the USA has quadrupled since 1987, and snow goose, seagull and cormorant populations have also increased (Eschenfedler, 2001). In the UK, flocks of Canada geese and greylag may number up to 1000 individuals.

The probability of finding a large flock of birds at high altitudes where aircraft speeds are higher enhances the risk of impact. As well, the majority of airports are positioned away from populated areas and closer to natural habitats of wild birds. These factors, combined with the increasing use of aircraft, have led to a rise in the number of recorded bird strikes during commercial flights, in particular during take-offs and landings, especially in recent years. Despite efforts to reduce the problem, the worldwide cost of bird strikes exceeds $USD 3 billion per year. An important part of these costs are associated with cancelling flights and making alternate travel arrangements for affected passengers (Short et al., 2000).

4.1 Experimental analysis of bird strike

Because of the importance of this problem, before aircrafts are put into operational use they must show certain elements of structural integrity following a bird strike. For certification, the aviation authorities specify that components should be struck by a standardised bird fired by a pneumatic launcher at operational velocity. For instance, turbofan blades must be able to withstand impacts from various sizes and numbers of birds (depending on the size of the engine) without immediate catastrophic failure; Boeing 777 engines are required to produce at least 75% of full rated thrust after simultaneously taking in four 1.125 kg birds (Lewis, 1995). Other aircraft parts (windshields, radome, fuselage and the leading edges of the wings and empennage) and helicopter parts (fuselage, windshield and rotor blades) are also vulnerable to damage from bird strikes, and therefore must be designed to be bird-proof.

For final certification, testing the use of real birds is mandatory. Individuals of the same mass, even of the same species, may present some differences in mass-density that may cause different effects upon impact. However the authorities only define the mass, not the species to be used for testing. It is also difficult to ensure that the target point is struck properly because the irregular shape of real birds means that results are not uniform. Eliminating the need to use animals for testing would therefore be useful. To deal with these problems, many companies use bird substitutes for pre-certification impact tests (Budgey, 2000) but there is no standardisation in the design of artificial birds. The design of bird substitutes attempts to reproduce the pressure loading obtained in a real bird impact, rather than reproducing the biometric parameters of real birds. Meat, foam, rubber, silicone, neoprene, wax, emulsion, and gelatin have been used as materials (Wilbeck, 1978; Nizampatnam, 2007; Baughn and Graham, 1988). Dynamic loading produced with gelatine provides the best agreement with the impact of a real bird. Gelatine is therefore the material most used in experimental research projects. Moreover, analysis techniques must be validated before they can be used in the design process.

4.2 Modeling the bird strike

The design/manufacture/test cycle for a bird-impact–proof aircraft component may be a long, expensive process (Nizampatnam, 2007). To reduce the number of sample tests, a reliable design tool to predict the structural behaviour of components under bird strike impact is necessary. Numerical simulations enable the use of various design approaches to minimise weight according to bird strike design requirements.

The design of bird-proof aircraft components requires the use of complex numerical tools that should be capable of dealing with the highly non-linear characteristics of the problem. However, a simplified description of the process by means of a theoretical model provides useful insight into the stages of the impact event. Wilbeck (1978) developed a theoretical model for the sequential stages that occur during an impact.

Theoretical model of Wilbeck (1978)

Wilbeck described the impact process by means of the following sequence of consecutive events (Figure 26).

Shock regime. During the first moments of the impact of a projectile against a rigid target, a compressive shock wave is generated at the contact surface and transmitted through the projectile as each succeeding layer of particles comes to rest (Figure 26a). This happens so quickly that the particles

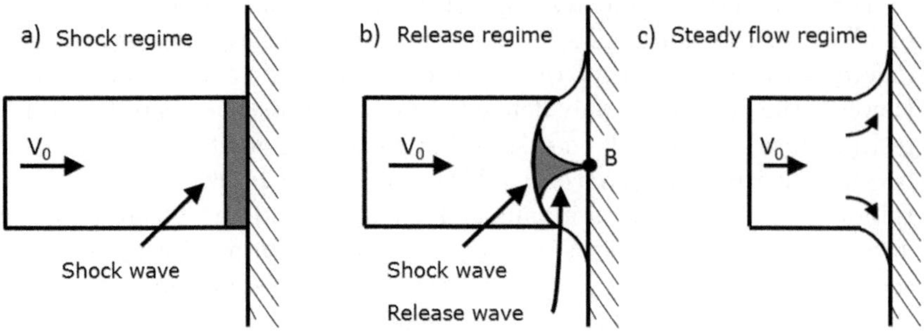

Figure 26. Schematic description of an impact process according to Wilbeck (1978).

do not have time to receive information from the lateral free surface, and deformation occurs under plane strain conditions. The pressure in this region is very high and can be considered to be constant. This process is adiabatic and the shock pressure pulse is well described by the characteristic Hugoniot relationship

$$P_H = \rho_0 V_0 V_{shock} \tag{2}$$

where ρ_0 is the initial density of the projectile, V_0 is the impact velocity of the soft body, and V_{shock} is the velocity of shock wave propagation in the material. The model uses an empirical linear relationship between the shock wave velocity and the impact velocity, valid in the range of pressure where there is no phase transition

$$V_{shock} = c_0 + kV_0 \tag{3}$$

where c_0 is the speed of sound in the uncompressed material and k is an experimental constant. Therefore, at low impact velocity the shock wave velocity tends toward the speed of sound in the material. However, at high impact velocities the shock wave velocity is higher than c_0, so the increase of Hugoniot pressure with impact velocity is non-linear. For water, Eq. (3) accurately fits results up to Mach=1.2 if $k = 2$ and $c_0 = 1482.9$ m/s, values obtained by Rice et al. (1958) at a temperature of 20 °C and a pressure of 1 atmosphere.

Release regime. As the shock wave propagates, the particles close to the side of the projectile's edge are subjected to a high pressure gradient due to shock stresses at the axis and the stress-free condition at the lateral surface. Thus the material particles are accelerated radially outward. From the lateral free surface, release waves propagate to the axis of the projectile, relieving the radial pressures (Figure 26b).

The time t_B needed for the release wave to reach the central point of the contact surface, point B in Figure 26b, determines the duration of the shock pressure pulse acting on the target. It can be calculated by $t_B = a/C_r$, where a is the radius of the projectile and C_r is the speed of sound in the shocked material, obtained from the slope of the pressure-density curve at the Hugoniot state

$$C_r = \sqrt{\frac{dP}{d\rho}} \tag{4}$$

In a shocked material, the slope of the pressure–density curve increases with density, and the release waves propagate faster than the shock wave. This makes it likely for t_B to be small (less than 1 μs for a typical bird projectile geometry), causing pressure applied on the target to decay rapidly (Iannucci, 1992, 1998). This effect can be seen in Figure 27, representative of a typical time–pressure curve for a soft body impacting on a rigid plate.

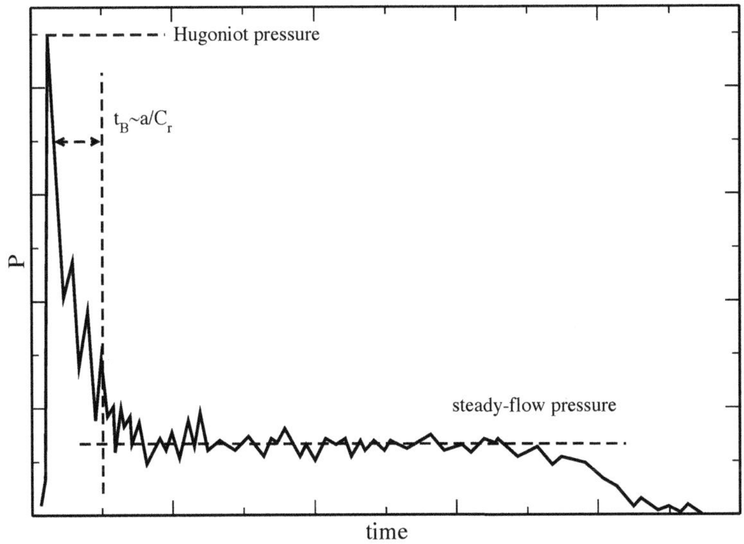

Figure 27. Typical time-pressure curve for a normal soft-body impact on a rigid plate.

Steady flow regime. This phase starts after several reflections of the release waves, contributing to a decrease of the shock pressure on the projectile. During this regime (Figure 27c), the pressure and velocity in the projectile can be considered constant and the particles flow along streamlines.

The existence of this regime depends on the length/diameter ratio (L/D) of the impactor. For a projectile with a ratio greater than a critical value, the steady flow condition should exist for some time. Typically, a ratio $L/D > 2$ is need for this regime to be clearly detected. For a very low L/D ratio the impact event could finish before the steady flow regime appears.

In the case of a cylindrical projectile impacting normally on a rigid plate, the point at the centre of the contact surface is the stagnation point with zero velocity. During this stage, the flow of the projectile can be assumed incompressible and the stagnation pressure P_{stg} can be obtained from the Bernouilli equation by

$$P_{stg} = \frac{1}{2}\rho_0 V_0^2 \tag{5}$$

To determine the radial pressure at any point of the contact surface, Wilbeck (1978) considered the exponential radial distribution obtained by Banks and Chandrasekhara (1963) in a water jet

$$P(r) = P_{stg} exp\left[-\frac{1}{2}\left(\frac{r}{a}\right)^2\right] \tag{6}$$

a being the radius of the projectile.

The simplified description of the sequence of regimes that appear during the impact process developed by Wilbeck (1978) enables the characteristic time–pressure curve due to perpendicular soft-body impact on a rigid plate to be determined, justifying the initial short peak in pressure and the long-lasting steady state.

Bird constitutive equations

Birds are mostly composed of water, and at the range of impact velocities at which real bird strikes take place, they behave like a fluid. Thus equations of state (EOS) are commonly used to define the pressure–volume relationship in a soft body since they provide reasonable accuracy.

From the the continuity equation, the Hugoniot linear relationships for density and pressure, and the linear relation for V_{shock} given by Eq. (3), the following EOS can be derived (Wilbeck, 1978)

$$P = \rho_0 c_0^2 \frac{\rho(\rho - \rho_0)}{((1-k)\rho + k\rho_0)^2} \tag{7}$$

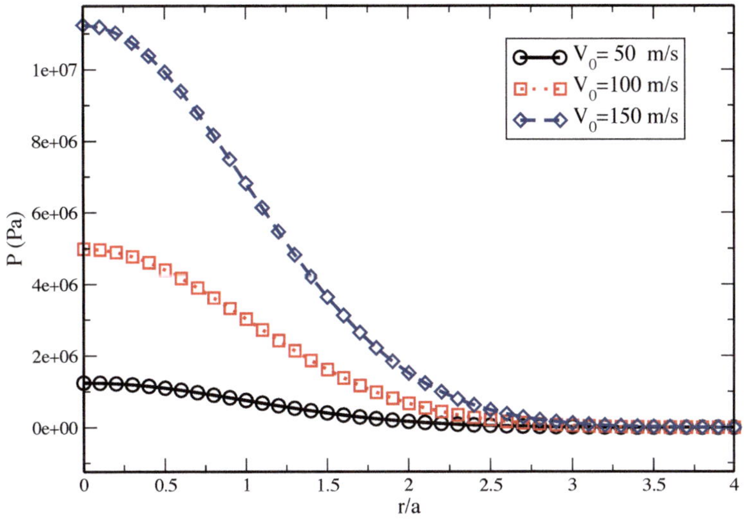

Figure 28. Radial pressure distribution during the steady flow regime for different impact velocities.

Using $\mu = \rho/\rho_0 - 1$, it can be rewritten as

$$P = \rho_0 c_0^2 \frac{\mu(1+\mu)}{((1-k)\mu - 1)^2} \tag{8}$$

For low values of μ, this expression may be expanded in a polynomial form (implemented in some commercial numerical codes) as

$$P = \beta_1 \mu + \beta_2 \mu^2 + \beta_3 \mu^3 + O(4) \tag{9}$$

where $\beta_1 = \rho_0 c_0^2$, $\beta_2 = (2k-1)\beta_1$, $\beta_3 = (k-1)(3k-1)\beta_1$.

Some authors (Murnaghan, 1944; Cogolev et al., 1963) established empirical EOSs that are easier to handle for theoretical models, which can be expressed in the form

$$P = D\left[\left(\frac{\rho}{\rho_0}\right)^B - 1\right] \tag{10}$$

Ruoff (1967) deduced the values of the parameters B and D for this expression to deal with the linear Hugoniot relationships, resulting in $B = 4k - 1 = 7$ and $D = \rho_0 c_0^2 / B = 321$ MPa for water.

The differences between the above EOSs (Eqs. 8–10) can be observed in Figure 29. Note that in range of pressures reached in a bird strike (below 200 m/s), the differences between them are negligible; therefore any of them is equally valid for simulation purposes.

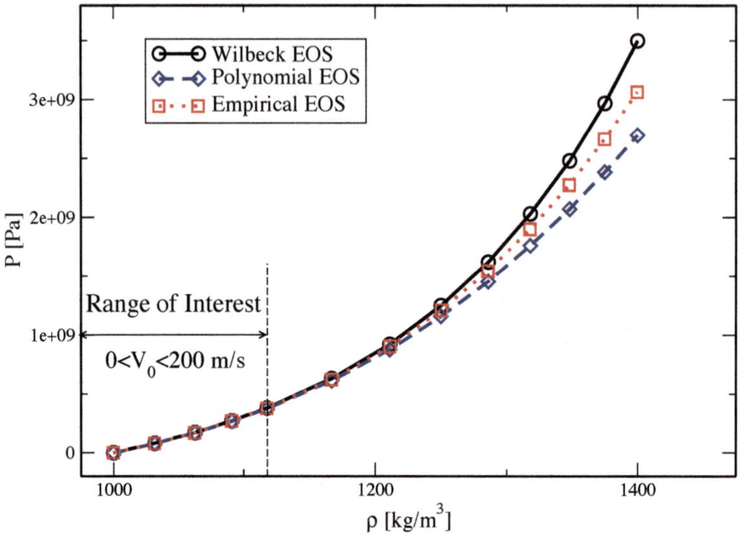

Figure 29. Shock pressure estimation by different EOS.

To consider the effects of the anatomical cavities in the birds' structure, the equation of state must include the void content (porosity). The presence of porosity in a soft body has a great effect on both the shock velocity and the compressibility of a material during impact. Wilbeck (1978) used the theory of Torvik (1970) to predict pressure–density relationships in a homogeneous mixture of water and air. The hypotheses assumed by the model are: i) a porous material that is macroscopically homogeneous and isotropic, ii) the density of each constituent is the same as for a homogeneous sample of the given constituent at the same pressure, and iii) the shock pressure is the same in each component.

Figure 30 shows the Hugoniot pressure as a function of the impact velocity for different values of initial porosity. It can be seen that porosity induces a sharp decrease in the pressure, and it therefore has to be considered during the shock stage. However, porosity barely changes the pressure during the steady flow stage, as can be seen in Figure 31.

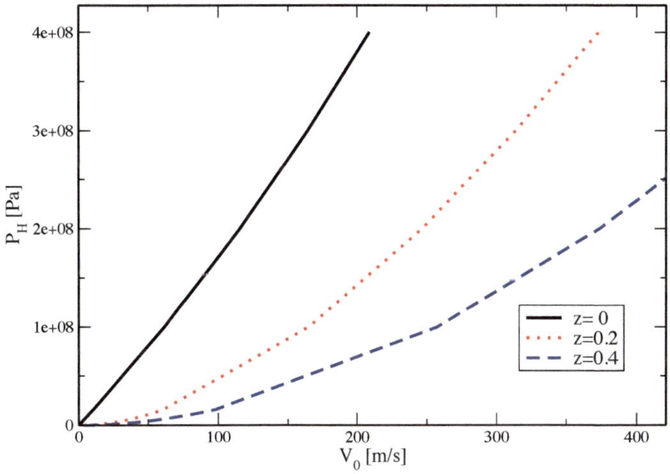

Figure 30. Hugoniot shock pressure versus impact velocity for water with different values of the porosity. z: air volume fraction.

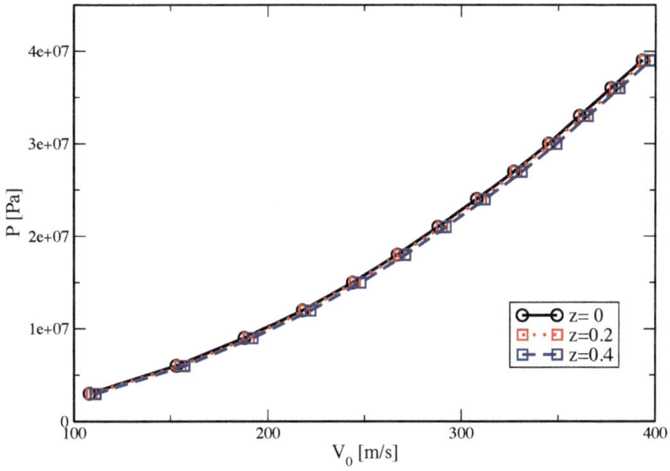

Figure 31. Steady flow pressure versus impact velocity for water with porosity. z: air volume fraction.

Numerical techniques

The bird strike problem is highly nonlinear and the numerical codes used to simulate it must be able to deal with large deformations, contact, and non-linear behaviour of materials. Additionally, they typically offer a variety of discretisation schemes, such as Lagrangian with remeshing, Eulerian, Arbitrary Lagrangian–Eulerian (ALE) and smoothed particle hydrodynamics (SPH). Numerical analyses of bird strike simulation available in the literature were done with codes using explicit time-integration schemes.

The features of the numerical schemes relating to the simulation of the bird strike problem will now be summarised (Figure 32a).

Lagrangian modeling with remeshing. The Lagrangian impactor model shows good agreement with experimental results in the absence of large distortions. However, in the case of problems with large deformation, the distortion of a Lagrangian mesh may lead to inaccurate results in the impact pressure curve (due to severe hourglassing), negative volume elements, and large computational time (Georgiadis et al., 2008; Castelletti and Anghileri, 2003).

To deal with mesh distortion in a a bird strike problem, adaptive remeshing can be used where needed, but the new mesh generation increases numerical errors and computational cost (Nizampatnam, 2007). Another option to prevent numerical problems is element deletion (Stoll and Brockman, 1997), but it can cause additional inaccuracy due to momentum and energy loss, and artificial oscillation in the contact forces, leading to a worse correlation with experimental results (Lavoie et al., 2007; Airoldi and Cacchione, 2006). Georgiadis et al. (2008) concluded that the use of Lagrangian modeling for a soft-body strike may be not the most recommendable technique .

Eulerian modeling. Due to the hydrodynamic behavior of the soft body during impact, some authors have used the Eulerian modeling technique, which is mainly applied in fluid mechanics problems and flow processes (Huertas, 2006; Nizampatnam, 2007). Here, the mesh is fixed in space and the material flows through it (Figure 32b). Boundaries are not clearly defined and element size needs to be small for accurate results. The entire domain where material may be present at any moment must be mapped, causing a very high computational cost compared with the Lagrangian model (Nizampatnam, 2007). However, there may be no stability problems due to the mesh deformation.

ALE formulation. This approach (Figure 32c) has a high computational cost compared to other methods and, according to Castelletti and Anghileri (2003) and to Anghileri et al. (2005c), the results obtained for bird strike modeling are not satisfactory due to the large deformation of the moving mesh.

SPH method. The SPH method has been used in bird strike simulation for the bird model (Lacome, 2004; Zammit et al., 2010; Anghileri et al., 2005b; Liu et al., 2008; Georgiadis et al., 2008; Wu et al., 2009; Salehi et al., 2010) (Figure 32d). According to these authors, this method as used for bird strike problems has a lower computational cost than Lagrangian soft-body impactors, has high stability, and gives good correlations with experimental results.

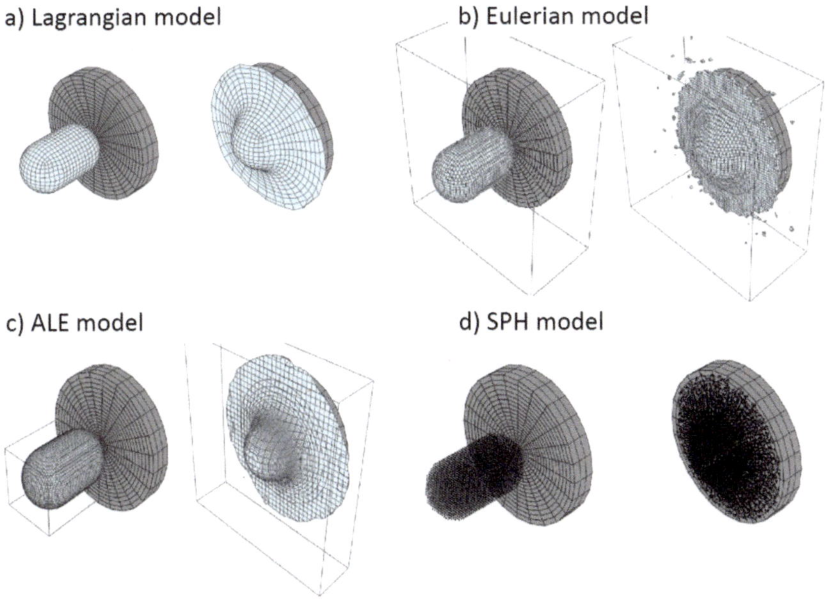

Figure 32. Meshes for soft-body impact on a rigid plate (based on Heimbs (2011b)).

All these approaches have pros and cons, and the choice often depends on the particular problem to be analysed, the numerical code, geometry and mass of the impactor, impact velocity, as well as other factors.

Impactor geometry

The development of numerical techniques and constitutive models has enabled the bird strike problem to be analysed in greater depth. Since artificial bird shape is not standardised, the influence of the impactor geometry is an important issue to analyse, and this has been done by several authors (McCallum and Constantinou, 2005; Airoldi and Cacchione, 2006; Zhu and Tong, 2008). Typically, bird substitutes have a simplified geometry, the torso of the bird being represented by a right cylinder, a cylinder with hemispherical ends, an ellipsoid, or a sphere (Figure 33). Nizampatnam (2007) studied the shock and steady-flow pressure with these four shapes, and the best correlation with experimental data is obtained by a cylinder with hemispherical ends. The geometry mainly influences the shock pressure. This is highest for the right cylinder due to its largest initial contact area, with a shock peak around 43% higher than for the cylinder with hemispherical ends, which in turn is about 30% higher than for the ellipsoidal impactor (Meguid et al., 2008). They also found that a change in the length to diameter ratio of the impactor has little influence on the normalized impact pressure and impulse. Changing the impactor mass from 4 to 8 lb also slightly increases the pressure peak. These conclusions are drawn from the theoretical approximation of the problem by Wilbeck (1978).

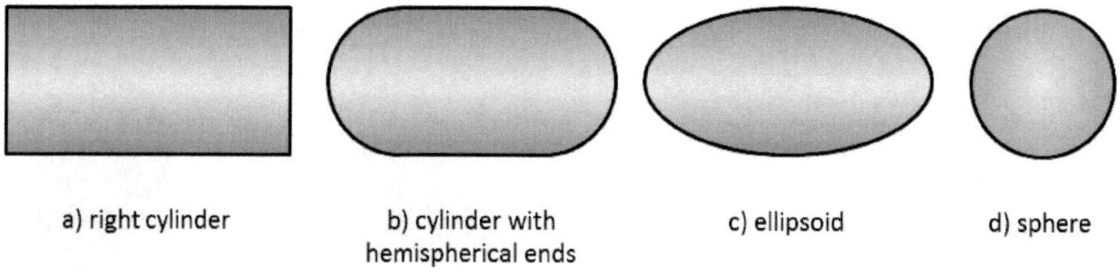

Figure 33. Different impactor geometries.

Some authors question whether the birds typically used in the certification tests are representative of a real bird strike during flight (Budgey, 2000). Therefore, more realistic shapes for the artificial bird, including neck, torso and wings (McCallum and Constantinou, 2005), and even bones and lungs (Nizampatnam, 2007), or considering different densities and equations of state in the model (Nizampatnam, 2007) have been proposed. Comparing the typical pressure–time curve obtained from a cylindrical impactor with hemispherical ends, some peaks appear to be attributable to the impact of the head and torso. McCallum and Constantinou (2005) concluded that the initial neck impact pre-stresses the target panel before the torso impacts, which may affect the level of damage predicted for the structure. Nevertheless, there are no available experimental results for a real bird strike with forward-facing head and extended wings for comparison with numerical predictions.

5 Ice Impact

Aircraft structures may be subjected to a variety of ice impact hazards. The radome, landing lights, canopy, engine intake, and leading edges of the wing and tail empennage are all vulnerable to high-velocity ice impact during takeoff, flight, and landing. In addition, ice flung from the edge of a propeller blade may impact the nacelle of the twin engine or the fuselage. In particular, the open-rotor engines being considered by the main aircraft manufacturers for their next generation of jet airliners for medium-range routes will be positioned on the empennage. Thus, the effect of impact on composite panels by ice slabs from propeller blades will have to be evaluated. For spacecraft, ice presented one of the most serious debris impact threats to thermal protection systems on the Space Shuttle Orbiter. Ice that forms on the fuel lines of the external tank, if dislodged during flight, can impact the tiles or the reinforced carbon-carbon wing leading edge of the spacecraft (Fasanella and Boitnott, 2006). All these structural elements must have tolerance to damage caused by ice impact, and their design should consider ice impact as a potential threat. However, commercial simulation codes do not have suitable constitutive equations to model mechanical behavior of ice, since it is not a structural material. Likewise, less attention has been paid to the experimental study of ice under impact conditions. Consequently, the mechanical behavior of ice must be studied through experimental tests, and numerical tool models suitable for simulating ice at high strain rates must be developed (Anghileri et al., 2005a).

The following section presents a review of experimental and modeling developments for the behavior of ice at high strain rates, with special attention to aircraft applications.

5.1 Experimental tests and mechanical behavior of ice at high strain rates

Most studies of the mechanical behavior of ice are of *ice Ih*, the most common on earth. This kind of ice is formed when liquid water is cooled below 0° C at ambient pressure. It can also be deposited directly from vapor with no intervening liquid phase, such as the accretion of glaze or rime ice on the airfoil of airplanes. Ice Ih possesses the hexagonal crystal structure seen in the shape of snowflakes. The other crystal forms of ice are thermodynamically stable only at high pressures.

The mechanical behavior of ice has been widely studied for arctic ship transportation, oil and gas facilities, and for cold ocean and glacier research. Deformation of ice in these fields occurs at low strain rates in the range 10^{-7} to 10^{-1} s^{-1}. Creep and quasi-static experimental tests have been undertaken to gain knowledge about its behavior (Haynes, 1978; Cole, 1988; Gold, 1988; Schulson, 1990, 2001; Petrovic, 2003). These tests show the noticeable influence of the conditions of ice formation (leading to single crystal, columnar or granular polycrystal structures) on its mechanical properties. The presence of inclusions or air bubbles, which act as stress concentrators, also influences the mechanical properties of ice. Some researchers therefore obtained standard material properties from ice-manufacturing companies, in order to minimize dispersion in the experimental values. Young's modulus has been reported to be in the range of 9.7 to 11.2 GPa and Poisson's ratio varies from 0.29 to 0.32. Tensile strength varies between 0.7 and 3.1 MPa and depends on the specimen volume, following a Weibull statistical distribution. Compressive strength ranges between 5 and 25 MPa. Some properties (such as compressive strength) are strongly affected by temperature as well, so experimental results are commonly provided for a given set of pre-defined temperatures (e.g. −10, −20, −30, −40°C). In tension, ice shows brittle behavior due to crack nucleation and cleavage. On the other side, compression increases ductility and strength, like in other brittle materials, the mechanism usually hypothesized being intergranular friction. However, a change to brittle compressive failure appears at strain rates higher than 10^{-2} s^{-1}, although this value may change depending on the temperature and the microstructure.

When it impacts aircraft, the strain rate in the ice exceeds 10^{1} s^{-1}, easily reaching values on the order of 10^{3} s^{-1}. Therefore specific tests have been used to analyze its behavior under these conditions. Various devices have been used to study the compressive behavior of ice at high strain rates. The *high-speed universal testing machine* ($\varepsilon \sim 10^{1}$ s^{-1}) and *split Hopkinson pressure bar* ($\varepsilon \sim 10^{3}$ s^{-1}) both aim to determine the compressive strength, and the *drop weight tower* ($\varepsilon \sim 10^{2}$ s^{-1}) and *impact on rigid target* ($\varepsilon \sim 10^{3}$ s^{-1}) are used to analyze the impact behavior of the threat, evaluate contact forces and validate simulation tools.

Various authors have performed dynamic tests with the high-speed universal testing machine. Following his results obtained in quasi-static conditions (Jones, 1982), Jones (1997) extended compression tests to the range 10 s^{-1}, which showed a continuous increase of strength with strain rate. This author found a characteristic brittle failure initiated by longitudinal vertical cleavage cracks and followed by the explosion of the specimen.

The split Hopkinson pressure bar was used by Dutta et al. (2003), Kim

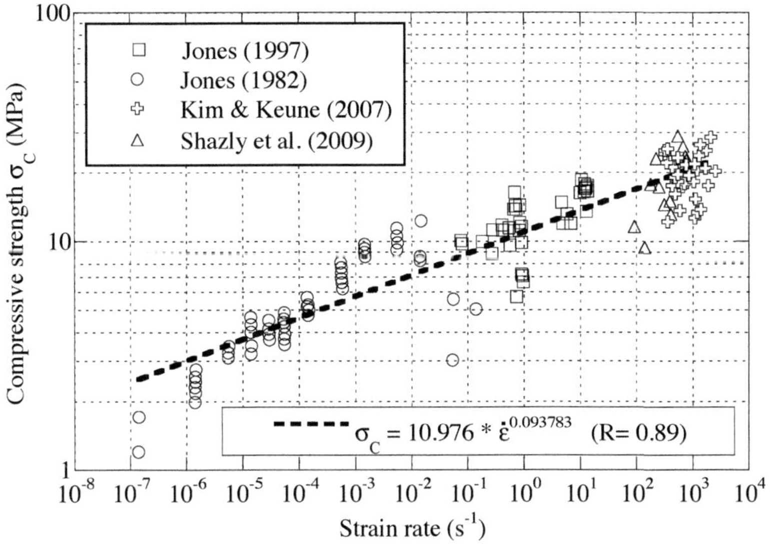

Figure 34. Strain rate sensitivity of compressive strength of ice (Jones, 1982, 1997; Shazly et al., 2009; Kim and Keune, 2007).

and Keune (2007) and Shazly et al. (2009). The results suggest that the response at high strain rates appears to be independent of microstructure. As in quasi-static conditions, strength increases as temperature decreases. The peak strength of ice increases monotonically with strain rate, and the relationship seems to be well approximated by a power law (Figure 34). An important observation found in these results is that the samples do not lose their load-carrying capacity catastrophically even after peak stress is reached during dynamic compression. This residual (tail) strength of the damaged/fragmented ice is sizable, and in some cases is larger than the quasi-static compression strength reported for ice (Shazly et al., 2009).

The results obtained by Fasanella and Boitnott (2006) using the drop weight tower were consistent with previously determined trends: contact force increased with increasing strain rate and the effect of internal crystalline structure was smaller at high impact velocities. High-velocity impact tests on a rigid plate showed the same trends (Kim et al., 2003; Pereira et al., 2006; Johnson et al., 2006; Chuzel, 2009; Combescure et al., 2011). These authors launched cylindrical or spherical ice projectiles at ballistic velocities onto a rigid target supported by a dynamic load cell to measure the contact force during impact events, and observed no discernible differences in force as a result of differences in the crystalline structure. The maximum load increased with increasing velocity as a result of the velocity-dependent Hugoniot peak stress, also observed in other soft body impacts such as those produced by birds (Johnson et al., 2006). A cleavage fracture mode associated with a small number of large fragments is observed at the earlier stage of impact, followed by comminution of the ice. Some of these results were later used by these and other authors to validate numerical models. To this end, Guégan et al. (2011) studied the kinematics of ice fragmentation, such as particle cloud velocity, launching spherical ice specimens against a transparent rigid plate.

In addition to the development and execution of tests intended to analyze the mechanical behavior of ice at high strain rates, the impact of ice against panels made of aeronautical materials is used to analyze the coupled response of threat and target. The increased use of lightweight materials in the aircraft industry has led to an increased probability of these materials being exposed to ice impacts. Composites are particularly susceptible to damage when subjected to out-of-plane dynamic loads. In addition, aluminum sheets are used as skins for structural elements and their thinness makes them liable to denting or even to perforation when exposed to hail threats. Various authors have studied the effect of single or multiple ice impacts on CFRP and GFRP laminates (Kim et al., 2003; Johnson et al., 2006; Juntikka and Olsson, 2009; Park and Kim, 2010; Appleby-Thomas et al., 2011). The main findings corroborate the sequence of damage observed in other types of dynamic tests on ice; cleavage cracks split the ice specimen into several large fragments, followed by a crushing process which converts the ice into a set of small particles that then behave like a fluid.

Chuzel and co-workers (Chuzel, 2009; Chuzel et al., 2010; Combescure et al., 2011) carried out normal impact tests on aeronautical aluminum sheets, using ice impactors with different nose shapes –flat and conical– and launching them at different incidence angles. The largest permanent deformation of the sheet was found for flat specimens and at normal impact, a result that could be predicted by the theory developed by Wilbeck (1978) for the impact of low strength projectiles. This work explains how the shock pressure generated upon impact decreases as the angle between contacting surfaces increases. A thorough comparison of the performance of various ice protection plate materials –2024-T3 aluminium sheets, aramid reinforced thermoplastic resin and fiber metal laminates–, under single and multiple impacts, may be found in Tambunan and Vlot (1995).

5.2 Modeling ice impact

Literature describing models for the simulation of ice impact is also scarce. Regarding constitutive equations for ice at high strain rates, some authors have used a Huber-Mises plasticity model, commonly available in finite element commercial codes, to describe the mechanical behavior of ice. In these studies, the plastic flow model is used in combination with a failure criterion. Kim and Kedward (1999) and Kim and Keune (2000) considered a critical value for the plastic strain and a critical value for the hydrostatic stress, both triggering the onset of failure, after which the material behaves like a fluid carrying only hydrostatic pressures. This material model was used to determine the forces reached during the impact of an ice sphere at high velocity hitting a rigid target or a CFRP panel. Despite the simplicity of the constitutive equation, appropriate for metals but not for brittle and pressure-dependent materials such as ice, the model was able to capture the trends related to peak force and failure threshold energy in the composite laminates. Later, Park and Kim (2010) used the same model to simulate and analyze transverse ice impact onto single-lap adhesive joints in CFRP composite panels. In addition, Anghileri et al. (2005a) used the Huber-Mises model to simulate high-speed ice impact onto metal sheets and to investigate the influence of the numerical scheme –Lagrangian FE, ALE and SPH were considered– on the reliability of the simulation. Park and Kim (2010) used a Huber-Mises elastoplastic model with a hydrostatic failure criterion, setting the deviatoric stress of the failed material to zero and keeping hydrostatic stress at the cutoff pressure.

The first constitutive equation specifically developed for ice at high strain rates was that of Carney et al. (2006). The hypotheses of the model rely on the experimental findings of several authors cited above (Schulson, 2001; Petrovic, 2003; Dutta et al., 2003; Shazly et al., 2006; Pereira et al., 2006; Fasanella and Boitnott, 2006). Strength is assumed to exhibit a logarithmic strain rate sensitivity and pressure dependence. Various failure modes are considered; the critical value of the plastic strain, pressure cut-off in compression, and pressure cut-off in tension. The model was validated with the test results of Pereira et al. (2006) and good agreement was found between the experimental and numerical results although somewhat high values of compressive and tensile strengths should be used.

Chuzel and colleagues (Chuzel, 2009; Chuzel et al., 2010) used a constitutive equation based on the damage model of Mazars (Mazars, 1984), and included viscous effects. Since the original model of Mazars results in an undesired increase in strength when applied to dynamic compression conditions (Chuzel et al., 2010), it was modified to allow degradation for compressive stress states.

Well beyond the range of impact velocities typical of aircraft applications, Sherburn and Horstemeyer (2010) proposed and validated a model for the hypervelocity range ($v \sim 10^3$ m/s), suitable for spacecraft applications, based on the Bammann–Chiesa–Johnson rate and temperature-dependent equation and on the Mie–Gruniesen equation of state.

Recently, Pérnas-Sánchez et al. (2012) proposed a constitutive model to simulate the behavior of ice under impact conditions. The model combines the essential features of the mechanical behavior of ice at high strain rates, as deduced from the test performed by other authors, and has the advantage that their parameters may be taken from the results of these experimental studies. The model properly reproduces the behavior of the ice in terms of the force induced during the impact and offers the advantage of easy implementation in numerical codes. This model is now briefly described.

Constitutive model by Pérnas-Sánchez et al. (2012)

Assuming that elastic strains (and rates) are very small compared to unity or to plastic strains (and rates) in ice impact, the additive decomposition of the rate of deformation tensor in its corresponding elastic and plastic components, generally assumed for hypoelastic-plastic materials (Nemat-Nasser, 1982; Khan and Huang, 1995), is assumed in the model

$$\mathbf{d} = \mathbf{d}^e + \mathbf{d}^p \tag{11}$$

The elastic strain rate is provided by the following expression of Hooke's law

$$\boldsymbol{\sigma}^\nabla = \mathbf{C} : \mathbf{d}^e = \mathbf{C} : (\mathbf{d} - \mathbf{d}^p) \tag{12}$$

where $\boldsymbol{\sigma}^\nabla$ is an objective rate of the Cauchy stress tensor and $\mathbf{C}$ is the Hooke stress-strain tensor defined by the elastic constants G and K.

For the description of the inelastic behaviour of ice, the model considers the experimental observations made by several authors (Schulson, 2001; Shazly et al., 2006, 2009), namely pressure dependence of strength, increase in compressive strength with strain rate, and residual strength after damage. Thus the model assumes a Drucker–Prager yield function (1952) to define the pressure dependence

$$f = \bar{\sigma} - (\sigma_{0y} + 3\alpha p) \tag{13}$$

where 3α is a parameter related to the internal friction angle of the material and σ_{0y} is the material cohesion (Figure 35). $\bar{\sigma}$ is the equivalent stress defined as

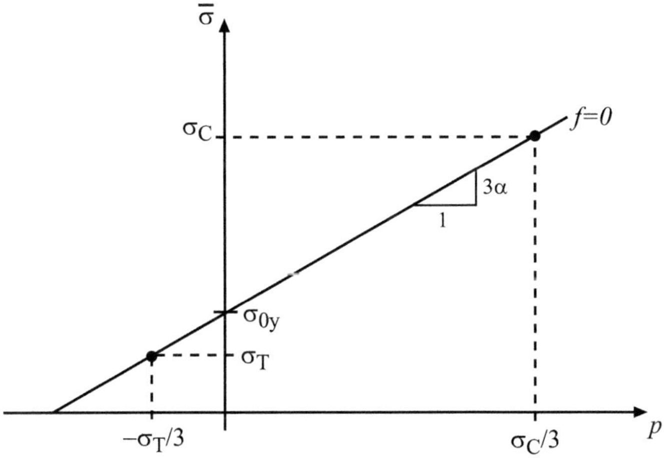

Figure 35. Yield locus in the $\{p, \bar{\sigma}\}$ space.

$$\bar{\sigma} = \sqrt{\frac{3}{2}\mathbf{s} : \mathbf{s}} \tag{14}$$

$\mathbf{s}$ being the deviatoric stress tensor, and p the hydrostatic pressure defined as

$$p = -\frac{\boldsymbol{\sigma} : \mathbf{1}}{3} \tag{15}$$

Both α and σ_{0y} parameters may be related to the yield stress in compression σ_C and tension σ_T

$$\alpha = \frac{\sigma_C - \sigma_T}{\sigma_C + \sigma_T} \tag{16}$$

$$\sigma_{0y} = \frac{2\sigma_C \sigma_T}{\sigma_C + \sigma_T} \tag{17}$$

thus the Drucker–Prager yield surface is completely defined once σ_C and σ_T are known. According to the experimental findings of several authors (Jones, 1982, 1997; Shazly et al., 2006; Kim and Keune, 2007), the compressive strength is dependent on strain rate; the same authors suggested that the viscous dependence is approximately linear in a log-log plot (see Figure 34), so a power law with strain rate sensitivity m was proposed

$$\sigma_C\left(\dot{\bar{\varepsilon}}^p\right) = \sigma_{C0}\left(\frac{\dot{\bar{\varepsilon}}^p}{\dot{\bar{\varepsilon}}_0}\right)^m \tag{18}$$

$\dot{\bar{\varepsilon}}^p$ being the equivalent plastic strain rate

$$\dot{\bar{\varepsilon}}^p = \sqrt{\frac{2}{3}\mathbf{d}^p : \mathbf{d}^p} \tag{19}$$

Regarding tensile behavior, as mentioned above, experimental results (Petrovic, 2003) show brittle failure and a negligible influence of strain rate in the strength of ice, so a constant value was chosen for σ_T.

To describe the inelastic flow, associative plasticity laws are often inappropriate since they overestimate the volumetric part of the plastic strain (Jirásek and Bazănt, 2002). Therefore a non-associated plastic flow rule is chosen

$$d^p = \dot{\lambda}\frac{\partial \Psi}{\partial \boldsymbol{\sigma}} = \dot{\lambda}\Psi_\sigma \tag{20}$$

where λ is the plastic multiplier, and the plastic potential Ψ is formally analogous to the yield function (13) but it has a lower slope with pressure α_Ψ

$$\Psi = \bar{\sigma} - (3\alpha_\Psi p + \Upsilon) \tag{21}$$

Υ being a dummy parameter and $\alpha_\Psi = k\alpha$ with $0 \leq k < 1$. The stress gradient of the plastic potential may be expressed as

$$\Psi_\sigma = \frac{3}{2}\frac{\mathbf{s}}{\bar{\sigma}} + \alpha_\Psi \mathbf{1} \tag{22}$$

and the flow rule leads to the following expression

$$d^p = \dot{\lambda}\left(\frac{3}{2}\frac{\mathbf{s}}{\bar{\sigma}} + \alpha_\Psi \mathbf{1}\right) \tag{23}$$

The above equations must be solved subject to the Kuhn–Tucker complementary conditions

$$\dot{\lambda} \geq 0, \quad f \leq 0, \quad \dot{\lambda} f = 0 \tag{24}$$

Property	Symbol	Value
Density	ρ	$897.6\ kg/m^3$
Young's Modulus	E	9.31 GPa
Poisson Rate	ν	0.33
Reference compressive strength	σ_{C0}	10.976 MPa
Compressive strain rate sensitivity	m	0.093783
Tensile strength	σ_{T0}	1.72 MPa
Internal friction factor	k	1.15

Table 1. Model parameters for ice (Pérnas-Sánchez et al., 2012)

and the consistency condition

$$\dot{\lambda}\dot{f} = 0 \tag{25}$$

As proposed by Carney et al. (2006), there are two different pressure cut-off limits; the first in tension P_T^{lim} and the other in compression P_C^{lim}. The ice is assumed to fail if

$$p < P_T^{lim} = -\frac{\sigma_T}{3} \tag{26}$$

or if

$$p > P_C^{lim} = \frac{\sigma_C}{3} \tag{27}$$

In both cases, the failure condition sets the deviatoric stress to zero and the pressure is only allowed to be greater than or equal to zero; it is supposed that after failure the broken ice can only withstand hydrostatic compressive stresses. The parameters for the ice, taken from the available literature, are given in Table 1.

The constitutive model can be incrementally integrated using a semi-implicit return mapping algorithm, in which the direction of inelastic flow is evaluated at time $n+1$, and the slope of both the yield function α and of the plastic potential α_Ψ are evaluated at time n. Since the yield surface remains constant during the plastic return, the first iteration enables the increment in the plastic multiplier to be determined and the stress updated. A complete description of the algorithm can be found in Pérnas-Sánchez et al. (2012).

In order to validate the proposed model, the experimental results obtained by Pereira et al. (2006) and presented by Carney et al. (2006) were chosen. In these tests, cylindrical ice projectiles were launched against a steel plate tied to a load cell. The main result of these tests was the load vs time recorded at the load cell. The ice material was modeled using the developed constitutive equation, which was implemented in the commercial finite element code LS-Dyna v971 (LSTC, 2010). Lagrangian, ALE and SPH solvers were used. In the Lagrangian mesh, an erosion criteria based on a maximum value of 1.5 for the equivalent strain was established to avoid numerical problems due to mesh distortion.

Experimental and numerical results are compared in Figure 36 which shows the results for ice impacts at 152 m/s. The figure depicts the time history of the force induced in the load cell for the three different methods analyzed. All of them find good agreement with the maximum force and pulse extension; only the SPH integration method slightly overestimates this value with respect to the experimental results.

6 Tyre impact

The design of aeronautic structures must take into account the possible impact that may occur as a result of tyre fragments being shed from wheels during takeoff and landing. This kind of impact has taken on particular importance in view of the repercussions of the accident involving the Concorde supersonic plane in Paris in the year 2000. The kinetic energy of the impact caused a strong pressure wave and large deformation of the fuel tank walls resulting in the catastrophic failure of the plane. This kind of problem, involving a transfer of energy between an impacting body and the impacted surface, has been studied in relation to the ballistic behaviour of rigid bodies and deformable bodies (Johnson et al., 1982,b). The great majority of these studies conclude that if the stiffness of the structure and the impacting body is comparable, the recovery of the elastic energy stored generates changes in the shape and trajectory of the projectile. In the case of the impact of a piece of rubber on a sheet of material which is more rigid (a metal or composite structure), a significant amount of the initial kinetic energy leads to the deformation of the projectile. Similarly, the kinetic energy transferred to the structure depends as much on the interaction between the impacting object and the surface impacted as on the behaviour of the projectile during the process (Karagiozova and Mines, 2007). In this way, the fractions of energy absorbed by the tyre fragment and the structure impacted are affected by the speed of impact, the angle of impact and the friction coefficient. The complexity of the process of tyre impact

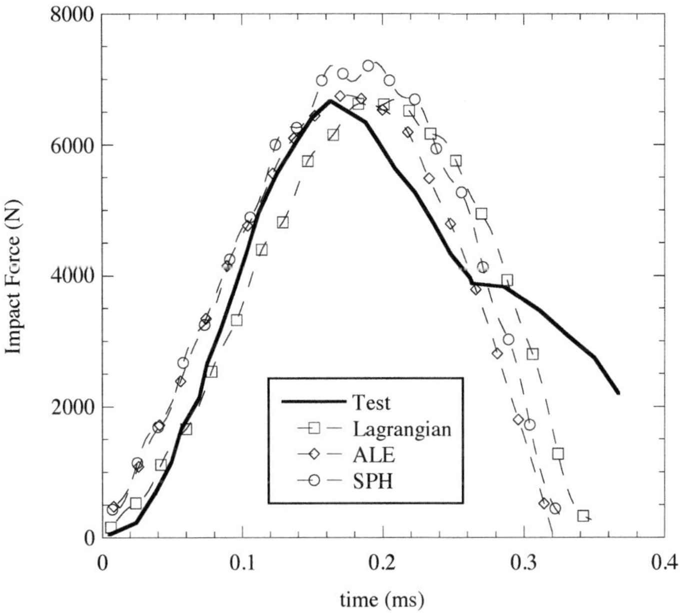

Figure 36. Impact force vs. time curves; experimental and numerical results for the three numerical solutions. Ice impact velocity 152 m/s (Pérnas-Sánchez et al., 2012).

(large deformation, non-linear behaviour, contact problems, inertial effects) means that experimental methods are needed both in terms of the mechanical characterisation of materials in impact conditions, as well as to validate the constitutive and simulation models developed.

6.1 Experimental analysis

Aircraft tyres are designed to withstand high loads during short periods and to guarantee stability in adverse conditions such as high pressure gradients due to crosswinds, hydroplaning, as well as high temperatures and brake friction. To guarantee this, aircraft tyres are made of a rubber matrix, usually natural rubber, with fabric reinforcements which tend to be Nylon. Figure 37 shows the configuration of a cross section of an aircraft tyre (Mines et al., 2007). The central part of the rubber matrix is reinforced with Nylon fabric plaited along its plane at a typical angle of $\pm 30^o$.

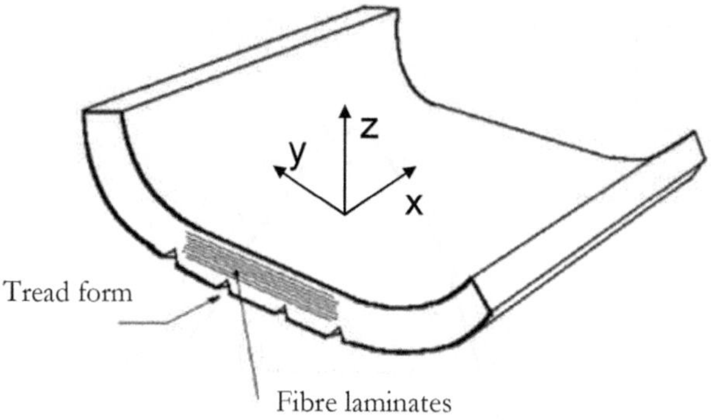

Figure 37. Cross section of a tyre (Mines et al., 2007).

x is the hoop direction of the tyre and y is the transverse direction, while the z direction corresponds to thickness. The outer area of the tyre which comes into contact with the runway has a higher concentration of rubber to provide greater wear resistance. To determine the appropriate mechanical properties of the polymer matrix on the one hand and the fabric reinforcement on the other, static and dynamic characterisation tests are needed (Mines et al., 2007).

Static behaviour

When a tyre fragment impacts against an aircraft, bending and crushing of the projectile itself occurs, resulting in significant deformations in the material related to stress states involving both tension and compression. Thus, with this kind of problem, a first estimate of the mechanical behaviour of the impacting object is usually carried out through uniaxial traction and compression tests (Johnson et al., 1982b). In this respect, the results obtained by Mines et al. (2007) on tyre samples of the Concorde supersonic plane are of particular interest. The experimental data corresponding to uniaxial compression (direction z) shows the behaviour of the material to be non-linear (Figure 38) as well as the presence of large strains before failure.

In addition, the results of the static tensile tests show anisotropic material properties, with a marked difference in behaviour being noted between directions x and y (Figure 39). The mechanical behaviour of the material is dominated by the matrix properties in a transverse direction to the reinforcement, in such a way that the failure stress is greater in the x direction

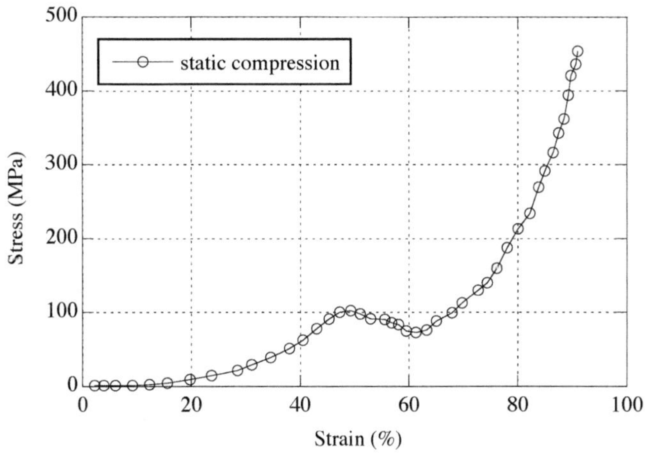

Figure 38. Nominal stress versus strain for static compression test of a Concorde tyre (Mines et al., 2007).

than the y, with the latter resulting in higher strain levels.

Dynamic behaviour

In order to consider the possible influences of strain rate on the behaviour of polymers in compression, tests can be carried out in different devices, such as drop weight towers for intermediate strain rates, and gas guns for high strain rates. Results of this kind of test carried out on tyre rubber (Mines et al., 2007) show that strain rate has no perceptible influence for static and dynamic compression stress states (Figure 40). With regards to the effect of strain rate in tension, tests carried out with a dynamic universal testing machine on tyre shows the effect of strain rate to be small and not to generate significant changes in the stress-strain curve (Figure 41). The differences in the stress value is due more to the inertial effects associated with the size of the sample than to any dependency of the mechanical properties of the material with the strain rate.

Impact tests

The performance of real impact tests, with tyre fragments propelled at speeds of hundreds of meters per second against metal sheets or composites, are essential in order to obtain the relevant information. Among the

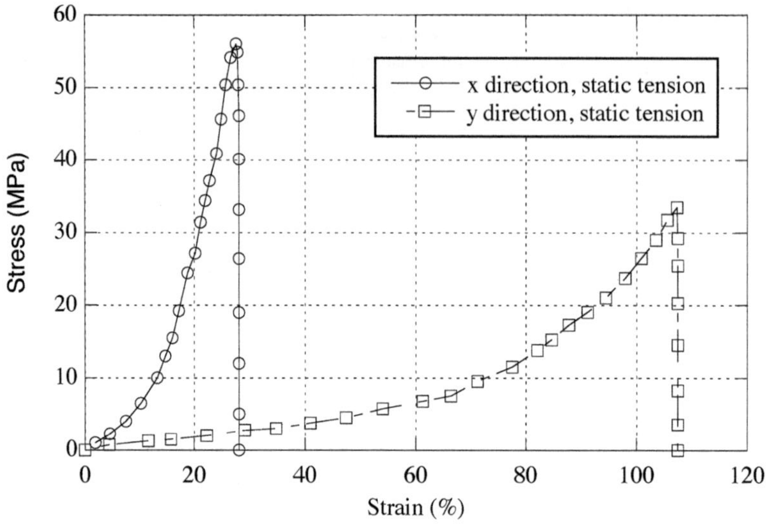

Figure 39. Nominal stress versus strain for static tensile test of a Concorde tyre (Mines et al., 2007).

influential variables of the problem, an analysis of the effect of angle and impact speed is of particular interest. Very few experimental tests can be found in the literature related to the impact of rubber or rubber fragments, aside from those of Mines et al. (2007) and Guégan et al. (2010). The work carried out by Mines et al. (2007) analyses the impact of cubic and parallelpiped-shaped tyre fragments that are launched at speeds of between 75 and 135 m/s and impact angles of 30°, 60° and 90° (normal impact) against aluminium sheets. The work carried out by Guégan et al. (2010) analyses the impact of rubber balls propelled at 135 m/s and impact angles from 0° to 90°. The results obtained demonstrates that in cases of impact at low speeds (75 m/s) and small angles of incidence (close to that of tangential impact) significant bending of the tyre fragment occurs and it slides along a large distance on the impacted surface. In this situation, a second impact against the aircraft structure may occur due to elastic rebound, which results in damage to a large area of the structure. By contrast, with high impact speeds (135 m/s) at a perpendicular angle to the impact surface, the possibility of rebound is reduced and deformations only occur at the first point hit by the projectile. The complexity involved in carrying out these kinds of tests, together with the scarcity of experimental data found in the

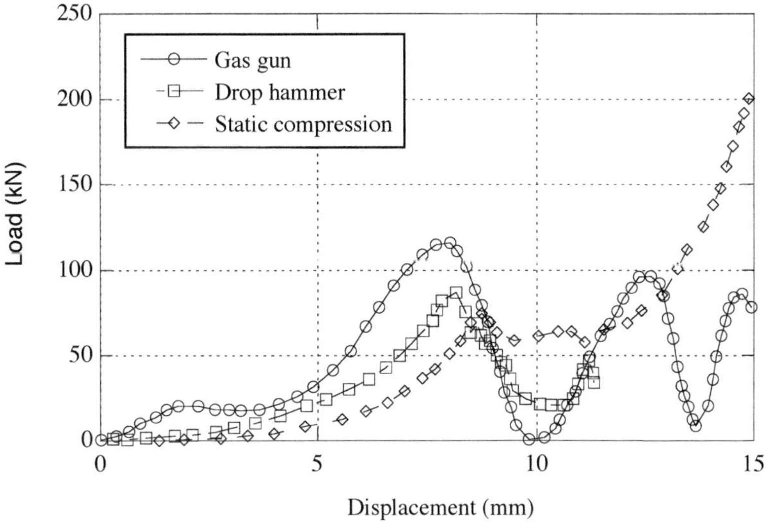

Figure 40. Comparison between static and dynamic compression data in terms of load and displacement of a specimen of tyre rubber (Mines et al., 2007).

scientific literature available, makes the use of numerical simulation tools necessary in order to advance knowledge of the mechanisms at work in this problem.

6.2 Modeling tyre fragment impact

As already indicated, the structure of the materials used in aircraft tyres is complex, consisting of a rubber matrix and the pre-established reinforcements. Reinforcements are made of nylon threads whose stiffness in compression is significantly less than in tension (Watanabe and Kaldjian, 1985; Mines et al., 2007). This structure of the material results in anisotropic properties and non-linear behaviour when the material is subjected to large deformations. As such, the structure needs to be adequately modelled to take into account the stiffness of the impacting object and the way it interacts with the impacted structure. The most commonly used model for tyre rubber takes into consideration the isotropic behaviour hypothesis and hyperelastic behaviour (Treloar, 1975; Ogden, 1998; Johnson et al., 2009), whereby large deformations take place without plastic deformations occur-

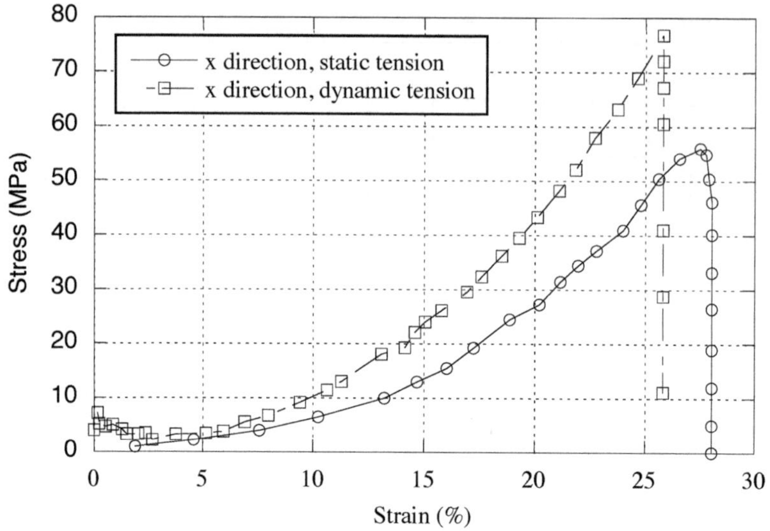

Figure 41. Comparison between static and dynamic tensile results in terms of nominal stress and strain of a Concorde tyre (Mines et al., 2007).

ring or any dissipation of internal energy (a reversible process). According to the hypothesis on compressible or incompressible material, different strain energy functions can be considered which result in different material models. For incompressible materials, the constitutive model due to Ogden (1998) can be formulated, whereby the energy density is expressed through a power function of the principal stretches. Particular values of the exponents in Ogden's power law leads to the Mooney-Rivlin model (Mooney, 1940; Rivlin, 1948), used in the numerical simulation of tyre impact by different authors (Mines et al., 2007; Johnson et al., 2009). If the material is compressible, a strain energy function can be considered that is dependent on the porosity of the material, and the Blatz and Ko (1962) model can be formulated. This model has been used in the modelling of rubber polymers by different authors (Feng et al., 2006).

In relation to numerical simulation, there are different ways of approaching the behaviour of the reinforced rubber. One of the traditional forms is based on the analysis techniques of fibrous composite materials (Noor and Tanner, 1985; Tabaddor and Stafford, 1985). With this kind of modelling, the material properties are derived from the average of the properties of the individual components of each layer, taking into account laminated shell-type elements and anisotropic behaviour laws. However, this approach requires a number of material constants related to the average layers and its application is limited in the case of materials exhibiting significant anisotropic behaviour. An alternative approach (Watanabe and Kaldjian, 1985; Reese et al., 2001) consists in the use of one-dimensional bar elements for modelling the reinforcement. The nodes of these elements are linked to those of the matrix elements, fulfilling the compatibility conditions. With regard to matrix, most models assume the hypotheses of the hyperelastic behaviour of rubber (Green and Adkins, 1960; Treloar, 1975; Karagiozova and Mines, 2007). The existing literature includes very few studies on numerical simulation that consider the problem of the impact of tyre fragments on composite panels, apart from the important work of Johnson et al. (2009). This work analyses the impact of tyre fragments at speeds of up to 93 m/s, using a Mooney-Rivlin model for the behaviour of rubber. The numerical results show that the failure of the composite is mainly determined by delamination –on top of fiber failure–, and that the projectile absorbs a significant part of the energy of the shock upon deformation. In any case, the problem of impact caused by tyre fragments is a specific and open field of research which still requires the development of suitable numerical simulation tools.

Bibliography

F.L. Addessio, M.W. Schraad, and M.W. Lewis. Physics-based damage predictions for simulating testing and evaluation experiments. Technical report, Los Alamos National Laboratory, 1997. LA-UR-97-4877.

A. Airoldi and B. Cacchione. Modelling of impact forces and pressures in lagrangian bird strike analyses. *International Journal of Impact Engineering*, 32:1651–1677, 2006.

M. Anghileri, L.M.L. Castelletti, F. Invernizz, and M. Mascheroni. A survey of numerical models for hail impact analysis using explicit finite element codes. *International Journal of Impact Engineering*, 31:929–944, 2005a.

M. Anghileri, L.M.L. Castelletti, F. Invernizzi, and M. Mascheroni. Birdstrike onto the composite intake of a turbofan engine. In *5th European LS-DYNA User's Conference*, Birmingham, UK, May 2005b.

M. Anghileri, L.M.L. Castelletti, and V. Mazza. Birdstrike: approaches to the analysis of impacts with penetration. In M. Alves and N. Jones, editors, *Impact loading of lightweight structures*, pages 63–74. WIT Press, 2005c.

M. Anghileri, L.M.L. Castelletti, and M. Tirelli. Fluid-structure interaction of water filled tanks during the impact with the ground. *International Journal of Impact Engineering*, 31:235–254, 2005d.

G.J. Appleby-Thomas, P.J. Hazell, and G. Dahini. On the response of two commercially-important CFRP structures to multiple ice impacts. *Composite Structures*, 93:2619–2627, 2011.

J.A. Artero-Guerrero, J. Pernas-Sánchez, D. Varas, and J. López-Puente. Numerical analysis of CFRP fluid-filled tubes subjected to high-velocity impact. *Composite Structures*, 96:286–297, 2013.

J.G. Avery. *Design Manual for Impact Damage Tolerant Aircraft Structure.* AGARD, 1981.

V.D. Azzi and S.W. Tsai. Anisotropic strength of composites. *Experimental Mechanics*, 5:283–288, 1965.

M.A. Badie, E. Mahdi, and A.M.S. Hamouda. An investigation into hybrid carbon/glass fiber reinforced epoxy composite automotive drive shaft. *Materials & Design*, 32:1485–1500, 2011.

R.E. Ball. *A computer program for the geometrically nonlinear static and dynamic analysis of arbitrarily loaded shells of revolution: theory and user's manual.* NASA, 1972.

R.E. Ball. Aircraft fuel tank vulnerability to hydraulic ram: Modification of the northrup finite element computer code BR-1 to include fluid-structure interaction. theory and user's manual for BR-1HR. Technical Report 57B p74071, Naval Postgraduate School, Monterey, CA, 1974.

R.B. Banks and D.V. Chandrasekhara. Experimental investigation of the penetration of a high-velocity gas jet through a liquid surface. *Journal of Fluid Mechanics*, 15:13–34, 1963.

K.S. Bates Jr. Aircraft fuel tank entry wall-projectile interaction studies. Master's thesis, Naval Postgraduate School, Monterey, CA, 1973.

T.V. Baughn and L.W. Graham. Simulation of a birdstrike impact on aircraft canopy material. *Journal of Aircraft*, 25:659–664, 1988.

P.J. Blatz and W.L. Ko. Application of finite elasticity to the deformation of rubbery materials. *Trans Soc Rheol*, 6:223–251, 1962.

S.J. Bless, J.P. Barber, P.F. Fry, and R.K. Newman. Studies of hydrodynamic ram induced by high velocity spherical fragment simulators. Technical Report AFML-TR-77-11, Dayton Univ OH Research Institute, 1977.

T. Børvik, A.G. Hanssen, M. Langseth, and L. Olovsson. Response of structures to planar blast loads. A finite element engineering approach. *Computers and Structures*, 87:507–520, 2009.

J.C. Brew and P.A. Lagace. Quadratic stress criterion for initiation of delamination. *Journal of Composite Materials*, 22:1141–1155, 1988.

R. Budgey. The development of a substitute artificial bird by the international birdstrike research group for use in aircraft component testing. In *International Bird Strike Committee ISBC25/WP-IE3*, Amsterdam, The Netherlands, 2000.

W.J. Cantwell. The influence of target geometry on the high-velocity impact response of CFRP. *Composite Structures*, 10:247–265, 1988a.

W.J. Cantwell. The influence of fiber stacking-sequence on the high-velocity impact response of CFRP. *Journal of Material Science Letters*, 7:756–758, 1988b.

W.J. Cantwell and J. Morton. Comparison of the low and high velocity impact response of CFRP. *Composites*, 20:545–551, 1989.

W.J. Cantwell and J. Morton. Impact perforation of carbon fibre reinforced plastic. *Composites Science and Technology*, 38:545–551, 1990.

W.J. Cantwell, P.T. Curtis, and J. Morton. An assessment of the impact performance of CFRP reinforced with high-strain carbon fibres. *Composites Science and Technology*, 25:133–148, 1986.

G. Caprino, I. Crivelli Visconti, and A. Di Ilio. Composite materials response under low-velocity impact. *Composite Structures*, 2:261–271, 1984.

G. Caprino, V. Lopresto, C. Scarponi, and G. Briotti. Influence of material thickness on the response of carbon-fabric/epoxy panels to low velocity impact. *Composites Science and Technology*, 59:2279–2286, 1999.

G. Caprino, A. Langella, and V. Lopresto. Indentation and penetration of carbon fibre reinforced plastic laminates. *Composites Part B: Engineering*, 34:319–325, 2003.

K. Carney, D. Benson, P. Dubois, and R. Lee. A phenomenological high strain rate model with failure for ice. *International Journal of Solids and Structures*, 43:7820–7839, 2006.

L.M.L. Castelletti and M. Anghileri. Multiple birdstrike analysis - a survey of feasible techniques. In *30th European Rotorcraft Forum*, pages 495–505, Marseille, France, 2003.

F. Chang and K.A. Chang. A progressive damage model for laminated composites containing stress concentrations. *Journal of Composite Materials*, 21:834–855, 1987.

J.K. Chen and D.F. Medina. The effects of projectile shape on laminated composite perforation. *Composite Science and Technology*, 58:1629–1639, 1998.

J.K. Chen, F.A. Allahdadi, and T.C. Carney. High-velocity impact of graphite/epoxi composite laminates. *Composite Science and Technology*, 57:1369–1379, 1997.

Y. Chuzel. *Caractérisation expérimentale et simulation numérique d'impacts de glace a haute vitesse*. PhD thesis, INSA, Lyon, 2009.

Y. Chuzel, A. Combescure, M. Nucci, R. Ortiz, and Y. Perrin. Development of hail material model for high speed. In *11th International LS-DYNA User's Conference*, pages 17–26, 2010.

V.N. Cogolev, V.G. Mirkin, and G.J. Yablokova. Approximate equation of state for solids. *Zhurnal Prikladnoi Mekhaniki i Teknickeskoi Fiziki*, 5: 93–98, 1963.

D. Cole. Crack nucleation in polycrystalline ice. *Cold Regions Science and Technology*, 15:79 – 87, 1988.

A. Combescure, Y. Chuzel-Marmot, and J. Fabis. Experimental study of high-velocity impact and fracture of ice. *International Journal of Solids and Structures*, 48:2779–2790, 2011.

M.F.S.F. de Moura and A.T. Marques. Prediction of low velocity impact damage in carbon-epoxy laminates. *Composites Part A: Applied Science and Manufacturing*, 33:361–368, 2002.

T. Gómez del Río, R. Zaera, E. Barbero, and C. Navarro. Damage in CFRPs due to low velocity impact at low temperature. *Composites Part B: Engineering*, 36:41–50, 2005.

P.J. Disimile, L.A. Swanson, and N. Toy. The hydrodynamic ram pressure generated by spherical projectiles. *International Journal of Impact Engineering*, 36:821–829, 2009.

D.C. Drucker and W. Prager. Soil mechanics and plastic analysis or limit design. *Quarterly of Applied Mathematics*, 10:157–165, 1952.

P.K. Dutta, D.M. Cole, E.M. Schulson, and D.S. Sodhi. A Fracture Study of Ice Under High Strain Rate Loading. In *International Offshore and Polar Engineering Conference*, pages 465–472, 2003.

P. Eschenfedler. Wildlife hazards to aviation. In *ICAO/ACI Airports Conference*, Miami, 2001.

E.L. Fasanella and R.L. Boitnott. Dynamic Crush Characterization of Ice. Technical report, NASA, February 2006.

E.L. Fasanella, R.L. Boinott, and S. Kellas. Test and analysis correlation of high speed impacts of ice cylinders. In *9th International LS-DYNA User's Conference*, Dearborn, Michigan, June 2006.

Z.Q. Feng, B. Magnain, and J.M. Cros. Solution of large deformation impact problems with friction between Blatz-Ko hyperelastic bodies. *International Journal of Engineering Science*, 44:113–126, 2006.

C.J. Freitas, C.E. Anderson Jr, J.D. Walker, and D.L. Littlefield. Hydrodynamic ram: A benchmark suite. In *ASME Pressure Vessel Piping Conference and Symposium on Structures Under Extreme Loading Conditions*, pages 63–74, 1996.

A.E. Fuhs and R.E. Ball. FY 73 hydraulic ram studies. Technical Report AD0776536, Monterey, CA, 1974.

K. Fujii, M. Aoki, N. Kiuchi, and E. Tasuda. Impact perforation behavior of CFRPs using high-velocity steel sphere. *International Journal of Impact Engineering*, 27:497–508, 2002.

S. Georgiadis, A.J. Gunnion, R.S. Thomson, and B.K. Cartwright. Bird-strike simulation for certification of the Boeing 787 composite moveable trailing edge. *Composite Structures*, 86:258–268, 2008.

R.A. Gingold and J.J. Monaghan. Smoothed particle hydrodynamics: Theory and application to non-spherical stars. *Mothly Notices of the Royal Astronomical Society*, 181:375–389, 1977.

L.W. Gold. On the elasticity of the ice plates. *Canadian Journal of Civil Engineering*, 15:1080–1084, 1988.

A.E. Green and J.E. Adkins. *Large elastic deformations and nonlinear continuum mechanics.* Oxford Clarendon Press, 1960.

A. Grimaldi, A. Solloa, M. Gudab, and F. Marulob. Parametric study of a SPH high velocity impact analysis. A bird strike windshield application. *Composite Structures*, 96:263–275, 2013.

P. Guégan, R. Othman, D. LeBreton, F. Pasco, N. Swiergiel, and P. Thevenet. Experimental investigation of rubber ball impacts on aluminium plates. *International Journal of Crashworthiness*, 15:391–399, 2010.

P. Guégan, R. Othman, D. LeBreton, F. Pasco, P. Villedieu, J. Meyssonnier, and S. Wintenberger. Experimental investigation of the kinematics of post-impact ice fragments. *International Journal of Impact Engineering*, 38:786–795, 2011.

A.G. Hanssen, Y. Girard, L. Olovsson, T. Berstad, and M. Langseth. A numerical model for bird strike of aluminium foam-based sandwich panels. *International Journal of Impact Engineering*, 32:1127–1144, 2006.

F.D. Haynes. Effect of Temperature on the Strength of Snow-ice. Technical Report 78:27, Cold Regions Research and Engineering Laboratory, U.S. Army, 1978.

S. Heimbs. Computational methods for bird strike simulations: A review. *Computers and Structures*, 89:2093–2112, 2011a.

S. Heimbs. Computational methods for bird strike simulations: A review. *Computers and Structures*, 89:2093–2112, 2011b.

W.M. Herlin and J.G. Avery. Hydraulic ram structural response computer program (HRSR). Technical report, Boeing Co., 1981. Prepared under Contract N60530-80-C-0242 for Naval Weapons Center, China Lake, California.

C.M. Holm. Hydraulic ram presure measurements. Master's thesis, Naval Postgraduate School, Monterey, CA, 1974.

D.P. Holm. Hydraulic ram shock wave and cavitation effects on aircraft fuell cell survivability. Master's thesis, Naval Postgraduate School, Monterey, CA, 1973.

M.V. Hosur, M. Adbullah, and S. Jeelani. Studies on the low-velocity impact response of woven hybrid composites. *Composite Structures*, 67:253–262, 2005.

J.P. Hou, N. Petrinic, C. Ruiz, and S.R. Hallett. Prediction of impact damage in composite plates. *Composites Science and Technology*, 60: 273–281, 1997.

C.A. Huertas. Robust bird-strike modeling using LS-DYNA. Master's thesis, University of Puerto Rico at Mayagüez, 2006.

L. Iannucci. Bird strike on composite panels. In *DYNA3D User's Conference*, Manchester, UK, 1992.

L. Iannucci. *Foreign Object Impact and Energy Absorbing Structure*, chapter Bird-strike impact modeling. John Wiley & Sons, London, England, 1998.

ICAO. In *Proposed Amendment to Annex 14. Unpublished.* 2001.

W.W. Jarzab, R. Chwalinski, W.E. Pfrang, and G. Tokar. Fluid-structure interaction effects in tank structures due to sloshing and hydrodynamic ram coupled lagrangian-eulerian simulations. In *International Conference: Spacecraft Structures and Mechanical Testing*, 1988.

S.T. Jenq, F.B. Hsiao, I.C. Lin, D.G. Zimcik, and M. Nejad Ensan. Simulation of a rigid plate hit by a cylindrical hemi-spherical tip-ended soft impactor. *Computational Materials Science*, 39:518–526, 2007.

M. Jirásek and Z.P. Bazănt. *Inelastic Analysis of Structures.* John Wiley & Sons, England, 2002.

A. Johnson, M. Holzapfel, H. Kraft, and A. Reiter. Measurement of Ice Mechanical Properties. Technical Report IB 435 2006/55, DLR, 2006.

A.F. Johnson, N. Toso-Pentecôte, and D. Schwinn. Modelling damage in composite aircraft panels under tyre rubber impact. In *Proceeding of 17th International Conference on Composite Materials*, 2009.

G.R. Johnson and W.H. Cook. A constitutive model and data for metals subjected to large strains, high strain rates, and temperatures. In *Proceedings of 7th International Symposium on Ballistics*, pages 1–7, The Hague, The Nederlands, 1983.

W. Johnson, A.K. Sengupta, and S.K. Ghosh. High velocity oblique and ricochet mainly of long rod projectiles: an overview. *International Journal Mechanical Sciences*, 24:425–436, 1982.

W. Johnson, A.K. Sengupta, and S.K. Ghosh. Plasticine modelled high velocity oblique impact and ricochet of long rods. *International Journal Mechanical Sciences*, 24:437–455, 1982b.

S.J. Jones. The confined compressive strength of polycrystalline ice. *Journal of Glaciology*, 28:171–177, 1982.

S.J. Jones. High Strain-Rate Compression Tests on Ice. *The Journal of Physical Chemistry B*, 101:6099–6101, 1997.

R. Juntikka and R. Olsson. Experimental and modelling study of hail impact on composite structures. In *11th International Conference on Composite Materials*, 2009.

L.C. Kappel. Hydraulic ram shock phase effects on fuel cell survivability. Master's thesis, Naval Postgraduate School, Monterey, CA, 1974.

D. Karagiozova and R.A.W. Mines. Impact of aircraft rubber tyre fragments on aluminium alloy plates: II - Numerical simulation using LS-DYNA. *International Journal Impact Engineering*, 34:647–667, 2007.

A.S. Khan and S. Huang. *Continuum Theory of Plasticity.* John Wiley & Sons, N.Y., 1995.

H. Kim and K.T. Kedward. Experimental and numerical analysis correlation of hail ice impacting composite structures. *Composite Structures*, 68:1–11, 1999.

H. Kim and J.N. Keune. Modeling Hail Ice Impacts and Predicting Impact Damage Initiation in Composite Structures. *AIAA Journal*, 38:1278–1288, 2000.

H. Kim and J.N. Keune. Compressive strength of ice at impact strain rates. *Journal of Materials Science*, 42:2802–2806, 2007.

H. Kim, D.A. Welch, and K.T. Kedward. Experimental investigation of high velocity ice impacts on woven carbon/epoxy composite panels. *Composites Part A: Applied Science and Manufacturing*, 34:25–41, 2003.

J.H. Kim and H.C. Shin. Application of the ALE technique for underwater explosion analysis of a submarine liquefied oxygen tank. *Ocean Engineering*, 35:812–822, 2008.

J.K. Kim, L.M. Leung, W.R. Lee, and Y. Hirai. Impact performance of a woven fabric CFRP laminate. *Polymers & Polymer Composites*, 4: 549–561, 1996.

K.D. Kimsey. Numerical simulation of hydrodynamic ram. Technical Report ARBRL-TR-02217, US Army Ballistic Research Laboratory, 1980.

N.F. Knight, N. Jaunky, R.E. Lawson, and D.R. Ambur. Penetration simulation for uncontained engine debris impact on fuselage-like panels using ls-dyna. *Finite Elements in Analysis and Design*, 36:99–133, 2000.

M. Koishi, T. Okano, L. Olovsson, H. Saito, and M. Makino. Hydroplaning simulation using fluid-structure interaction in ls-dyna. In *9th International LS-DYNA User's Conference*, Dearborn, Michigan, June 2006.

J.L. Lacome. Smoothed particle hydrodynamics method in LS-DYNA. In *3rd German LS-DYNA forum*, Bamberg, Germany, October 2004.

F. Larsson. Damage tolerance of a stitched carbon/epoxy laminate. *Composites Part A: Applied Science and Manufacturing*, 28:923–934, 1997.

M.A. Lavoie, A. Gakwaya, M. Nejad-Ensan, and D.G. Zimcik. Validation of available approaches for numerical bird strike modeling tools. *International Review of Mechanical Engineering*, 1:380–389, 2007.

C.M. Lewis. Engine bird ingestion. *Airliner*, 1:17–19, 1995.

J. Liu, Y.L. Li, and F. Xu. The numerical simulation of a bird-impact on an aircraft windshield by using the SPH method. *Advanced Materials Research*, 33-37:851–856, 2008.

J. López-Puente, R. Zaera, and C. Navarro. The effect of low temperatures on the intermediate and high velocity impact response of CFRPs. *Composites Part B: Engineering*, 33:559–566, 2002.

J. López-Puente, R. Zaera, and C. Navarro. High energy impact on woven laminates. *Journal de Physique IV*, 110:639–644, 2003.

J. López-Puente, R. Zaera, and C. Navarro. Experimental and numerical analysis of normal and oblique ballistic impacts on thin carbon/epoxy woven laminates. *Composites Part A: Applied Science and Manufacturing*, 39:374–387, 2008.

LSTC. *LS-DYNA Keyword User's Manual V.971.* Livermore, California, 2007.

LSTC. *LS-DYNA User's Manual.* Livermore, California, 2010.

Y. Lu and Z. Wang. Characterization of structural effects from above-ground explosion using coupled numerical simulation. *Computers and Structures*, 84:1729–1742, 2006.

L.B. Lucy. A numerical approach to the testing of the fission hypothesis. *The Astronomical Journal*, 82:1013–1024, 1977.

E.A. Lundstrom. Fluid/structure interaction in hydraulic ram. In *Proceedings of the Hydrodynamic Ram Seminar*, pages 223–230, 1977.

E.A. Lundstrom and T. Anderson. Hydraulic ram model for high explosive ammunition. In *Symposium on Shock and Wave Propagation, Fluid-Structure Interaction and Structural Responses. ASME Pressure Vessels and Piping Conference*, 1989.

E.A. Lundstrom and E. Stull. Fluid dynamic analysis of hydraulic ram II (results of experiments). Technical Report JTCG/AS 73-T-291, Joint Technical Coordinating Group/Aircraft Survivability, 1973.

B. MacKinnon. *Sharing the skies: an aviation industry guide to the management of wildlife hazards.* Civil Aviation, Transport Canada, 2004.

P.O. Marklund and L. Nilsson. Simulation of airbag deployment using a coupled fluid-structure approach. In *7th International LS-DYNA User's Conference*, Dearborn, Michigan, May 2002.

J. Mazars. *Application de la mécanique de l'endommagement au comportement non linéaire et à la rupture du béton de structure.* PhD thesis, Université Pierre et Marie Curie - Paris 6, 1984.

S.C. McCallum and C. Constantinou. The influence of bird-shape in bird-strike analysis. 5th European LS-DYNA users conference, Birmingham, UK, 2005.

D.F. Medina and J.K. Chen. Three-dimensional simulations of impact induced damage in composite structures using the parallelized SPH method. *Composites Part A: Applied Science and Manufacturing*, 31: 853–860, 2000.

S.A. Meguid, R.H. Mao, and T.Y. Ng. FE analysis of geometry effects of an artificial bird striking an aeroengine fan blade. *International Journal of Impact Engineering*, 35:487–498, 2008.

R.A.W. Mines, S. McKown, and R.S. Birch. Impact of aircraft rubber tyre fragments on aluminium alloy plates: I-experimental. *International Journal of Impact Engineering*, 34:627–646, 2007.

M. Mooney. A theory of large elastic deformation. *J Applied Physics*, 11: 582–592, 1940.

L.S. Mueller. Experiments in hydraulic ram. Master's thesis, Naval Postgraduate School, Monterey, CA, 1974.

F.D. Murnaghan. The compressibility of media under extreme pressures. *Proceedings of the National Academy of Sciences of the United States of America*, 30:244–247, 1944.

S. Nemat-Nasser. On finite deformation elastoplasticity. *International Journal of Solids and Structures*, 18:857–872, 1982.

R.R.V. Neves, G.B. Micheli, and M. Alves. An experimental and numerical investigation on tyre impact. *International Journal of Impact Engineering*, 37:685–693, 2010.

L.S. Nizampatnam. *Models and methods for bird strike load predictions.* PhD thesis, Wichita State University, 2007.

A.K. Noor and J.A. Tanner. Advances and trends in the development of computational models for tires. *Composite Structures*, 20:517–533, 1985.

R.W. Ogden. *Nonlinear Elastic Deformations.* Dover, 1998.

B. Page. Entry wall strain measurements during hydraulic ram. Master's thesis, Naval Postgraduate School, Monterey, CA, 1975.

H. Park and H. Kim. Damage resistance of single lap adhesive composite joints by transverse ice impact. *International Journal of Impact Engineering*, 37:177–184, 2010.

J.W. Patterson. Fuel cell pressure loading during hydraulic ram. Master's thesis, Naval Postgraduate School, Monterey, CA, 1975.

M.J. Pavier and M.P. Clarke. Experimental techniques for the investigation of the effects of impact damage on carbon-fibre composites. *Composites Science and Technology*, 55:545–551, 1995.

N. Pentecôte and D. Kohlgrueber. Crash on water: a highly multi-physics problem. In *EUROPAM2004, 14th European Conference and Exhibition on Digital Simulation for Virtual Engineering*, Paris, October 2004.

J.M. Pereira, S.A. Padula, D.M. Revilock, and M.E. Melis. Forces generated by high velocity impact of ice on a rigid structure. Technical Report TM-2066-214263, NASA, 2006.

J. Pérnas-Sánchez, D.A. Pedroche, D. Varas, J. López-Puente, and R. Zaera. Numerical modeling of ice behavior under high velocity impacts. *International Journal of Solids and Strutures*, 49:1919–1927, 2012.

J.J. Petrovic. Mechanical properties of ice and snow. *Journal of Material Science*, 38:1–6, 2003.

F. Peyraut. Loading restrictions for the Blatz-Ko hyperelastic model: application to a finite element analysis. *International Journal of Non-Linear Mechanics*, 39:969–976, 2004.

F. Poehlmann-Martins, J. Gabrys, and M. Souli. Hydrodynamic ram analysis of non-exploding projectile impacting water. In *Proceedings of the 2005 ASME Pressure Vessels and Piping Division Conference*, 2005.

S. Reese, T. Raible, and P. Wriggers. Finite element modelling of orthotropic material behaviour in pneumatic members. *International Journal Solids Structures*, 38:9525–9544, 2001.

M.H. Rice, R.G. McQueen, and J.M. Walsh. *Solid State Physics*, chapter Compression of Solids by Strong Shock Waves, pages 1–63. Academic Press, New York, 1958.

R.S. Rivlin. Large elastic deformations of isotropic materials. IV. Further developments of the general theory. *Philosophical Transactions of the Royal Society*, 241:379–397, 1948.

A.L. Ruoff. Linear shock-velocity-particle-velocity relationship. *Journal Applied Physics*, 38:4976–4980, 1967.

G. Sala. Post-impact behaviour of aerospace composites for high-temperature applications: experiments and simulations. *Composites Part B: Engineering*, 28:651–665, 1997.

H. Salehi, S. Ziaei-Rad, and M.A. Vaziri-Zanjani. Bird impact effects on different types of aircraft bubble windows using numerical and experimental methods. *International Journal of Crashworthiness*, 15:93–106, 2010.

P. Santini, D. Palmieri, and M. Marchetti. Numerical simulation of fluid-structure interaction in aircraft fuel tanks subjected to hydrodynamic ram penetration. In *Proceedings of the 21st ICAS Congress*, 1998.

M. Sayer, N.B. Bektas, and H. Callioglu. Impact behavior of hybrid composite plates. *Journal of Applied Polymer Science*, 118:580–587, 2010a.

M. Sayer, N.B. Bektas, and O. Sayman. An experimental investigation on the impact behavior of hybrid composite plates. *Composite Structures*, 92:1256–1262, 2010b.

E. Schulson. The brittle compressive fracture of ice. *Acta Metallurgica et Materialia*, 38:1963–1976, 1990.

E. Schulson. Brittle failure of ice. *Engineering Fracture Mechanics*, 68: 1839–1887, 2001.

L.E. Schwer. Preliminary assesment of non-lagrangian methods for penetration simulation. In *8th International LS-DYNA User's Conference*, Dearborn, Michigan, May 2004.

C.M. Seddon, K. Moodie, A.M. Thyer, and M. Moatamedi. Preliminary analysis of fuel tank impact. *International Journal of Crashworthiness*, 9:237–244, 2004.

M. Shazly, V. Prakash, and B.A. Lerch. High strain-rate compression testing of ice. Technical report, NASA Glenn Research Center, 2006.

M. Shazly, V. Prakash, and B.A. Lerch. High strain-rate behavior of ice under uniaxial compression. *International Journal of Solids and Structures*, 46:1499–1515, 2009.

J.A. Sherburn and M.F. Horstemeyer. Hydrodynamic modeling of impact craters in ice. *International Journal of Impact Engineering*, 37:27–36, 2010.

K. Shintate and H. Sekine. Numerical simulation of hypervelocity impacts of a projectile on laminated composite plate targets by means of improved SPH method. *Composites Part A: Applied Science and Manufacturing*, 35:683–692, 2004.

J.J. Short, M.E. Kelley, R.J. Speelman, and R.E. McCarty. Birdstrike prevention: applying aeroscience and bio-science. In *International Bird Strike Committee, IBSC25/WP-RS4*, Amsterdam, April 2000.

Y.P. Siow and V.P.W. Shim. An experimental study of low velocity impact damage in woven fiber composites. *Journal of Composite Materials*, 32: 1178–1202, 1998.

W.R. Soper. Hydraulic ram studies. Master's thesis, Naval Postgraduate School, Monterey, CA, 1973.

M. Souli, L. Olovsson, and I. Do. ALE and fluid-structure interaction capabilities in LS-DYNA. In *7th International LS-DYNA User's Conference*, Dearborn, Michigan, May 2002.

C.E. Sparks, R.L. Hinrichsen, and D. Friedmann. Comparisson and validation of smooth particle hydrodynamic (SPH) and coupled euler lagrange (CEL) techniques for modeling hydrodynamic ram. In *Proceedings of the 46th AIAA/ASME/ASCE/AHS/ASC Structures, Structural Dynamics & Materials Conference*, 2005.

F. Stoll and R.A. Brockman. Finite element simulation of high-speed soft-body impacts. In *Proceedings of the 38th AIAA/ASME/ASCE/AHS/ASC Structures, Structural Dynamics & Materials Conference*, pages 334–344, Kissimmee, FL, April 1997.

C.T. Sun and V. Potti. A simple model to predict residual velocities of thick composite laminates subjected to high velocity impact. *International Journal of Impact Engineering*, 18:339–353, 1996.

F. Tabaddor and J.R. Stafford. Some aspects of the rubber composite finite element analysis. *Composite Structures*, 21:327–339, 1985.

P. Tambunan and A. Vlot. Ice ball impact on aircraft fuselage protection plates. Technical report, TU Delft, 1995.

Y. Tanabe, M. Aoki, K. Fujii, H. Kasano, and E. Yasuda. Fracture behavior of CFRPs impacted by relatively high-velocity steel sphere. *International Journal of Impact Engineering*, 28:627–642, 2003.

P.J. Torvik. A simple theory for shock propagation in homogeneous mixtures. Technical Report AFIT-TR-70-3, Air Force Institute of Technology, Wright-Patterson Air Force Base, Ohio, 1970.

D. Townsend, N. Park, and PM. Devall. Failure of fluid filled structures due to high velocity fragment impactl. *International Journal of Impact Engineering*, 29:723–733, 2003.

L.R.G. Treloar. *The physics of rubber elasticity.* Oxford Clarendon Press, 1975.

S.W. Tsai and E.M. Wu. A general theory of strength for anisotropic materials. *Journal of Composite Materials*, 5:58–80, 1971.

D. Varas, J. López-Puente, and R. Zaera. Experimental analysis of fluid filled aluminium tubes subjected to high velocity impact. *International Journal of Impact Engineering*, 36:81–91, 2009a.

D. Varas, R. Zaera, and J. López-Puente. Numerical modelling of the hydrodynamic ram phenomenon. *International Journal of Impact Engineering*, 36:363–374, 2009b.

D. Varas, R. Zaera, and J. López-Puente. Experimental study of CFRP fluid-filled tubes subjected to high-velocity impact. *Composite Structures*, 93:2598–2609, 2011.

D. Varas, J. López-Puente, and R. Zaera. Numerical analysis of the hydrodynamic ram phenomenon in aircraft fuel tanks. *American Institute of Aeronautics and Astronautics Journal*, 50:1621–1630, 2012.

M. Vesenjak, M. Matthae, H. Mullerschon, and Z. Ren. Fluid models in LS-DYNA and their interaction with a structure in dynamic simulations. In *Proceedings of PVP2005. ASME Pressure Vessels and Piping Division Conference*, Denver, 2005.

R. Vignjevic, T. De Vuyst, J. Campbell, and L. Libersky. Modelling of hydrodynamic ram using smoothed particle hydrodynamics. In *Proceedings of the 5th International Conference on Dynamics and Control of Systems and Structures in Space*, Cambridge, UK, 2002.

H. Wang and T. Vukhanh. Damage extension in carbon fiber/peek crossply laminates under low velocity impact. *Journal of Composite Material*, 28: 545–551, 1994.

H. Wang and T. Vukhanh. Low-velocity impact damage in laminated composites materials. *Key Engineering Materials*, 141-1:277–304, 1998.

Y. Watanabe and M.J. Kaldjian. Modelling and analysis of bias-ply motorcycle tires. *Mathematical Modelling*, 6:80, 1985.

B. Whittingham, I.H. Marshall, T. Mitrevski, and R. Jones. The response of composite structures with pre-stress subject to low velocity impact damage. *Composite Structures*, 66:685–698, 2004.

J.S. Wilbeck. Impact behavior of low strength projectiles. Technical Report AFML-TR-77-134, Air Force Materials Laboratory, 1978.

L. Wu, Y.N. Guo, and Y.L. Li. Bird strike simulation on sandwich composite structure of aircraft radome. *Explosion and Shock Waves*, 29:642–647, 2009.

A. Zammit, M. Kim, and J. Bayandor. Bird-strike damage tolerance analysis of composite turbofan engines. In *ICAS 2010, 27th international Congress of the Aeronautical Sciences*, Nice, France, September 2010.

A. Zhang and K. Suzuki. A comparative study of numerical simulations for fluid-structure interaction of liquid-filled tank during ship collision. *Ocean Engineering*, 34:645–652, 2007.

G. Zhou, J.C. Lloyd, and J.J. McGuirk. Experimental evaluation of geometric factors affecting damage mechanisms in carbon/epoxy plates. *Composites Part A: Applied Science and Manufacturing*, 32:2279–2286, 2001.

S. Zhu and M. Tong. Study on bird shape sensitivity to dynamic response of bird strike on aircraft windshield. *J Nanjing Univ Aeron Astronaut*, 40:551–555, 2008.

Computer estimation of plastic strain localization and failure for large strain rates using viscoplasticity

Tomasz Łodygowski* and Wojciech Sumelka*
* Institute of Structural Engineering,
Poznan University of Technology, Poznań, Poland
{*tomasz.lodygowski, wojciech.sumelka*}@*put.poznan.pl*

Abstract The problem of modelling extreme dynamic events for metallic materials including strain rates over 10^7 s^{-1} and temperatures reaching melting point is still vivid in theoretical, applied and computational mechanics. Such thermomechanical processes are highly influenced by elasto-viscoplastic wave effects (their propagation and interaction) and varying initial anisotropy caused by existing defects in metals structure like microcracks, microvoids, mobile and immobile dislocations densities being together a cause of overall induced anisotropy during deformation (from the point of view of meso-macro continuum mechanics approach). It should be emphasised, that the most reliable way for estimation of such processes needs nowadays a complex phenomenological models due to limitations of current experimental techniques (it is still not possible to measure the evolution of crucial quantities e.g. temperature for extreme dynamic processes) and computational capabilities.

Within this document we consider recent achievements of Perzyna's type viscoplasticity theory for metallic materials accounting for anisotropic description of damage suitable for modelling plastic strain localization and failure for large strain rates.

1 Introduction

Qualitative and quantitative mathematical modelling of extreme dynamic events in metals is still an open question in mechanics. One should emphasise the embarrassing situation, that the experiment, can not still give a detailed/unique answer for extremely fast thermomechanical processes about evolution of crucial phenomena (e.g. strain or temperature evolution). Hence we face a situation in which the theory must extrapolate/forecast experimental results.

In the discussion of correctness of a damage (simultaneously strains localisation and temperatures) description by a constitutive model for phenomena including large strain rates we can distinguish two levels: global and local Sumelka and Łodygowski (2011). We have good *global damage approximation* (GDA) if (global) strain-stress curves from experiment and mathematical model are close to each other. We have good *local damage approximation* (LDA) if apart from the global we have particularly good coincidence in: macrodamage initiation time, velocity of macrodamage evolution and the geometry of macrodamage pattern.

Under continuum damage mechanics it can be shown that the first class of problems (GDA) can be covered by scalar damage models, while for the second class of problems (LDA) one needs higher order tensor in the model describing damage Glema et al. (2010b); Łodygowski et al. (2012). However due to the fact, that the identification of the material functions and parameters (necessary for industrial applications) for models with directional damage is awkward, hence they are still on the level of scientific considerations.

This document deals with recent achievements of Perzyna's type viscoplasticity theory for metallic materials under adiabatic conditions being a standard assumption for extreme dynamics. Special attention is devoted for metals anisotropy (initial and induced) which highly influences the overall process of dynamic deformation. The discussed constitutive structure has a deep physical interpretation derived from the analysis of a single crystal and a polycryslal behaviours Perzyna (2005, 2008); Sumelka (2009).

The remaining part of this paper is organised as follows.

In Section 2 we deal with experimental observations concerning metals anisotropy and its influence on behaviour of metallic materials under dynamic conditions.

Section 3 governs the fundamental results of Perzyna's type viscoplasticity accounting for anisotropic damage description and identification of assumed material functions and parameters Sumelka (2009).

In Section 4 computer implementation of the model in Abaqus/Explicit program utilising VUMAT subroutine is discussed.

Finally, in Section 5 verification of the model based on numerical modelling double chamber pipe under dynamic compression, showing real industrial application concerning designing of energy absorbers, is presented.

Section 6 concludes the document.

2 Experimental Motivation

Let us focus the attention on experimental observations of metals behaviour concerning especially microdamage anisotropy in metals being a source of the overall metal anisotropy. Other sources of anisotropy like different sizes and shapes of adjacent grains Narayanasamy et al. (2009), presence of different phases like pearlite or ferrite Pęcherski et al. (2009) are not discussed (cf. Fig. 1). Also on purpose we omit the description of well recognised and described in literature phenomena like e.g. kinematical hardening, rate sensitivity, length scale sensitivity or plastic non-normality (they are included into the model of course). For detailed information, please see the review reports, e.g. Łodygowski (1996); Perzyna (1998, 2005); Glema (2004).

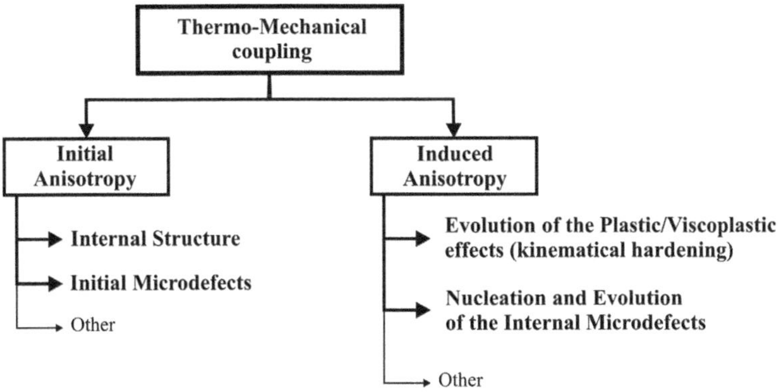

Figure 1. Anisotropy in thermo-mechanics of metals

So, the always existing defects in metals structure like microcracks, microvoids, mobile and immobile dislocations densities Abu Al-Rub and Voyiadjis (2006); Voyiadjis and Abu Al-Rub (2006) (cf. Fig. 2) cause anisotropy of metals. It is then clear, that for a proper mathematical modelling of metals behaviour one should include this type of anisotropy into the formal description. The frequently used isotropic simplification for metals should be thought of as a first approximation which carries not enough information for modern applications Glema et al. (2010a) (though it certainly does not disavow such an approach in many applications cf. Klepaczko (2007); Rusinek and Klepaczko (2009)).

Coming back to the experimental results, being the crux of the matter of this section, we propose the following two statements Sumelka (2009):

(i) intrinsic microdefects are anisotropic,

(ii) evolution of microdamages is directional.

Figure 2. The anisotropy of the HSLA-65 steel microstructure Narayanasamy et al. (2009).

Statement (i) confirms the experimental results, that metals anisotropy, caused by the intrinsic defects, comes not only from its existence but especially from its inhomogeneous structure.

As an example, let us consider the effects of flat plate impact experiment in 1145 aluminium Seaman et al. (1976). The separation observed is preceded by the evolution of microdamages (microvoids), consisting for undamaged material of three stages: nucleation, growth and coalescence. Notice in Fig. 3 that all of the microdefects are elongated perpendicularly to impact direction, thus to maximal tensile stresses. In this experiment they have approximately an ellipsoidal shape. So, intrinsic defects have directional nature. Their anisotropy influences the whole deformation process, having a considerable impact on it.

Statement (ii) expresses explicitly experimental fact that the anisotropic properties of the continuum body evolve directionally during the deformation process (cf. experimental results presented in Grebe et al. (1985)). Notice, that it is a consequence of structure rearrangement itself but especially by directional evolution of intrinsic defects. As an example in Fig. 4, the evolution of microvoids in the region of forming shear band is presented. It is clearly seen, that the evolution is directional, microvoids are being elongated through the shear band. So, the existing or nucleating microdamages growth is directional according to the imposed deformation process, induc-

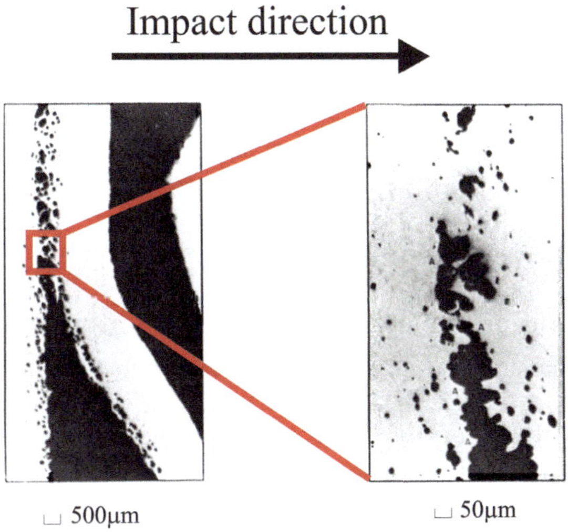

Figure 3. Cracks anisotropy in 1145 aluminium after flat plate impact experiment Seaman et al. (1976).

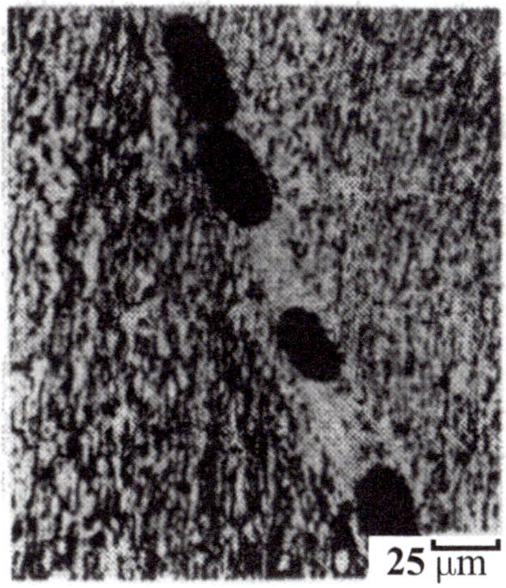

Figure 4. Anisotropic microcracks in the shear band region in Ti-6 pct Al-4 pct V alloy (after Grebe et al. (1985)).

ing the anisotropic evolution of material properties.

As concluding remark of this section let us recall, that the microdamage evolution mechanism in metals generally has three stages: defects nucleation, their growth and coalescence. All of them are anisotropic and should be described by the material model. These observations will be transferred into the model utilising the concept of microdamage tensor.

3 Mathematical Modelling

3.1 The Concept of Microdamage Tensor

The microdamage tensor plays an important role in the presented formulation. Notice that although we use the prefix "micro", it should not be confused with micro-scale observations. Namely, we use the continuum description, so we are between macro and meso scales ($10^{-4} \div 10^{-3}$m), but we "transfer" the nomenclature from micro-scale Longere et al. (2005). An analogous problem was discussed in Tikhomirov et al. (2001) according to the notion *microcracks*, from which we repeat the following Tab. 1 being a conclusion.

Table 1. Representation of materials defects

Type of defects	Mathematical description	Science
Microcracks, microvoids	Stochastical distributions, overall variables, and fractal mechanics	Micromechanics
Microcracks, microvoids	**Damage variables**	**Continuum damage mechanics**
Macrocraks	Embedded displacement discontinuities	Fracture mechanics

It is obvious that the concept of microdamage tensor should have its clear physical interpretation giving a straight answer concerning how to measure its components experimentally. To understand the whole concept let us follow the results discussed in Sumelka (2009); Glema et al. (2010b)

The Physical Interpretation of Microdamage Tensor Suppose, that for the selected points P_i in the material body $\mathcal{B}$, on three perpendicular planes, the ratio of the damaged area to the assumed characteristic area of the representative volume element (RVE) can be measured i.e.

$$\frac{A_i^p}{A}, \tag{1}$$

where A_i^p is a damaged area and A denotes assumed characteristic area of the RVE - cf. Fig. (5). Based on the calculated ratios $(\frac{A_i^p}{A})$ three vectors are obtained. Theirs modules are equal to those ratios and are normal to RVE's planes (see Fig. (5)).

Measurements presented above can be repeated in any different configuration by rotating these three planes about point O - Fig. (5). So, for every measurement configuration, from those three vectors, we compose the resultant. Afterwards, we choose that configuration, in which the resultant module is largest. Such resultant is called the *main microdamage vector* and is denoted by $\hat{\boldsymbol{\xi}}^{(m)}$ Sumelka and Glema (2007) i.e.

$$\hat{\boldsymbol{\xi}}^{(m)} = \frac{A_1^p}{A}\hat{\mathbf{e}}_1 + \frac{A_2^p}{A}\hat{\mathbf{e}}_2 + \frac{A_3^p}{A}\hat{\mathbf{e}}_3, \tag{2}$$

where $\hat{(\cdot)}$ denotes the principal directions of microdamage with $A_1^p \geq A_2^p \geq A_3^p$.

In the following step, based on the main microdamage vector, we build the *microdamage vector*, denoted by $\hat{\boldsymbol{\xi}}^{(n)}$ Sumelka and Glema (2007)

$$\hat{\boldsymbol{\xi}}^{(n)} = \frac{1}{\left\|\hat{\boldsymbol{\xi}}^{(m)}\right\|}\left(\left(\frac{A_1^p}{A}\right)^2\hat{\mathbf{e}}_1 + \left(\frac{A_2^p}{A}\right)^2\hat{\mathbf{e}}_2 + \left(\frac{A_3^p}{A}\right)^2\hat{\mathbf{e}}_3\right). \tag{3}$$

Finally, we postulate the existence of microdamage tensorial field $\boldsymbol{\xi}$

$$\boldsymbol{\xi} = \begin{bmatrix} \xi_{11} & \xi_{12} & \xi_{13} \\ \xi_{21} & \xi_{22} & \xi_{23} \\ \xi_{31} & \xi_{32} & \xi_{33} \end{bmatrix}, \tag{4}$$

which we define in its principal directions by applying the formula combining the microdamage vector and microdamage tensor Sumelka and Glema (2007)

$$\hat{\boldsymbol{\xi}}^{(n)} = \hat{\boldsymbol{\xi}}\mathbf{n}, \tag{5}$$

where

$$\mathbf{n} = \sqrt{3}\left\|\hat{\boldsymbol{\xi}}^{(m)}\right\|^{-1}\left(\hat{\xi}_1^{(m)}\hat{\mathbf{e}}_1 + \hat{\xi}_2^{(m)}\hat{\mathbf{e}}_2 + \hat{\xi}_3^{(m)}\hat{\mathbf{e}}_3\right), \tag{6}$$

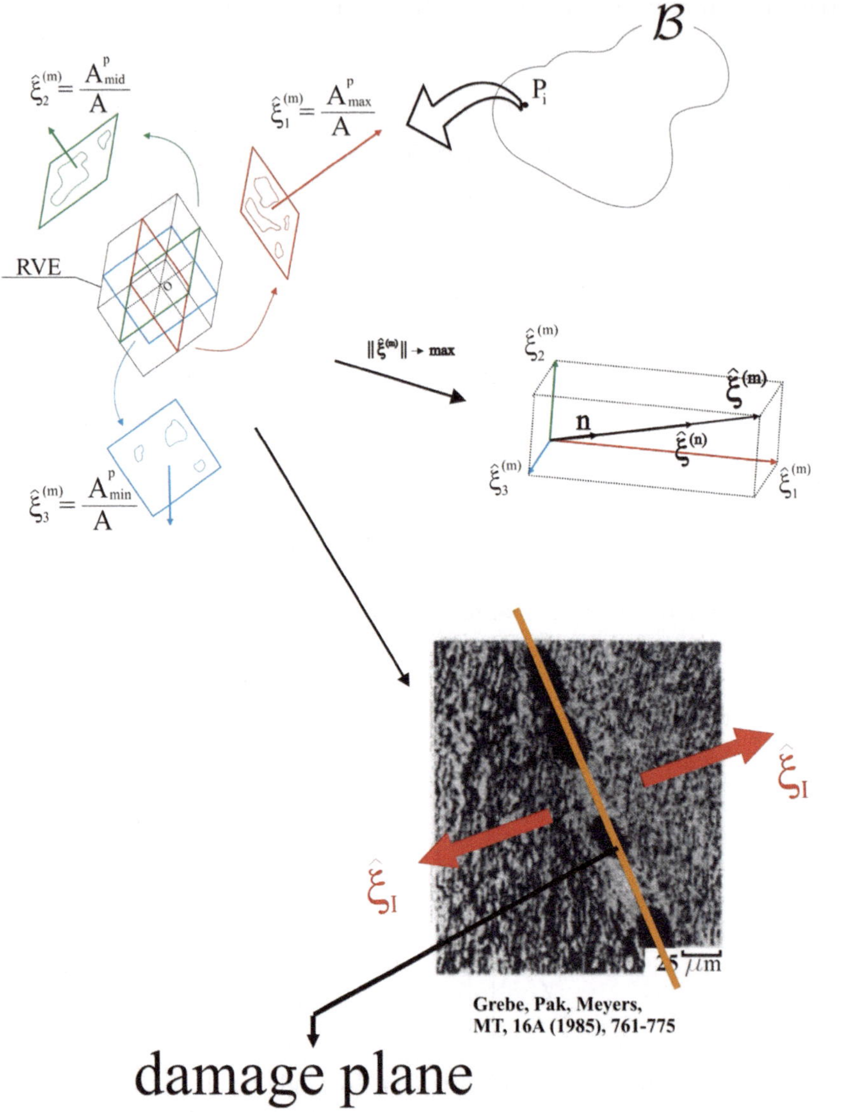

Figure 5. The concept of microdamage tensor

leading to the fundamental result

$$\hat{\boldsymbol{\xi}} = \frac{\sqrt{3}}{3} \begin{bmatrix} \hat{\xi}_1^{(m)} & 0 & 0 \\ 0 & \hat{\xi}_2^{(m)} & 0 \\ 0 & 0 & \hat{\xi}_3^{(m)} \end{bmatrix}. \tag{7}$$

Hence, we obtain the following physical interpretation of the microdamage tensor components: *The diagonal components ξ_{ii} of the microdamage tensor $\boldsymbol{\xi}$, in its principal directions, are proportional to the components of the main microdamage vector $\xi_i^{(m)}$ which defines the ratio of the damaged area to the assumed characteristic area of the RVE, on the plane perpendicular to the i direction.*

Such interpretation clearly states, that the damage plane is the one perpendicular to the maximal principal value of $\boldsymbol{\xi}$. Simultaneously, it gives a tool for a graphical presentation of the anisotropy evolution during post processing of the numerical results. Namely, by tracing the principal directions of $\boldsymbol{\xi}$ we also trace the softening directions and can predict macrodamage path(s) Glema et al. (2010a,b).

Moreover, taking the Euclidean norm from the microdamage field $\hat{\boldsymbol{\xi}}$, we obtain

$$\sqrt{\boldsymbol{\xi} : \boldsymbol{\xi}} = \frac{\sqrt{3}}{3} \sqrt{\left(\frac{A_1^p}{A}\right)^2 + \left(\frac{A_2^p}{A}\right)^2 + \left(\frac{A_3^p}{A}\right)^2}. \tag{8}$$

Now, assuming that the characteristic length of RVE cube is l we can rewrite Eq. (8) as

$$\sqrt{\boldsymbol{\xi} : \boldsymbol{\xi}} = \frac{\frac{\sqrt{3}\, l}{3} \sqrt{(A_1^p)^2 + (A_2^p)^2 + (A_3^p)^2}}{l^3}. \tag{9}$$

Eq. (8) results in additional physical interpretation for microdamage tensor appears, namely, the Euclidean norm of the microdamage field defines the scalar quantity called the *volume fraction porosity* or simply *porosity* Perzyna (2008)

$$\sqrt{\boldsymbol{\xi} : \boldsymbol{\xi}} = \xi = \frac{V - V_s}{V} = \frac{V_p}{V}, \tag{10}$$

where ξ denotes porosity (scalar damage parameter), V is the volume of a material element, V_s is the volume of the solid constituent of that material element and V_p denotes void volume

$$V_p = \frac{\sqrt{3}\, l}{3} \sqrt{(A_1^p)^2 + (A_2^p)^2 + (A_3^p)^2}. \tag{11}$$

The Limit Values of Microdamage Evolution The interpretations of the microdamage tensorial field impose the mathematical bounds for a microdamage evolution, as

$$\xi \in< 0,1 >, \quad \text{and} \quad \hat{\boldsymbol{\xi}}_{ii} \in< 0,1 > . \tag{12}$$

However, the physical bounds are different. This fact is crucial for further numerical analyses and shows the influence of the initial microdamage field on a macrodamage mode.

There are unwavering experimental evidence showing the existence of the initial porosity in metals (denoted by ξ_0) which is of order $\xi_0 \cong 10^{-4} \div 10^{-3}$ Nemes and Eftis (1991). That porosity however, can not reach the theoretical full saturation, i.e. $\xi = 1$, during the deformation process. Real maximal fracture porosity in metals depends on the tested material and the process characteristics (strain rates) and is of the order $0.09 \div 0.35$ Dornowski and Perzyna (2002, 2006).

The only information regarding the initial microdamage tensorial field is its norm (porosity). It is crucial to notice that current experimental approach can not give the answer about damage directions (particularly during extremely fast thermomechanical processes - crucial for this paper). However, as shown in Sumelka and Łodygowski (2011) we can theoretically test the model for possible scenarios of the initial microdamage tensorial field distributions, to show its meaningful role in the deformation process. This initial state (note that the porosity must increase its initial value by several orders to reach the fracture porosity) will influence in particular: macrodamage initiation time, the velocity of macrodamage evolution and the geometry of macrodamage.

3.2 Theory of Perzyna's Type Viscoplasticity Accounting for Anisotropic Damage

We start with the description of the kinematics of the body, than we will postulate fundamental constitutive axioms and finally by fulfilling standard set of balance principles we will obtain constitutive model.

Kinematics Let us consider two descriptions of the material body motion, namely Lagrangean (material, referential) and Eulerian (spatial, current). Those descriptions span two manifolds denoted by $\mathcal{B}$ and $\mathcal{S}$ respectively Marsden and Hughes (1983).

Points in $\mathcal{B}$ are denoted by $\mathbf{X}$ while in $\mathcal{S}$ by $\mathbf{x}$. Coordinate system for $\mathcal{B}$ is denoted by $\{X^A\}$ with base $\mathbf{E}_A$ and for $\mathcal{S}$ we have $\{x^a\}$ with base $\mathbf{e}_a$. Dual bases in those coordinate systems are denoted by $\mathbf{E}^A$ and $\mathbf{e}^a$, respectively.

The tangent spaces in $\mathcal{B}$ and $\mathcal{S}$ are written as $T_{\mathbf{X}}\mathcal{B} = \{\mathbf{X}\} \times V^3$ and $T_{\mathbf{x}}\mathcal{S} = \{\mathbf{x}\} \times V^3$. It is understood as Euclidean vector space V^3, regarded as vectors emanating from points $\mathbf{X}$ and $\mathbf{x}$, respectively Marsden and Hughes (1983).

Taking Riemannian space on manifolds $\mathcal{B}$ and $\mathcal{S}$, i.e. $\{\mathcal{B}, \mathbf{G}\}$ and $\{\mathcal{S}, \mathbf{g}\}$ we define the metric tensors $\mathbf{G} : T\mathcal{B} \to T^*\mathcal{B}$ and $\mathbf{g} : T\mathcal{S} \to T^*\mathcal{S}$ where $T\mathcal{B}$ and $T\mathcal{S}$ denotes the tangent bundles of $\mathcal{B}$ and $\mathcal{S}$, respectively, while $T^*\mathcal{B}$ and $T^*\mathcal{S}$ denotes their dual tangent bundles. Explicit definitions for metric tensors are then $G_{AB}(\mathbf{X}) = (\mathbf{E}_A, \mathbf{E}_B)_{\mathbf{X}}$ and $g_{ab}(\mathbf{x}) = (\mathbf{e}_a, \mathbf{e}_b)_{\mathbf{x}}$ where $(,)_{\mathbf{X}}$ and $(,)_{\mathbf{x}}$ denote inner product in $\mathcal{B}$ and $\mathcal{S}$, respectively.

The regular motion of the material body, treated as a series of the immersing of the abstract body $\mathcal{B}$ in the Euclidean point space E^3 Rymarz (1993), can be written as

$$\mathbf{x} = \phi(\mathbf{X}, t), \tag{13}$$

thus $\phi_t : \mathcal{B} \to \mathcal{S}$ is a C^1 actual configuration of $\mathcal{B}$ in $\mathcal{S}$, at time t. The tangent of ϕ, defines the two point tensor field $\mathbf{F}$, called deformation gradient, which describes all local deformation properties and is the primary measure of deformation Perzyna (1978); Holzapfel (2000). Thus we have

$$\mathbf{F}(\mathbf{X}, t) = T\phi = \frac{\partial \phi(\mathbf{X}, t)}{\partial \mathbf{X}}, \tag{14}$$

and using the notion of tangent space

$$\mathbf{F}(\mathbf{X}, t) = T_{\mathbf{X}}\mathcal{B} \to T_{\mathbf{x}=\phi(\mathbf{X},t)}\mathcal{S}, \tag{15}$$

so $\mathbf{F}$ is a linear transformation for each $\mathbf{X} \in \mathcal{B}$ and $t \in I \subset \mathbb{R}^1$.

The map ϕ is assumed to be uniquely invertible (smooth homeomorphism) ($\mathbf{X} = \phi^{-1}(\mathbf{x}, t)$), hence there exists the inverse of deformation gradient

$$\mathbf{F}^{-1}(\mathbf{x}, t) = \frac{\partial \phi^{-1}(\mathbf{x}, t)}{\partial \mathbf{x}}, \tag{16}$$

and the tensor field $\mathbf{F}$ is non-singular ($\det(\mathbf{F}) \neq 0$), and because of the impenetrability of matter $\det(\mathbf{F}) > 0$. So we have the following the important decomposition, called polar decomposition

$$\mathbf{F} = \mathbf{R}\mathbf{U} = \mathbf{v}\mathbf{R}, \tag{17}$$

where the $\mathbf{R}$ is rotation tensor (unique, proper orthogonal) which measures local orientation and $\mathbf{U}$ and $\mathbf{v}$ define unique, positive define, symmetric tensors called the right (or material) stretch tensor and the left (or spatial) stretch tensor, respectively (stretch tensors measure the local shape). Using the notion of tangent space we have for each $\mathbf{X} \in \mathcal{B}$, $\mathbf{U}(\mathbf{X}) : T_{\mathbf{X}}\mathcal{B} \to T_{\mathbf{X}}\mathcal{B}$, and for each $\mathbf{x} \in \mathcal{S}$, $\mathbf{v}(\mathbf{x}) : T_{\mathbf{x}}\mathcal{S} \to T_{\mathbf{x}}\mathcal{S}$. Thus the local motion characterised by $\mathbf{F}$ can be decomposed into pure stretch and pure rotation.

Let us now define the general class of Lagrangean and Eulerian strain measures, defined through one single scale function given by (cf. Hill (1978); Xiao et al. (1998)),

$$\mathbf{E} = g(\mathbf{C}) = \sum_{i=1}^{3} g(\breve{\chi}_i)\mathbf{C}_i,$$

and

$$\mathbf{e} = g(\mathbf{B}) = \sum_{i=1}^{3} g(\breve{\chi}_i)\mathbf{B}_i,$$

where the scale function $g(\cdot)$ is a smooth increasing function with the normalised property $g(1) = g'(1) - 1 = 0$, $\breve{\chi}_i$ are used to denote distinct eigenvalues of the right and left Cauchy-Green tensors $\mathbf{C}$ and $\mathbf{B}$ respectively, and $\mathbf{C}_i$ and $\mathbf{B}_i$ are the corresponding subordinate eigenprojections. From this class we choose different strain measures for Lie and Logarithmic formulations.

The Green-Lagrange strain tensor has been chosen Perzyna (2005) ($\mathbf{E} : T_{\mathbf{X}}\mathcal{B} \to T_{\mathbf{X}}\mathcal{B}$)

$$2\mathbf{E} = \mathbf{C} - \mathbf{I}, \tag{18}$$

where $\mathbf{E}$ stands for the Green-Lagrange strain tensor, $\mathbf{I}$ denotes the identity on $T_{\mathbf{X}}\mathcal{B}$ and

$$\mathbf{C} = \mathbf{F}^T \cdot \mathbf{F} = \mathbf{U}^2 = \mathbf{B}^{-1}. \tag{19}$$

By analogy for the spatial (Eulerian) strain measure the Euler-Almansi strain tensor has been accepted ($\mathbf{e} : T_{\mathbf{x}}\mathcal{S} \to T_{\mathbf{x}}\mathcal{S}$)

$$2\mathbf{e} = \mathbf{i} - \mathbf{c}, \tag{20}$$

where $\mathbf{e}$ stands for the Euler-Almansi strain tensor, $\mathbf{i}$ denotes the identity on $T_{\mathbf{x}}\mathcal{S}$. We have also

$$\mathbf{c} = \mathbf{b}^{-1} \quad \text{and} \quad \mathbf{b} = \mathbf{F} \cdot \mathbf{F}^T = \mathbf{v}^2, \tag{21}$$

where tensor $\mathbf{b}$ is sometimes referred to as the Finger deformation tensor.

Using push-forward and pull-back operation we obtain for covariant material and spatial strain measures the following relation

$$\mathbf{e}^\flat = \phi_*(\mathbf{E}^\flat) = \mathbf{F}^{-T}\mathbf{E}^\flat\mathbf{F}^{-1}, \tag{22}$$

and

$$\mathbf{E}^\flat = \phi^*(\mathbf{e}^\flat) = \mathbf{F}^T\mathbf{e}^\flat\mathbf{F}, \tag{23}$$

where

$$\phi_*((\cdot)^\flat) = \mathbf{F}^{-T}(\cdot)^\flat \mathbf{F}^{-1} \tag{24}$$

stands for push-forward, and

$$\phi^*((\cdot)^\flat) = \mathbf{F}^T(\cdot)^\flat \mathbf{F}, \tag{25}$$

for pull-back.

To describe the finite elasto-viscoplastic deformation the multiplicative decomposition of the total deformation gradient has been accepted

$$\mathbf{F}(\mathbf{X}, t) = \mathbf{F}^e(\mathbf{X}, t) \cdot \mathbf{F}^p(\mathbf{X}, t). \tag{26}$$

This decomposition justified by the micromechanics of single crystal plasticity Perzyna (1998) states that the component $\mathbf{F}^e$ is a lattice contribution to $\mathbf{F}$, while $\mathbf{F}^p$ describes the deformation solely due to plastic shearing on crystallographic slip systems.

The inverse of the local elastic deformation $\mathbf{F}^{e^{-1}}$ releases from the stress state in every surrounding $(\mathcal{N}(\mathbf{x}) \subset \phi(\mathcal{B}))$ in an actual configuration. The configuration obtained by the linear map $\mathbf{F}^{e^{-1}}$ from actual configuration $\mathcal{S}$ is called unstressed configuration and is denoted by $\mathcal{S}'$. Thus one can write (see Fig. 6)

$$\mathbf{F}^e : T_\mathbf{y}\mathcal{S}' \to T_\mathbf{x}\mathcal{S}, \quad \mathbf{F}^p : T_\mathbf{X}\mathcal{B} \to T_\mathbf{y}\mathcal{S}', \tag{27}$$

where the material point in the configuration $\mathcal{S}'$ is characterised by $\mathbf{y}$.

The introduced decomposition of $\mathbf{F}$ allows us to define the fundamental strain tensors in both formulations considered.

The viscoplastic strain tensor $\mathbf{E}^p : T_\mathbf{X}\mathcal{B} \to T_\mathbf{X}\mathcal{B}$ can be written as

$$2\mathbf{E}^p = \mathbf{C}^p - \mathbf{I}, \tag{28}$$

where

$$\mathbf{C}^p = \mathbf{F}^{p^T} \cdot \mathbf{F}^p = \mathbf{U}^{p^2} = \mathbf{B}^{p^{-1}} \quad \text{and} \quad \mathbf{E}^e = \mathbf{E} - \mathbf{E}^p, \tag{29}$$

while the elastic strain tensor $\mathbf{e}^e : T_\mathbf{x}\mathcal{S} \to T_\mathbf{x}\mathcal{S}$ is

$$2\mathbf{e}^e = \mathbf{i} - \mathbf{c}^e, \tag{30}$$

where

$$\mathbf{c}^e = \mathbf{b}^{e^{-1}} \quad \text{and} \quad \mathbf{b}^e = \mathbf{F}^e \cdot \mathbf{F}^{e^T} = \mathbf{v}^{e2} \quad \text{and} \quad \mathbf{e}^p = \mathbf{e} - \mathbf{e}^e. \tag{31}$$

At the end of material body kinematics description it is fundamental for further discussion of the rate type constitutive relations, to define the rate

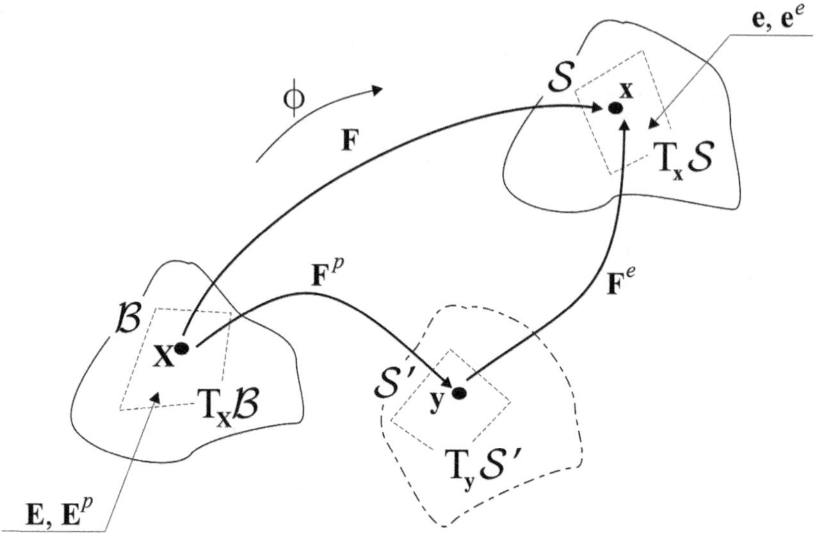

Figure 6. The interpretation of the multiplicative decomposition of **F**

of fields that describe the change of shape, position and orientation of a continuum body.

Using motion defined by Eq. (13), the spatial velocity $\boldsymbol{v}$ is simply

$$\boldsymbol{v}(\mathbf{x}, t) = \dot{\mathbf{x}} = \frac{\partial \phi}{\partial t}. \tag{32}$$

Taking the gradient of $\boldsymbol{v}$, the tensor field (non-symmetric, second order), called spatial velocity gradient, is obtained Holzapfel (2000)

$$\mathbf{l}(\mathbf{x}, t) = \frac{\partial \boldsymbol{v}(\mathbf{x}, t)}{\partial \mathbf{x}}, \tag{33}$$

where $\mathbf{l}$ stands for spatial velocity gradient. Using the notion of deformation gradient and multiplicative decomposition of deformation gradient, the Eq. (33) is rewritten to the form Perzyna (2005)

$$\mathbf{l} = \dot{\mathbf{F}} \cdot \mathbf{F}^{-1} = \dot{\mathbf{F}}^{e} \cdot \mathbf{F}^{e-1} + \mathbf{F}^{e} \cdot (\dot{\mathbf{F}}^{p} \cdot \mathbf{F}^{p-1}) \cdot \mathbf{F}^{e-1} = \mathbf{l}^{e} + \mathbf{l}^{p}, \tag{34}$$

which introduces the elastic $\mathbf{l}^e$ and plastic $\mathbf{l}^p$ parts of spatial velocity gradient. On the other hand the additive decomposition of spatial velocity gradient to symmetric and antisymmetric parts generates covariant tensor field $\mathbf{d}$ called rate of deformation tensor and also covariant tensor field $\mathbf{w}$ called spin tensor, with the definitions

$$\mathbf{l} = \mathbf{d} + \mathbf{w} = \mathbf{d}^e + \mathbf{w}^e + \mathbf{d}^p + \mathbf{w}^p, \tag{35}$$

$$\mathbf{d} = \frac{1}{2}(\mathbf{l} + \mathbf{l}^T), \tag{36}$$

$$\mathbf{w} = \frac{1}{2}(\mathbf{l} - \mathbf{l}^T). \tag{37}$$

Let us first introduce the notion of objective Lie derivative (assuring diffeomorphisms). The Lie derivative of an arbitrary spatial tensor field φ is obtained using the following concept:

(i) compute the pull-back operation of φ - the material field Φ is obtained,

(ii) take the material time derivative of Φ,

(iii) carry out the push-forward operation of the result field from (ii).

The scheme can be summarised as

$$\mathrm{L}_{\boldsymbol{\upsilon}}(\varphi) = \phi_* \left(\frac{\mathrm{D}}{\mathrm{D}t} \phi^*(\varphi) \right), \tag{38}$$

where $\mathrm{L}_{\boldsymbol{\upsilon}}$ stands for Lie derivative.

So, applying Lie derivative to strain measure we obtain the following fundamental result

$$\mathbf{d}^\flat = \mathrm{L}_{\boldsymbol{\upsilon}}(\mathbf{e}^\flat). \tag{39}$$

Thus, Lie derivative states a direct relationship between the stretching $\mathbf{d}$, which is a direct natural measure of the rate of length of any line element and the rate of change of the angle between any two intersecting line elements in a deforming body, and the Eulerian strain $\mathbf{e}$ which measures a change of the length of any line element and change of the angle between any two intersecting line elements.

By analogy to Eq. (39) one can write

$$\mathbf{d}^{e\flat} = \mathrm{L}_{\boldsymbol{\upsilon}}(\mathbf{e}^{e\flat}), \quad \mathbf{d}^{p\flat} = \mathrm{L}_{\boldsymbol{\upsilon}}(\mathbf{e}^{p\flat}). \tag{40}$$

Constitutive Postulates Assuming that the balance principles hold, namely: conservation of mass, balance of momentum, balance of moment of momentum and balance of energy and entropy production, we define four constitutive postulates Perzyna (1986a, 2005) (below $\mathbf{e}$ depends on formulation of course):

(i) Existence of the free energy function ψ. Formally we apply it in the following form

$$\psi = \hat{\psi}(\mathbf{e}, \mathbf{F}, \vartheta; \boldsymbol{\mu}), \tag{41}$$

where $\boldsymbol{\mu}$ denotes a set of internal state variables governing the description of dissipation effects and ϑ represents temperature. It is important to notice, that we have used semicolon to separate the last variable due to its different nature (it introduces a dissipation to the model), without $\boldsymbol{\mu}$ the presented model describes thermoelasticity.

(ii) Axiom of objectivity.The material model should be invariant with respect to diffeomorphism (any superposed motion).

(iii) The axiom of the entropy production. For every regular process the constitutive functions should satisfy the second law of thermodynamics.

(iv) The evolution equation for the internal state variables vector $\boldsymbol{\mu}$ should be of the form

$$\mathrm{L}_{\boldsymbol{\upsilon}}\boldsymbol{\mu} = \hat{\mathbf{m}}(\mathbf{e}, \mathbf{F}, \vartheta, \boldsymbol{\mu}). \tag{42}$$

where evolution function $\hat{\mathbf{m}}$ has to be determined based on the experimental observations.

Constitutive Relations - General Form We can write the so called reduced dissipation inequality in the form Marsden and Hughes (1983); Sumelka (2009)

$$\frac{1}{\rho_{Ref}}\boldsymbol{\tau} : \mathbf{d} - (\eta\dot{\vartheta} + \dot{\psi}) - \frac{1}{\rho\vartheta}\mathbf{q} \cdot \mathrm{grad}\vartheta \geq 0, \tag{43}$$

where where ρ denotes actual and ρ_{Ref} reference densities, $\boldsymbol{\tau}$ denotes Kirchhoff stress, ψ is the free energy function, ϑ is absolute temperature, η denotes the specific (per unit mass) entropy and $\mathbf{q}$ is the heat flux. Using postulate (i), Eq. (43) can be rewritten to the form:

$$\left(\frac{1}{\rho_{Ref}}\boldsymbol{\tau} - \frac{\partial\hat{\psi}}{\partial\mathbf{e}}\right) : \mathbf{d} - \left(\eta + \frac{\partial\hat{\psi}}{\partial\vartheta}\right)\dot{\vartheta} - \frac{\partial\hat{\psi}}{\partial\boldsymbol{\mu}}\mathrm{L}_{\boldsymbol{\upsilon}}\boldsymbol{\mu} - \frac{1}{\rho\vartheta}\mathbf{q} \cdot \mathrm{grad}\vartheta \geq 0, \tag{44}$$

so because of arbitrariness

$$\boldsymbol{\tau} = \rho_{Ref}\frac{\partial\hat{\psi}}{\partial\mathbf{e}}, \tag{45}$$

$$\eta = -\frac{\partial\hat{\psi}}{\partial\vartheta}, \tag{46}$$

hence Eq. (44) reduces to

$$-\frac{\partial\hat{\psi}}{\partial\boldsymbol{\mu}}\mathrm{L}_{\boldsymbol{\upsilon}}\boldsymbol{\mu} - \frac{1}{\rho\vartheta}\mathbf{q}\cdot\mathrm{grad}\vartheta \geq 0. \tag{47}$$

Assuming now that internal state vector consists of two variables (discussed in details in next section), namely Perzyna (2008); Glema et al. (2009); Sumelka and Łodygowski (2011)

$$\boldsymbol{\mu} = (\in^p, \boldsymbol{\xi}), \tag{48}$$

where $\in^p$ is the equivalent plastic deformation $\dot{\in}^p = \left(\frac{2}{3}\mathbf{d}^p : \mathbf{d}^p\right)^{\frac{1}{2}}$, which describes the dissipation effects generated by viscoplastic deformation, and $\boldsymbol{\xi}$ is microdamage tensor which takes into account the anisotropic microdamage effects we can write general form of rate type constitutive equations for thermomechanical process under consideration.

Applying Lie derivative to formula Eq. (45), with internal state vector constant, or in other words keeping the history constant (thermoelastic process), we obtain the evolution equation for Kirchhoff stress tensor in the form Duszek–Perzyna and Perzyna (1994)

$$(\mathrm{L}_{\boldsymbol{\upsilon}}\boldsymbol{\tau})^e = \mathcal{L}^e : \mathbf{d}^e - \mathcal{L}^{th}\dot{\vartheta}, \tag{49}$$

where

$$\mathcal{L}^e = \rho_{Ref}\frac{\partial^2\hat{\psi}}{\partial\mathbf{e}^2}, \tag{50}$$

$$\mathcal{L}^{th} = -\rho_{Ref}\frac{\partial^2\hat{\psi}}{\partial\mathbf{e}\partial\vartheta}, \tag{51}$$

in above $\mathcal{L}^e$ denotes elastic constitutive tensor and $\mathcal{L}^{th}$ is thermal operator. Using the relations

$$(\mathrm{L}_{\boldsymbol{\upsilon}}\boldsymbol{\tau})^e = \dot{\boldsymbol{\tau}} - \boldsymbol{\tau}\cdot\mathbf{d}^e - \mathbf{d}^e\cdot\boldsymbol{\tau}, \tag{52}$$

and

$$\mathbf{d} = \mathbf{d}^e + \mathbf{d}^p, \tag{53}$$

we obtain final form of rate of Kirhchoff stress as

$$\mathrm{L}_{\boldsymbol{\upsilon}}\boldsymbol{\tau} = \mathcal{L}^e : \mathbf{d} - \mathcal{L}^{th}\dot{\vartheta} - (\mathcal{L}^e + \mathbf{g}\boldsymbol{\tau} + \boldsymbol{\tau}\mathbf{g}) : \mathbf{d}^p. \tag{54}$$

Using the energy balance in the form Perzyna (2005); Sumelka (2009)

$$\rho\vartheta\dot{\eta} = -\mathrm{div}\mathbf{q} - \rho\frac{\partial\hat{\psi}}{\partial\boldsymbol{\mu}}\cdot\mathrm{L}_{\boldsymbol{\upsilon}}\boldsymbol{\mu}, \tag{55}$$

and taking the rate of entropy, defined by the Eq. (46), we obtain the evolution equation for temperature as

$$\rho c_p \dot{\vartheta} = -\mathrm{div}\mathbf{q} + \vartheta \frac{\rho}{\rho_{Ref}} \frac{\partial \boldsymbol{\tau}}{\partial \vartheta} : \mathbf{d} + \chi^* \boldsymbol{\tau} : \mathbf{d}^p + \chi^{**} \mathbf{k} : \mathrm{L}_{\boldsymbol{\upsilon}} \boldsymbol{\xi}, \tag{56}$$

where the specific heat

$$c_p = -\vartheta \frac{\partial^2 \hat{\psi}}{\partial \vartheta^2}, \tag{57}$$

and the irreversibility coefficients χ^* and χ^{**} are determined by ($\mathbf{p}$ defines viscoplastic flow direction - discussed in details in next section)

$$\begin{aligned} \chi^* &= -\rho \left(\frac{\partial \hat{\psi}}{\partial \in^p} - \vartheta \frac{\partial^2 \hat{\psi}}{\partial \vartheta \partial \in^p} \right) \sqrt{\frac{2}{3}} \frac{1}{\boldsymbol{\tau} : \mathbf{p}}, \\ \chi^{**} &= -\rho \left(\frac{\partial \hat{\psi}}{\partial \boldsymbol{\xi}} - \vartheta \frac{\partial^2 \hat{\psi}}{\partial \vartheta \partial \boldsymbol{\xi}} \right) : \frac{1}{\mathbf{k}}. \end{aligned} \tag{58}$$

Next, we need to: specify an explicit definitions for assumed internal state variables ($\in^p, \boldsymbol{\xi}$), define materials functions and finally identify material parameters.

3.3 Adiabatic Process

Let us postulate the evolution equations for internal state variables as follows

$$\mathbf{d}^p = \Lambda \mathbf{p}, \tag{59}$$

$$\mathrm{L}_{\boldsymbol{\upsilon}} \boldsymbol{\xi} = \Lambda^h \frac{\partial h^*}{\partial \boldsymbol{\tau}} + \Lambda^g \frac{\partial g^*}{\partial \boldsymbol{\tau}}, \tag{60}$$

where Λ, Λ^h, Λ^g define the intensity of viscoplastic flow, microdamage nucleation and microdamage growth, while $\mathbf{p}$, $\frac{\partial h^*}{\partial \boldsymbol{\tau}}$, $\frac{\partial g^*}{\partial \boldsymbol{\tau}}$ define viscoplastic flow direction, microdamage nucleation direction and microdamage growth direction, respectively. It should be emphasised that there exists fracture microdamage state (ξ^F) for which catastrophe takes place, namely

$$\kappa = \hat{\kappa}(\in^p, \vartheta, \boldsymbol{\xi})|_{\xi=\xi^F} = 0. \tag{61}$$

Having defined evolution of $\in^p$ and $\boldsymbol{\xi}$ we can state the initial boundary value problem (IBVP) as follows.

Find $\phi, \boldsymbol{\upsilon}, \rho, \boldsymbol{\tau}, \boldsymbol{\xi}, \vartheta$ as functions of t and position $\mathbf{x}$ such that the following equations are satisfied Perzyna (1994); Łodygowski (1996); Łodygowski and Perzyna (1997a,b):

(i) the field equations

$$
\begin{aligned}
\dot{\phi} &= \boldsymbol{\upsilon}, \\
\dot{\boldsymbol{\upsilon}} &= \frac{1}{\rho_{Ref}}\left(\mathrm{div}\boldsymbol{\tau} + \frac{\boldsymbol{\tau}}{\rho}\cdot\mathrm{grad}\rho - \frac{\boldsymbol{\tau}}{1-(\boldsymbol{\xi}:\boldsymbol{\xi})^{\frac{1}{2}}}\mathrm{grad}(\boldsymbol{\xi}:\boldsymbol{\xi})^{\frac{1}{2}}\right), \\
\dot{\rho} &= -\rho\mathrm{div}\boldsymbol{\upsilon} + \frac{\rho}{1\quad(\boldsymbol{\xi}:\boldsymbol{\xi})^{\frac{1}{2}}}(\mathrm{L}_{\boldsymbol{\upsilon}}\boldsymbol{\xi}:\mathrm{L}_{\boldsymbol{\upsilon}}\boldsymbol{\xi})^{\frac{1}{2}}, \\
\dot{\boldsymbol{\tau}} &= \mathcal{L}^e:\mathbf{d} + 2\boldsymbol{\tau}\cdot\mathbf{d} - \mathcal{L}^{th}\dot{\vartheta} - (\mathcal{L}^e + \mathbf{g}\boldsymbol{\tau} + \boldsymbol{\tau}\mathbf{g}):\mathbf{d}^p, \qquad (62)\\
\dot{\boldsymbol{\xi}} &= 2\boldsymbol{\xi}\cdot\mathbf{d} + \frac{\partial g^*}{\partial\boldsymbol{\tau}}\frac{1}{T_m}\langle\Phi^g[\frac{I_g}{\tau_{eq}(\boldsymbol{\xi},\vartheta,\in^p)} - 1]\rangle, \\
\dot{\vartheta} &= \frac{\chi^*}{\rho c_p}\boldsymbol{\tau}:\mathbf{d}^p + \frac{\chi^{**}}{\rho c_p}\mathbf{k}:\mathrm{L}_{\boldsymbol{\upsilon}}\boldsymbol{\xi},
\end{aligned}
$$

(ii) the boundary conditions
   (a) displacement ϕ is prescribed on a part Γ_ϕ of $\Gamma(\mathcal{B})$ and tractions $(\boldsymbol{\tau}\cdot\mathbf{n})^a$ are prescribed on a part $\Gamma_{\boldsymbol{\tau}}$ of $\Gamma(\mathcal{B})$, where $\Gamma_\phi \cap \Gamma_{\boldsymbol{\tau}} = 0$ and $\Gamma_\phi \cup \Gamma_{\boldsymbol{\tau}} = \Gamma(\mathcal{B})$,
   (b) heat flux $\mathbf{q}\cdot\mathbf{n} = 0$ is prescribed on $\Gamma(\mathcal{B})$,

(iii) the initial conditions $\phi, \boldsymbol{\upsilon}, \rho, \boldsymbol{\tau}, \boldsymbol{\xi}, \vartheta$ are given for each particle $\mathbf{X} \in \mathcal{B}$ at $t = 0$,

are satisfied.

In Eq. (62) we have assumed that in the evolution of microdamage, nucleation term is omitted, we will assume appropriate initial microdamage state in computations (cf. Sumelka and Łodygowski (2011) to detailed discussion of such consequences) and because of adiabatic regime assumptions first two terms in temperature evolution in Eq. (56) are omitted.

3.4 Material Functions

For the evolution problem (62) we assume the following:

1. For elastic constitutive tensor $\mathcal{L}^e$

$$\mathcal{L}^e = 2\mu\mathcal{I} + \lambda(\mathbf{g}\otimes\mathbf{g}), \qquad (63)$$

   where μ, λ are Lamé constants.

2. For thermal operator $\mathcal{L}^{th}$

$$\mathcal{L}^{th} = (2\mu + 3\lambda)\theta\mathbf{g}, \qquad (64)$$

   where θ is thermal expansion coefficient.

3. For viscoplastic flow phenomenon $\mathbf{d}^p$ Perzyna (1963, 1966)

$$\mathbf{d}^p = \Lambda^{vp}\mathbf{p}, \tag{65}$$

where

$$\Lambda^{vp} = \frac{1}{T_m}\langle\Phi^{vp}(\frac{f}{\kappa} - 1)\rangle = \frac{1}{T_m}\langle(\frac{f}{\kappa} - 1)^{m_{pl}}\rangle, \tag{66}$$

$$f = \left\{J_2^{'} + \left[n_1(\vartheta) + n_2(\vartheta)(\boldsymbol{\xi} : \boldsymbol{\xi})^{\frac{1}{2}}\right] J_1^2\right\}^{\frac{1}{2}}, \tag{67}$$

$$n_1(\vartheta) = 0, \qquad n_2(\vartheta) = n = \text{const.}, \tag{68}$$

$$\kappa = \{\kappa_s(\vartheta) - [\kappa_s(\vartheta) - \kappa_0(\vartheta)] \exp\left[-\delta(\vartheta) \in^p\right]\} \left[1 - \left(\frac{(\boldsymbol{\xi} : \boldsymbol{\xi})^{\frac{1}{2}}}{\xi^F}\right)^{\beta(\vartheta)}\right], \tag{69}$$

$$\overline{\vartheta} = \frac{\vartheta - \vartheta_0}{\vartheta_0}, \quad \kappa_s(\vartheta) = \kappa_s^* - \kappa_s^{**}\overline{\vartheta}, \quad \kappa_0(\vartheta) = \kappa_0^* - \kappa_0^{**}\overline{\vartheta},$$
$$\delta(\vartheta) = \delta^* - \delta^{**}\overline{\vartheta}, \quad \beta(\vartheta) = \beta^* - \beta^{**}\overline{\vartheta}, \tag{70}$$

$$\xi^F = \xi^{F^*} - \xi^{F^{**}}\langle(\frac{||\mathrm{L}_{\boldsymbol{\upsilon}}\boldsymbol{\xi}|| - ||\mathrm{L}_{\boldsymbol{\upsilon}}\boldsymbol{\xi}_c||}{||\mathrm{L}_{\boldsymbol{\upsilon}}\boldsymbol{\xi}_c||})^{m_F}\rangle, \tag{71}$$

$$\mathbf{p} = \left.\frac{\partial f}{\partial \boldsymbol{\tau}}\right|_{\boldsymbol{\xi}=const} \left(\left|\left|\frac{\partial f}{\partial \boldsymbol{\tau}}\right|\right|\right)^{-1} = \frac{1}{[2J_2^{'} + 3A^2(\mathrm{tr}\boldsymbol{\tau})^2]^{\frac{1}{2}}}[\boldsymbol{\tau}^{'} + A\mathrm{tr}\boldsymbol{\tau}\boldsymbol{\delta}], \tag{72}$$

and f denotes the potential function Shima and Oyane (1976); Perzyna (1986a,b); Glema et al. (2009), κ is the isotropic work hardening-softening function Perzyna (1986a); Nemes and Eftis (1993), $\boldsymbol{\tau}^{'}$ represents the stress deviator, J_1, $J_2^{'}$ are the first and the second invariants of Kirchhoff stress tensor and deviatoric part of the Kirchhoff stress tensor, respectively, $A = 2(n_1 + n_2(\boldsymbol{\xi} : \boldsymbol{\xi})^{\frac{1}{2}})$, ξ^{F^*} can be thought as a quasi-static fracture porosity and $||\mathrm{L}_{\boldsymbol{\upsilon}}\boldsymbol{\xi}_c||$ denotes equivalent critical velocity of microdamage.

4. For microdamage mechanism we take the additional assumptions Dornowski (1999); Glema et al. (2009):
   - velocity of the microdamage growth is coaxial with the principal directions of stress state,
   - only positive (tension) principal stresses induces the growth of the microdamage,

so we have

$$\frac{\partial g^*}{\partial \boldsymbol{\tau}} = \langle \frac{\partial \hat{g}}{\partial \boldsymbol{\tau}} \rangle \| \langle \frac{\partial \hat{g}}{\partial \boldsymbol{\tau}} \rangle \|^{-1}, \quad \text{and} \quad \hat{g} = \frac{1}{2} \boldsymbol{\tau} : \mathcal{G} : \boldsymbol{\tau}, \tag{73}$$

$$\Phi^g \left(\frac{I_g}{\tau_{eq}(\boldsymbol{\xi}, \vartheta, \in^p)} - 1 \right) = \left(\frac{I_g}{\tau_{eq}} - 1 \right)^{m_g}, \tag{74}$$

where

$$\tau_{eq} = c(\vartheta)(1 - (\boldsymbol{\xi} : \boldsymbol{\xi})^{\frac{1}{2}}) \ln \frac{1}{(\boldsymbol{\xi} : \boldsymbol{\xi})^{\frac{1}{2}}} \left\{ 2\kappa_s(\vartheta) - [\kappa_s(\vartheta) - \kappa_0(\vartheta)] \, F(\xi_0, \boldsymbol{\xi}, \vartheta) \right\}, \tag{75}$$

$$c(\vartheta) = \text{const.},$$

$$F = \left(\frac{\xi_0}{1 - \xi_0} \frac{1 - (\boldsymbol{\xi} : \boldsymbol{\xi})^{\frac{1}{2}}}{(\boldsymbol{\xi} : \boldsymbol{\xi})^{\frac{1}{2}}} \right)^{\frac{2}{3}\delta} + \left(\frac{1 - (\boldsymbol{\xi} : \boldsymbol{\xi})^{\frac{1}{2}}}{1 - \xi_0} \right)^{\frac{2}{3}\delta}, \tag{76}$$

and

$$I_g = \bar{b}_1 J_1 + \bar{b}_2 (J_2')^{\frac{1}{2}} + \bar{b}_3 (J_3')^{\frac{1}{3}}. \tag{77}$$

$\bar{b}_i$ $(i = 1, 2, 3)$ are the material parameters, J_3' is the third invariant of deviatoric part of the Kirchhoff stress tensor.

Now, taking into account the postulates for microdamage evolution, and assuming that tensor $\mathcal{G}$ can be written as a symmetric part of the fourth order unity tensor $\mathcal{I}$ Łodygowski et al. (2008)

$$\mathcal{G} = \mathcal{I}^s, \quad \mathcal{G}_{ijkl} = \frac{1}{2} \left(\delta_{ik}\delta_{jl} + \delta_{il}\delta_{jk} \right), \tag{78}$$

we can write the explicit form of the growth function $\hat{g}$ as

$$\hat{g} = \frac{1}{2} \left(\tau_I^2 + \tau_{II}^2 + \tau_{III}^2 \right). \tag{79}$$

The gradient of $\hat{g}$ with respect to the stress field gives us the following matrix representation of a tensor describing the anisotropic evolution of microdamage

$$\frac{\partial \hat{g}}{\partial \boldsymbol{\tau}} = \begin{bmatrix} g_{11}\tau_I & 0 & 0 \\ 0 & g_{22}\tau_{II} & 0 \\ 0 & 0 & g_{33}\tau_{III} \end{bmatrix}. \tag{80}$$

In (80) $\tau_I, \tau_{II}, \tau_{III}$ are the principal values of Kirchhoff stress tensor.

5. For temperature evolution we consider the following relation

$$\mathbf{k} = \boldsymbol{\tau}. \tag{81}$$

4 Numerical Examples

4.1 Introductory Remarks

Up to this point we have defined a constitutive structure of Perzyna's type including anisotropic description for damage. The important features of this formulation can be summarised as: (i) the description is invariant with respect to any diffeomorphism (covariant material model), (ii) the obtained evolution problem is well-posed, (iii) sensitivity to the rate of deformation, (iv) finite elasto-viscoplastic deformations, (v) plastic non-normality, (vi) dissipation effects, (vii) thermo-mechanical couplings and (viii) length scale sensitivity.

For the moment let us put the attention to the feature (i) which is a result of consequent use of Lie derivative and is connected with objectivity of the description. The importance of the subject of matter lays in fact that we used to define spatial tensor rates in terms of Zaremba-Jaumann Zaremba (1903) or Green-Naghdi Green and Naghdi (1965) rates in most of popular constitutive models (and commercial programs like Abaqus Abaqus (2012)) although in some situations they can lead to nonphysical solutions Dienes (1979); Lehmann (1972); Nagtegaal and de Jong (1982); Xiao et al. (1997b). The situation described is most strongly vivid in extreme dynamics. It is than crucial to choose appropriately the objective rate definition from the set of all objective rates which is infinite, unfortunately. This problem was discussed in Sumelka (2013) were the importance of using Lie derivative was pointed out. Summarising using different objective rate one can observe differences in: global strain-stress space response, geometry and intensity of localised deformation zones, initiation time of macrodamage its direction and final fracture pattern. We conclude that covariant material model, being one of the most general in continuum mechanics, presents most robust and stable solution.

The proposed theory has been tested for many different types of IBVPs to indicate different aspects of the formulation. So, dynamic tension ans twisting was discussed in Sumelka (2009); Glema et al. (2010a); dynamic shearing was shown in Glema et al. (2010b); Sumelka and Łodygowski (2013b) while machining was a subject of Łodygowski et al. (2012) (cf. Fig. 7). In every process mention different type of combined loading was captured (e.g. different triaxiality, local temperatures or strains s̃train rates) and robustness of formulation was proved.

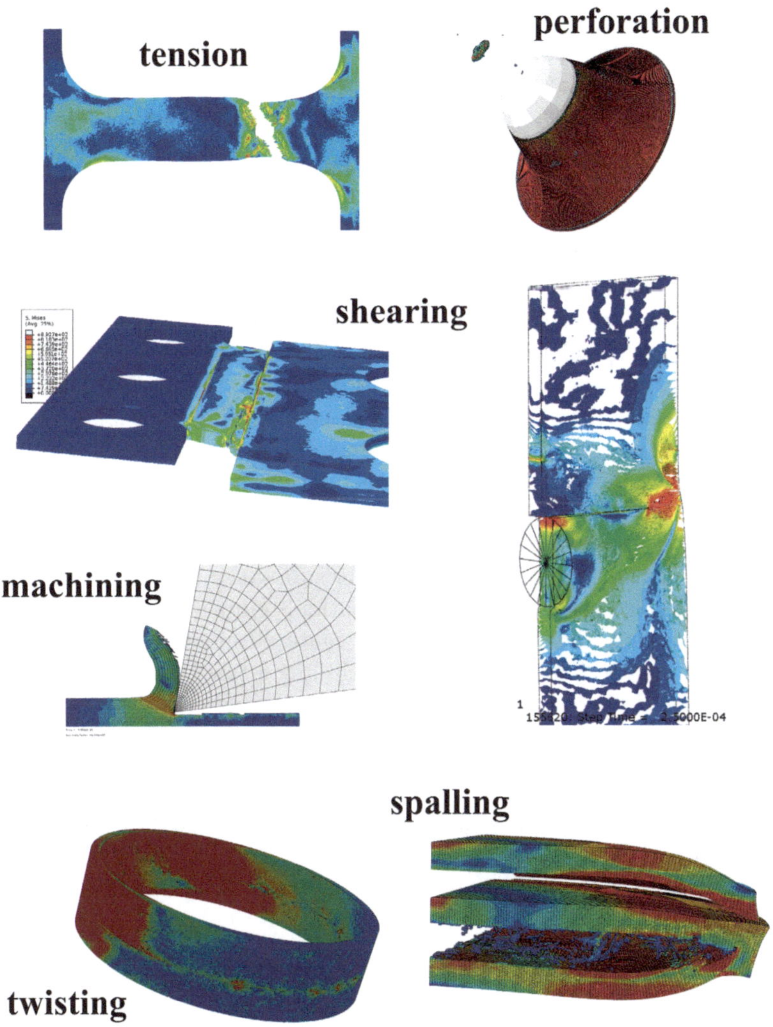

Figure 7. The processes solved under Perzyna's type viscoplasticity accounting for anisotropic damage

In the reaming part of this section the modelling of double chamber pipe under dynamic loading is shown as a example of real industrial application concerning designing of energy absorbers. But first we start with information about computer implementation of the theory in Abaqus/Explicit finite element program utilising Lie derivative.

4.2 Computer Implementation in Abaqus/Explicit

The solution of the IBVP defined by Eqs (62) has been obtained with the use of finite element method. The Abaqus/Explicit commercial finite element code has been adapted as a solver. The model, has been implemented in the software, by taking advantage of a user subroutine VUMAT, which is coupled with Abaqus system Abaqus (2012). The Abaqus/Explicit utilises central-difference time integration rule along with the diagonal ("lumped") element mass matrices. To remove damaged elements from the mesh (elements in which for every integration point fracture porosity was reached) the so called element deletion method is applied Song et al. (2008).

The stress update in VUMAT user subroutine needs some additional comments. During computations, the user subroutine VUMAT controls the evolution of stresses, viscoplastic deformation, temperature and microdamage fields. Recall, that in the presented material model Lie derivative has been taken into account for all rates, including stress rate.

Hence for the stress rate we enforce

$$\mathrm{L}_{\boldsymbol{\upsilon}}\boldsymbol{\tau} = \dot{\boldsymbol{\tau}} - \mathbf{l}^T \cdot \boldsymbol{\tau} - \boldsymbol{\tau} \cdot \mathbf{l}, \tag{82}$$

in opposition to the Green-Naghdi rate calculated by default in Abaqus/Explicit VUMAT user subroutine according to the following formula Abaqus (2012)

$$\boldsymbol{\tau}^{(G-N)^\circ} = \dot{\boldsymbol{\tau}} + \boldsymbol{\tau} \cdot \boldsymbol{\Omega} - \boldsymbol{\Omega} \cdot \boldsymbol{\tau}, \tag{83}$$

where $\boldsymbol{\Omega} = \boldsymbol{\Omega}^{(G-N)} = \dot{\mathbf{R}} \cdot \mathbf{R}^T$ represents the angular velocity of the material Dienes (1979) (or spin tensor Xiao et al. (1997a)) and $\mathbf{R}$ denotes the rotation tensor. It is also worth mentioning that the material model in Abaqus/Explicit VUMAT user subroutine is defined in corotational coordinate system, being described by the spin tensor $\boldsymbol{\Omega}$ (see Fig. 8).

To keep the VUMAT algorithm objective in Lie sense we apply the following approach. In the iterative procedure, we take the forward difference scheme as the material derivative of the second order tensor. Thus, for the material derivative of the Kirchhoff stress tensor we have

$$\dot{\boldsymbol{\tau}} \mid_i = \frac{\boldsymbol{\tau} \mid_{i+1} - \boldsymbol{\tau} \mid_i}{\Delta t}. \tag{84}$$

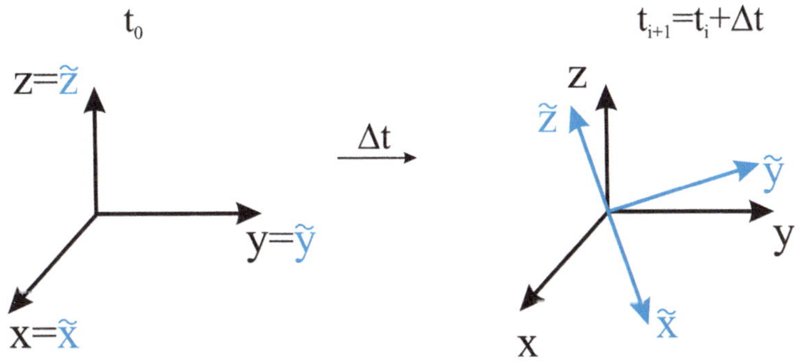

Figure 8. Initial (XYZ), current (xyz), corotational $(\tilde{x}\tilde{y}\tilde{z})$ and 'damage' $(\hat{x}\hat{y}\hat{x})$ coordinate systems

Using Eqns (82) and (83) we can write in the corotational coordinate system respectively

$$\tilde{\boldsymbol{\tau}} \mid_{i+1} = \mathbf{R}^T \mid_{i+1} \left[\boldsymbol{\tau} \mid_i + \Delta t \mathrm{L}_{\boldsymbol{\upsilon}} \boldsymbol{\tau} \mid_i + \Delta t (\mathbf{l}^T \mid_i \cdot \boldsymbol{\tau} \mid_i + \boldsymbol{\tau} \mid_i \cdot \mathbf{l} \mid_i) \right] \mathbf{R} \mid_{i+1}, \quad (85)$$

and

$$\tilde{\boldsymbol{\tau}} \mid_{i+1} = \mathbf{R}^T \mid_{i+1} \left[\boldsymbol{\tau} \mid_i + \Delta t \boldsymbol{\tau}^{(G-N)^\circ} \mid_i + \Delta t (\boldsymbol{\Omega} \mid_i \cdot \boldsymbol{\tau} \mid_i - \boldsymbol{\tau} \mid_i \cdot \boldsymbol{\Omega} \mid_i) \right] \mathbf{R} \mid_{i+1}. \quad (86)$$

Thus we see, that the Green-Naghdi rate, produces an additional term

$$\Delta t \;\; (\boldsymbol{\Omega} \mid_i \cdot \boldsymbol{\tau} \mid_i - \boldsymbol{\tau} \mid_i \cdot \boldsymbol{\Omega} \mid_i). \quad (87)$$

That is why one has to subtract this term since different is proposed. Hence, in the presented formulation for the stress update in VUMAT we have Sumelka (2009)

$$\tilde{\boldsymbol{\tau}} \mid_{i+1} = \mathbf{R}^T \mid_{i+1} \left[\boldsymbol{\tau} \mid_i + \Delta t \left(2\boldsymbol{\tau} \mid_i \cdot \mathbf{d} \mid_i + \mathrm{L}_{\boldsymbol{\upsilon}} \boldsymbol{\tau} \mid_i \right) + \boldsymbol{\Upsilon} \mid_i \right] \mathbf{R} \mid_{i+1}, \quad (88)$$

where $\boldsymbol{\Upsilon} \mid_i = -\Delta t \;\; (\boldsymbol{\Omega} \mid_i \cdot \boldsymbol{\tau} \mid_i - \boldsymbol{\tau} \mid_i \cdot \boldsymbol{\Omega} \mid_i)$ and $\boldsymbol{\tau} \mid_i = \mathbf{R} \mid_i \tilde{\boldsymbol{\tau}} \mid_i \mathbf{R}^T \mid_i$.

Such approach is necessary only for stresses, since other variables are kept as scalars. The detailed algorithm for the whole process can be found in Sumelka (2009).

4.3 Energy Absorption by Double Chamber Pipe Under Dynamic Loading

Material Parameters For identification we have used experimental data for HSLA-65 steel presented in Nemat-Nasser and Guo (2005). The HSLA-65 steel belongs to the class of HSLA steels (High-Strength Low-Alloy) which were developed in 1960s. The high strength of this steel (flow stresses are in the rage $400 \div 1200 MPa$ dependently on temperature) connected with good weldability, formability, thoughtness, elevated service life, less weight to the traditional highstrength steel, cause the broad range of its nowadays applications e.g. cars, trucks, cranes, bridges, naval surface vessels, submarines and other structures that are designed to handle large amounts of material efforts, frequently subjected to wide range of temperatures. The steel has the characteristics of the bcc structure, hence belongs to so called ferritic steels. As a consequence this metal has high temperature and strain rate sensitivity, and displays good ductility and plasticity (true strain $> 60\%$). The mechanical properties of the HSLA-65 steel are strongly affected by impurities in its internal structure. It is important that the processing (rolling) of the HSLA-65 steel can induce the anisotropy of its structure.

The set of material parameters is presented in Tables 2.

It should be emphasised however that the identification of such grate number of material parameters is awkward. From one point of view the results presented in Nemat-Nasser and Guo (2005) are not enough to calibrate the presented material model in which all variables (e.g. temperature, viscoplastic strain, microdamage) are coupled. On the other hand recall that current experimental techniques can not still give a detailed/unique answer for extremely fast thermomechanical processes. Thus parameters from Table 2 should be thought as a compromise, hence small fluctuations of them are possible (dependently on detailed experimental results showing competition of fundamental variables e.g. temperature, viscoplastic strain, microdamage). For proposition of reduction of the number of material parameters using soft computing methods cf. Sumelka and Łodygowski (2013a).

Table 2. Material parameters for HSLA-65 steel

$\lambda = 121.15$ GPa	$\mu = 80.769$ GPa	$T_m = 2.5\ \mu$s	$m_{pl} = 0.14$
$n_1 = 0$	$n_2 = 0.25$	$\chi^* = 0.8$	$\chi^{**} = 0.1$
$\kappa_s^* = 570$ MPa	$\kappa_s^{**} = 129$ MPa	$\kappa_0^* = 457$ MPa	$\kappa_0^{**} = 103$ MPa
$\delta^* = 6.0$	$\delta^{**} = 1.4$	$\beta^* = 11.0$	$\beta^{**} = 2.5$
$c = 0.067$	$\theta = 10^{-6}\ \mathrm{K}^{-1}$	$m_{md} = 1$	m_F –
$b_1 = 0$	$b_2 = 0.5$	$b_3 = 0$	$\|\|L_{\upsilon}\boldsymbol{\xi}_c\|\|$ – s^{-1}
$\xi^{F^*} = 0.36$	$\xi^{F^{**}} = 0$	$c_p = 470\ \frac{\mathrm{J}}{\mathrm{kgK}}$	$\rho_{Ref} = 7800\ \frac{\mathrm{kg}}{\mathrm{m}^3}$

In Fig. 9 the comparison of experimental and numerical results are presented. Notice that the numerical solution is obtained from full 3D thermomechanical analysis accounting for the previously mentioned anisotropic intrinsic microdamage process (microdamage state was initially isotropic during calibration procedure, however, the obtained parameters are valid for other initial microdamage states since its influence on GDA scatter is in experimentally acceptable bounds Sumelka and Łodygowski (2011)). In other words, the presented numerical results take into account the whole local processes. The curve fitting shows that using presented material model one can obtain the numerical simulations in a very good agreement with the experimental observations.

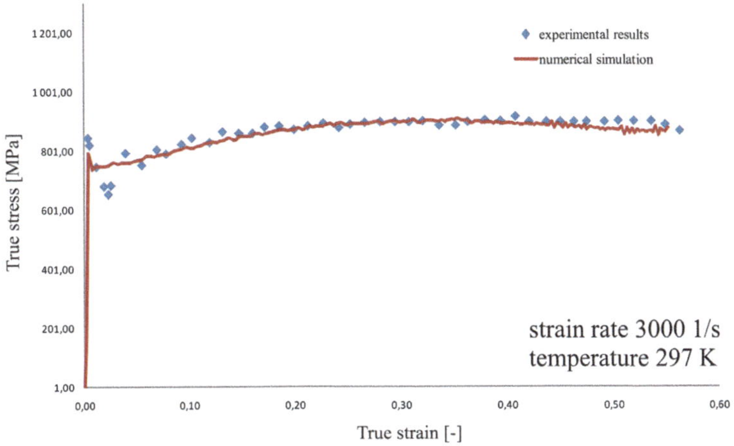

Figure 9. The comparison of strain-stress curve from experiment and numerical simulation

Computational Model The geometry of double chamber pipe is presented in Fig. 10. The dimensions of cross sections are $97.5 \times 70.5mm$, the wall thickness is $t = 2.5mm$ and the initial length is $l_0 = 400mm$. To obtain expected damage mode the top of the pipe was chamfer with $7deg$.

Because of anisotropy full spatial modelling was enforced for computational model. Thus, for the pipe mesh continuum elements were used - the C3D8R finite elements (8-node linear brick, reduced integration element). Totally $3Mdof$ were applied giving $9M$ of variables.

The remaining part of the computational model consists of rigid base which was fixed, and moving rigid surface causing compression of the pipe with initial velocity $v_0 = 100m/s$ and mass $500kg$ cf. Fig. 11. Other initial condition were. The initial temperature 296K was assumed. Because of lack of the experimental data concerning initial microdamage distribution in the

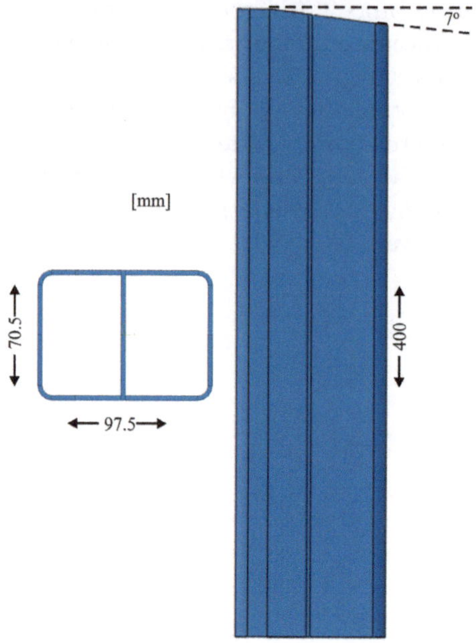

Figure 10. The geometry of double chamber pipe

specimen we have assumed it equal in every material point in the body and simultaneously we have assumed its initial isotropy. This simplification is crucial concerning the fact that the way of mapping of initial microdamage state has a strong impact on a final failure mode and the global answer from the specimen Sumelka and Łodygowski (2011). The components of microdamage tensor were chosen in a way to obtain initial porosity equal to $6 \cdot 10^{-4}$, namely

$$\boldsymbol{\xi}_0 = \begin{bmatrix} 34.64 \cdot 10^{-5} & 0 & 0 \\ 0 & 34.64 \cdot 10^{-5} & 0 \\ 0 & 0 & 34.64 \cdot 10^{-5} \end{bmatrix}.$$

To assure proper contact conditions the general contact in Abaqus/Explicit was applied, which includes the self contact conditions. The properties of a contact were: the hard normal contact (without penetration and unlimited contact stresses) and the tangential contact with Coulomb friction model (friction coefficient equal 0.05).

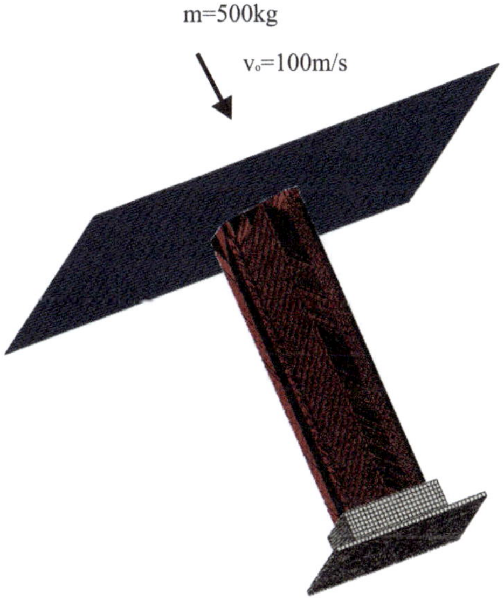

Figure 11. The computational model for double chamber pipe compression test

Results Discussion The impact of pipe by rigid surface had induced the elasto-viscoplastic wave propagation. The waves interactions during the process determine the strain localization zones Glema (2004). The further process evolution induces the intensified damage evolution in those zones, leading finally to failure - loss of continuity as result of macrodamage propagation. The process time up to full crush was $3.1e-3s$.

The deformation of the pipe is presented in Fig. 12. It is clear that the phenomena of buckling of pipe walls connected with the appearance of plastic hinges in the zones of severe viscoplastic deformation is responsible for energy absorption. Finally one pipe can absorb $0.15MJ$.

Let us now analyse selected results concerning Huber-Mises-Hencky (HMH) stress, equivalent viscoplastic deformation $\in^p$, temperature ϑ and porosity ξ evolution cf. Figs 13 and 14 - please notice that in both figures the maps of equivalent viscoplastic deformation, temperature and porosity are plotted on undeformed pipe geometry for easier interpretation.

Thus, soon after impact viscoplastic deformation occurs. In the zones of localised deformation the equivalent viscoplastic deformation locally exceeded 100%. The adiabatic temperature due to viscoplastic deformation and microdamage evolution in those zones has the level $800 \div 900K$. Finally,

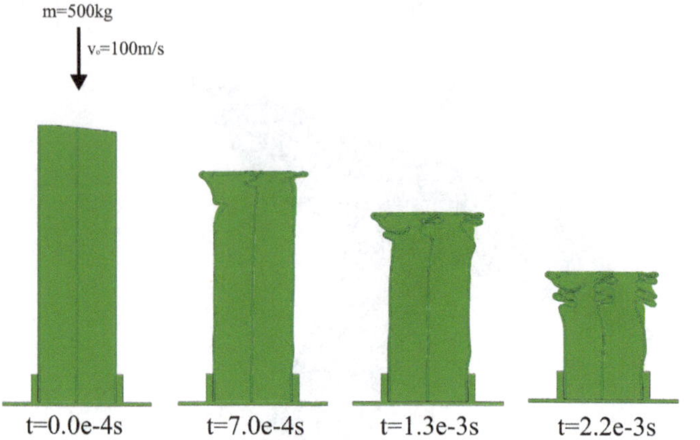

Figure 12. The deformation of the double chamber pipe for selected time points

the porosity level is generally below the fracture level $\xi^{F^*} = 0.36$, so the impact is absorbed homogeneously during the process of compression.

At the end let us mention that up to the end of analysis the adiabatic conditions assumption was hold by FEM model. Namely, according to general requirement the maximal time of analysis t_{max} should fulfil the condition Abaqus (2012)

$$t_{max} \leq \frac{\rho c_p}{6k} \Delta l^2, \tag{89}$$

where k denotes conductivity. In our case t_{max} was approximately $6.5e-3s$ - average finite element dimension for pipe was $0.5mm$.

5 Conclusions

This document presents the complexity of computer estimation of plastic strain localization and failure for large strain rates using Perzyna's type viscoplasticity accounting for anisotropic microdamage. Is is shown that presented constitutive model is well-posed, hence the obtained results are unique. Due to the relaxation time parameter (which implicitly introduces length scale) the solution is in general not mesh dependant. Detailed discussion on model implementation in Abaqus/Explicit programme is considered also. Formal results are illustrated with the numerical one showing the analysis of real industrial application concerning designing of energy absorbers for which proper description of viscoplastic deformation is crucial.

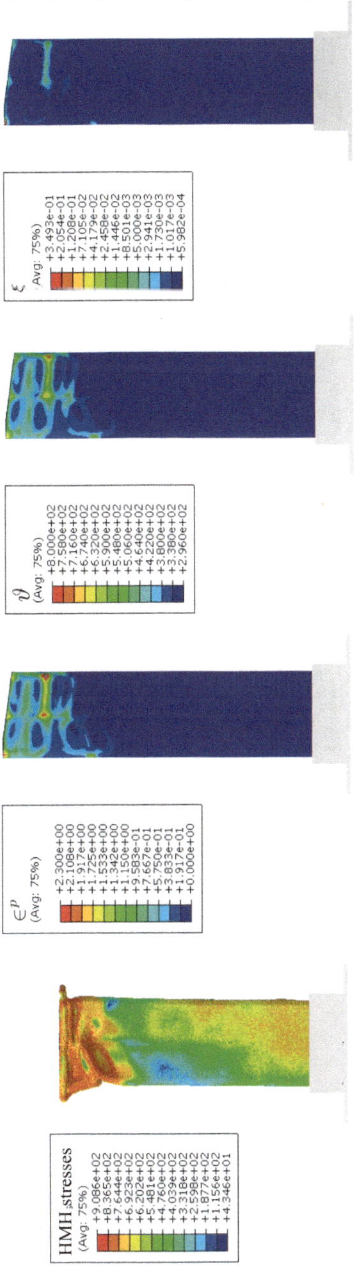

Figure 13. HMH stresses [MPa], equivalent viscoplastic deformation, temperature [K], porosity for $t = 7.0e - 4s$

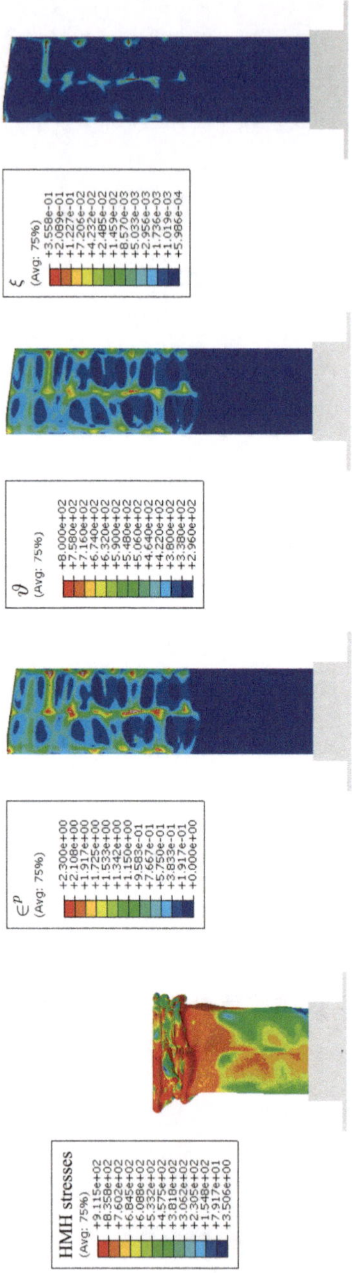

Figure 14. HMH stresses [MPa], equivalent viscoplastic deformation, temperature [K], porosity for $t = 1.8e - 3s$

Bibliography

Abaqus. *Abaqus Version 6.12 Theory Manual.* 2012.

R.K. Abu Al-Rub and G.Z. Voyiadjis. A finite strain plastic-damage model for high velocity impact using combined viscosity and gradient localization limiters: Part I - theoretical formulation. *International Journal of Damage Mechanics*, **15**(4):293–334, 2006.

J.K. Dienes. On the analysis of rotation and stress rate in deforming bodies. *Acta Mechanica*, **32**:217–232, 1979.

W. Dornowski. Influence of finite deformations on the growth mechanism of microvoids contained in structural metals. *Archives of Mechanics*, **51** (1):71–86, 1999.

W. Dornowski and P. Perzyna. Analysis of the influence of various effects on cycle fatigue damage in dynamic process. *Archive of Applied Mechanics*, **72**:418–438, 2002.

W. Dornowski and P. Perzyna. Numerical investigation of localized fracture phenomena in inelastic solids. *Foundations of Civil and Environmental Engineering*, **7**:79–116, 2006.

M.K. Duszek–Perzyna and P. Perzyna. *Analysis of the influence of different effects on criteria for adiabatic shear band localization in inelastic solids*, volume **50**, chapter Material Instabilities: Theory and Applications. ASME, New York, 1994.

A. Glema. *Analysis of wave nature in plastic strain localization in solids*, volume **379** of *Rozprawy*. Publishing House of Poznan University of Technology, 2004. (in Polish).

A. Glema, T. Łodygowski, W. Sumelka, and P. Perzyna. The numerical analysis of the intrinsic anisotropic microdamage evolution in elasto-viscoplastic solids. *International Journal of Damage Mechanics*, **18**(3): 205–231, 2009.

A. Glema, T. Łodygowski, and W. Sumelka. Towards the modelling of an anisotropic solids. *Computational Methods in Science and Technology*, **16**(1):73–84, 2010a.

A. Glema, T. Łodygowski, and W. Sumelka. Nowacki's double shear test in the framework of the anisotropic thermo-elasto-vicsoplastic material model. *Journal of Theoretical and Applied Mechanics*, **48**(4):973–1001, 2010b.

H.A. Grebe, H.-R. Pak, and Meyers M.A. Adiabatic shear localization in titanium and Ti-6 pct Al-4 pct V alloy. *Metallurgical and Materials Transactions A*, **16**(5):761–775, 1985.

A.E. Green and P.M. Naghdi. A general theory of an elastic–plastic continuum. *Archive for Rational Mechanics and Analysis*, **18**:251–281, 1965.

R. Hill. Aspects of invariance in solid mechanics. *Advances in Applied Mechanics*, **18**:1–75, 1978.

G.A. Holzapfel. *Nonlinear Solid Mechanics - A Continuum Approach for Engineering*. Wiley, 2000.

J.R. Klepaczko. *Constitutive relations in dynamic plasticity, pure metals and alloys*. Advances in constitutive relations applied in computer codes. CISM, Udine, Italy, July 23–27 2007.

Th. Lehmann. Anisotrope plastische Formänderungen. *Romanian J. Tech. Sci. Appl. Mech.*, **17**:1077–1086, 1972.

P. Longere, A. Dragon, H. Trumel, and X. Deprince. Adiabatic shear banding-induced degradation in thermo-elastic/viscoplastic material under dynamic loading. *International Journal of Impact Engineering*, **32**: 285–320, 2005.

J.E. Marsden and T.J.H Hughes. *Mathematical Foundations of Elasticity*. Prentice-Hall, New Jersey, 1983.

J.C. Nagtegaal and J.E. de Jong. Some aspects of non-isotropic work-hardening in finite strain plasticity. In Lee E.H. and Mallet R.L., editors, *Proceedings of the workshop on plasticity of metals at finite strain: theory, experiment and computation*, pages 65–102. Stanford University, 1982.

R. Narayanasamy, N.L. Parthasarathi, and C.S. Narayanan. Effect of microstructure on void nucleation and coalescence during forming of three different HSLA steel sheets under different stress conditions. *Materials and Design*, **30**:1310–1324, 2009.

S. Nemat-Nasser and W.-G. Guo. Thermomechanical response of HSLA-65 steel plates: experiments and modeling. *Mechanics of Materials*, **37**: 379–405, 2005.

J.A. Nemes and J. Eftis. Several features of a viscoplastic study of plate-impact spallation with multidimensional strain. *Computers and Structures*, **38**(3):317–328, 1991.

J.A. Nemes and J. Eftis. Constitutive modelling of the dynamic fracture of smooth tensile bars. *International Journal of Plasticity*, **9**(2):243–270, 1993.

T. Łodygowski. *Theoretical and numerical aspects of plastic strain localization*, volume **312** of *D.Sc. Thesis*. Publishing House of Poznan University of Technology, 1996.

T. Łodygowski and P. Perzyna. Localized fracture of inelastic polycrystalline solids under dynamic loading process. *International Journal Damage Mechanics*, **6**:364–407, 1997a.

T. Łodygowski and P. Perzyna. Numerical modelling of localized fracture of inelastic solids in dynamic loading process. *International Journal for Numerical Methods in Engineering*, **40**:4137–4158, 1997b.

T. Łodygowski, A. Glema, and W. Sumelka. Anisotropy induced by evolution of microstructure in ductile material. In *8th World Congress on Computational Mechanics (WCCM8), 5th European Congress on Computational Methods in Applied Sciences and Engineering (ECCOMAS 2008)*, Venice, Italy, June 30 - July 5 2008.

T. Łodygowski, A. Rusinek, T. Jankowiak, and W. Sumelka. Selected topics of high speed machining analysis. *Engineering Transactions*, **60**(1):69–96, 2012.

R.B. Pęcherski, W.K. Nowacki, Z. Nowak, and P Perzyna. Effect of strain rate on ductile fracture. a new methodology. In *Workshop, Dynamic Behaviour of Materials, In memory of our Friend and Colleague Prof. J.R. Klepaczko*, pages 65–73, Metz, France, May 13-15 2009.

P. Perzyna. The constitutive equations for rate sensitive plastic materials. *Quarterly of Applied Mathematics*, **20**:321–332, 1963.

P. Perzyna. Fundamental problems in viscoplasticity. *Advances in Applied Mechanics*, **9**:243–377, 1966.

P. Perzyna. *Termodynamika materiałów niesprężystych*. PWN, Warszawa, 1978. (in Polish).

P. Perzyna. Internal state variable description of dynamic fracture of ductile solids. *International Journal of Solids and Structures*, **22**:797–818, 1986a.

P. Perzyna. Constitutive modelling for brittle dynamic fracture in dissipative solids. *Archives of Mechanics*, **38**:725–738, 1986b.

P. Perzyna. Instability phenomena and adiabatic shear band localization in thermoplastic flow process. *Acta Mechanica*, **106**:173–205, 1994.

P. Perzyna. Constitutive modelling of dissipative solids for localization and fracture. In Perzyna P., editor, *Localization and fracture phenomena in inelastic solids*, chapter 3, pages 99–241. Springer, 1998. (CISM course and lectures - No.386).

P. Perzyna. The thermodynamical theory of elasto-viscoplasticity. *Engineering Transactions*, **53**:235–316, 2005.

P. Perzyna. The thermodynamical theory of elasto-viscoplasticity accounting for microshear banding and induced anisotropy effects. *Mechanics*, **27**(1):25–42, 2008.

A. Rusinek and J.R. Klepaczko. Experiments on heat generated during plastic deformation and stored energy for trip steels. *Materials and Design*, **30**(1):35–48, 2009.

Cz. Rymarz. *Mechanika ośrodków ciągłych*. PWN, Warszawa, 1993. (in Polish).

L. Seaman, D.R. Curran, and D.A. Shockey. Computational models for ductile and brittle fracture. *Journal of Applied Physics*, **47**(11):4814–4826, 1976.

S. Shima and M. Oyane. Plasticity for porous solids. *International Journal of Mechanical Sciences*, **18**:285–291, 1976.

J-H Song, H. Wang, and T. Belytschko. A comparative study on finite element methods for dynamic fracture. *Computational Mechanics*, **42**: 239–250, 2008.

W. Sumelka. *The Constitutive Model of the Anisotropy Evolution for Metals with Microstructural Defects*. Publishing House of Poznan University of Technology, Poznań, Poland, 2009.

W. Sumelka. The role of the covariance in continuum damage mechanics. *ASCE Journal of Engineering Mechanics*, 2013. (DOI: 10.1061/(ASCE)EM.1943-7889.0000600).

W. Sumelka and A. Glema. The evolution of microvoids in elastic solids. In *17th International Conference on Computer Methods in Mechanics CMM-2007*, pages 347–348, Łódź-Spała, Poland, June 19-22 2007.

W. Sumelka and T. Łodygowski. The influence of the initial microdamage anisotropy on macrodamage mode during extremely fast thermomechanical processes. *Archive of Applied Mechanics*, **81**(12):1973–1992, 2011.

W. Sumelka and T. Łodygowski. Reduction of the number of material parameters by ann approximation. *Computational Mechanics*, 2013a. (DOI: 10.1007/s00466-012-0812-9).

W. Sumelka and T. Łodygowski. Thermal stresses in metallic materials due to extreme loading conditions. *ASME Journal of Engineering Materials and Technology*, 2013b. (DOI: 10.1115/1.4023777).

D. Tikhomirov, R. Niekamp, and E. Stein. On three-dimensional microcrack density distribution. *ZAMM - Journal of Applied Mathematics and Mechanics*, **81**(1):3–16, 2001.

G.Z. Voyiadjis and R.K. Abu Al-Rub. A finite strain plastic-damage model for high velocity impacts using combined viscosity and gradient localization limiters: Part II - numerical aspects and simulations. *International Journal of Damage Mechanics*, **15**(4):335–373, 2006.

H. Xiao, O.T. Bruhns, and A. Meyers. Logarithmic strain, logarithmic spin and logarithmic rare. *Acta Mechanica*, **124**:89–105, 1997a.

H. Xiao, O.T. Bruhns, and A. Meyers. Hypo-elasticity model based upon the logarithmic stress rate. *Journal of Elasticity*, **47**:51–68, 1997b.

H. Xiao, O.T. Bruhns, and A. Meyers. Strain rates and material spin. *Journal of Elasticity*, **52**:1–41, 1998.

S. Zaremba. Sur une forme perfectionée de la théorie de la relaxation. *Bull. Int. Acad. Sci. Cracovie*, pages 594–614, 1903.

Inelastic Flow and Failure of Metallic Solids. Material Effort: Study Across Scales

Ryszard B. Pęcherski[‡], Kinga Nalepka[‡‡], Teresa Frąś[‡] and Marcin Nowak[‡]
[‡] Institute of Fundamental Technological Research, Polish Academy of Sciences, IPPT PAN, Warsaw
[‡‡] AGH University of Science and Technology, Krakow

Abstract The multiscale physical foundations of the concept of material effort in isotropic solid body are studied, in particular for solids revealing the strength differential effect. Various yield criteria result from this hypothesis: ellipsoid, paraboloid or hyperpoloid ones. The examples are discussed and visualized in the principal axes of stress or in the plane defined by coordinates: equivalent stress and mean stress. The numerical implementation of paraboloid yield surface forming plastic potential in an associated flow law is presented and an example of the identification of the strength differential effect ratio is discussed. Finally, some problems related with an account for the third invariant of stress deviator in failure criteria for isotropic solids and energy-based failure criteria for anisotropic solids are shortly discussed. Also the possibility for an account of two basic mechanisms responsible for inelastic flow: crystallographic slip and shear banding in modelling the inelastic flow law is outlined.

1 Introduction

Studying problems related with constitutive theory and modelling of materials that in particular are subjected to dynamic loading conditions we encounter the question of proper description of elastic range, which in practical applications is related with determining yield strength and yield surface in the space of stresses. Proper formulation of yield surface is important for the description of inelastic deformation of materials. The function describing the yield surface becomes then, assuming associated flow law, the plastic potential function. Therefore, adequate formulating the yield surface appears crucial for proper description of inelastic flow and failure of solids. On the other hand, there is a vast amount of experimental observations, in particular related with new materials, which show that the measured yield strength in tension is different from the yield strength in compression. We focus our attention to the discussion how to formulate yield criteria for isotropic materials that reveal the strength differential effect. Although this effect is not very strong in case of many metals and alloys, the strength differential ratio $\kappa = k_c/k_t$ often does not exceed 1.3, it can be important in some special problems. For instance, in high speed machining processes, where the pressure in the neighborhood of the tool material contact exceeds 1 GPa. Such an amount of pressure may remarkably influence plastic yielding and produce the strength differential effect. Also impact problems can provide examples of the wave interaction phenomena while the different values of yield strength in compression and tension may play a crucial role in adequate prediction of material failure. This explanation could give the answer for the apparent question about the application of presented theory for analysis and simulation of dynamic processes and impact phenomena. Assumptions of the energy-based hypothesis of material effort and multiscale character of the measure of material effort are discussed in chapter 2. Chapter 3 is devoted to the presentation of the Burzyński hypothesis of variable limit energy of volume change and distortion. The resulting failure criteria for isotropic solids are also discussed and their original graphical presentations on the plane equivalent stress mean stress are displayed. Chapter 4 contains examples and visualisations of the particular Burzyński failure criteria applied for many materials. There are presented applications of failure criteria, resulting from energy measure of material effort, to metallic solids, in particular nanocrystalline metals and to metal-ceramic composites. The possible extension of the study over polymers is also discussed. In chapter 5 the numerical implementation of elasto-plasticity equations associated with paraboloid yield condition is discussed. Finally, in chapter 6 some remarks about possible applications of energy-based approach in formulations of yield criteria for anisotropic solids are presented. Also an outline of viscoplasticity theory accounting for multiscale mechanism of shear banding is given. Possible applications in the analysis of deformation and failure processes are discussed and final conclusions are formulated. The contribution of the authors is distributed in the following way: Ryszard B. Pęcherski is the editor of the whole work and the author of chapters 1, 3, and 6. Chapter 2 is co-authored by Kinga Nalepka and Ryszard B. Pęcherski, while chapter 4 was prepared by Teresa Frąś and Ryszard B. Pęcherski and chapter 5 by Marcin Nowak and Ryszard B. Pęcherski. The help of Marcin Nowak by typesetting is thankfully acknowledged.

2 Electronic basis and multiscale character of material effort

The aim of this paragraph is to show in what way analysis of deformation processes at an atomistic level enables a new derivation of energy-based hypothesis of material effort. In a result, we obtain a reliable description of material strength that foundations constitute the achievements of solid state physics. The presented approach to the assessment of material strength gives a basis for examination of existing energy-based hypotheses of material effort and the resulting failure criteria for isotropic materials. We study in detail the concepts proposed by James Clerk Maxwell (Maxwell, 1936) and Maksymilian Tytus Huber (Huber, 1904) together with further generalization over materials revealing strength differential effect, which was derived by Włodzimierz Burzyński (Burzyński, 1928, 1929). The mentioned failure criteria can be related with the limit of proportionality, i.e. the limit of linear elasticity, plastic yield limit, the limit of strength that is usually related with brittle fracture or even with a structural transformation. The assumed definition depends on considered material or deformation processes. The presented results contradict the common opinion about the straightforward empirical correlation between the plastic resistance and elastic constants. Our conclusion is that the concept of resilience or more generally material effort provides the additional information that is requisite for the description of the material behaviour in a neighbourhood of a limit state.

2.1 The concept of material effort and its origin in the atomic theory of solid

The concept of material effort was clarified originally by (Burzyński, 1928) who additionally explained the subtle otherness to the term material strength:

"The main subject of the theory of elasticity is to mathematically determine the state of strain or stress in a solid body being under the conditions determined by the action of a system of external forces, the specific shape of the body and its elastic properties. The solution of this question exhausts the role of the elasticity theory and, next, the theory of strength of materials comes into play. Its equally important task is to give the dimensions of the considered body with definite exactness with respect to the states unwanted regarding the body safety on the one hand and most advantageous economical conditions on the other hand. This problem, very simple in the case of a uniaxial state of stress, becomes so complicated in a general case that from the beginning of the mentioned theories special attention has been paid to this question and an intermediate chapter, being at the same time the

final part of the theory of elasticity and the introduction to the strength of materials theory, has been introduced. This new passage deals with material effort and different hypotheses related with this concept". p. 129 (of English translation), (Burzyński, 1928), and further:

"Generally, under material effort we understand the physical state of a body, comprehended in the sense of elasticity or plasticity or material strength, and generated by a system of stresses, and related with them strains, in the body" p. 34 (of English translation), (Burzyński, 1928). The above mentioned citation, more widely discussed in (Frąś and Pęcherski, 2010), is illustrated in Fig. 1. This scheme reveals the confrontation with the unified theory of material behaviour, which introduces the complete constitutive relation holding from the moment of load application up to the limit state. It is visible that only in a simplified phenomenological theory the notion of material effort and the idea of material effort hypothesis is necessary. In the direct approach based on *ab initio* and molecular dynamics calculations (Hayes et al., 2004; He and Li, 2012) this intermediate state, recognized earlier by (Burzyński, 1928), can be omitted. On the other hand, the direct approach substantiates the phenomenological notion of *material effort*. Let

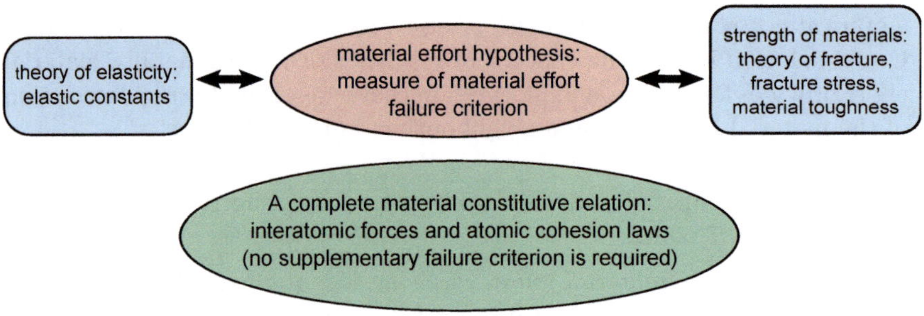

Figure 1. The scheme of the three-stage phenomenological description versus the unified theory of material behaviour which determines a complete constitutive theory (Hayes et al., 2004; He and Li, 2012).

us consider a deformable continuous body. Its material point corresponds to a Representative Volume Element (RVE) of the considered condensed matter, cf. (Pęcherski, 1983, 1998), where the pertinent discussion and basic references on this issue can be found. Any external loading applied to the deformable body changes on the atomic level the relative positions of the constituents of matter: nuclei and electrons or ions and valence electrons. Usually the Born-Oppenheimer approximation is assumed, which means that electrons follow the motion of the nuclei adiabatically and hence the total energy of the system becomes a unique function of the positions of the nuclei. As a consequence, the strength of chemical bonds in the RVE changes. It has gradually become clear that detailed knowledge of the nature of the chemical bonding is crucial to an adequate understanding the load bearing capacity, which characterizes the considered material. Therefore, it is possible to coin the following definition:

Material effort is a state of a material point of the loaded deformable body determined by the change in the energy of inter-atomic interactions in the RVE.

Let us note that this definition is a result of an analysis across scales. The notion of material effort commonly used in mechanics of materials on the macroscopic level of continuous deformable body is defined on the level of chemical bonds by the change in the energy of interatomic interactions and averaged over the RVE. According to basics of the solid-state physics, (Phillips, 2001), (Gilman, 2003), (Finnis, 2003), the density of valence (bonding) electrons and their distribution control cohesive forces, while the spatial distribution of valence electrons defines the elastic stiffness as well as the strength of a body. The values of these quantities are evaluated by means of the system energy. One of the more accurate ways of calculating the energy is solving Kohn-Sham equations formulated in the frame of Density Functional Theory (DFT), (Phillips, 2001), (Finnis, 2003), (Burke, 2012). This approach can be applied to solid bodies with different bonds: ionic, covalent as well as metallic ones. According to DFT, the system of electrons interacting with each other and with fixed ions is replaced by the system of non-interacting electrons with the same density as in the real system. The introduced system is subjected to acting of the effective potential. In this way, determination of the distribution of the electron density as well as evaluation of the total energy undergoes simplification. The presented approach is one of ab initio methods, which accurately describes interactions at the atomistic level. The computational cost of the methods is so large that they cannot be applied to the RVE, which is adequate for real material microstructures containing several hundred thousand of atoms. Therefore, semi-empirical potentials are introduced that determine the energy of interactions of an individual atom with the rest. In this way, we obtain the total energy as an average over the whole considered RVE. This energy determined on the atomic level is related with the elastic energy density of deformed body considered from the continuum perspective,

(Phillips, 2001):

$$\Phi = \frac{1}{\Omega}\left(E^{tot}_{def} - E^{tot}_{undef}\right) \tag{1}$$

where

$$\Phi = \frac{1}{2}C_{ijkl}\varepsilon_{kl} \tag{2}$$

is the elastic strain energy density of crystalline solid, ε_{ij} denote the components of small strain tensor and C_{ijkl} are the components of the elastic stiffness tensor in a Cartesian coordinate system. The quantity E^{tot}_{undef} constitutes the energy of the perfect crystal per atom. E^{tot}_{def} is similarly defined but for the deformed crystal. Additionally, Ω denotes the crystal volume per atom in the non-deformed state. It is worth to underline that for a crystal subjected to homogeneous strain ε, the RVE is the elementary cell. Metal/metal oxide interfaces constitute one of the types of microstructures observed experimentally, in which metal crystallites undergo some visible tension, what provides an excellent tool visualizing the concept of material effort or resilience. The reason is that the crystallites have to adjust to the stiff ceramic layer in order to form a coherent or semi-coherent structure (Wolf and Yip, 1992), (Nalepka and Pęcherski, 2009), (Nalepka, 2012). The state, in which the metal layer remains, is registered by means of the High Resolution Electron Microscopy (HRTEM). The observations have been performed in cooperation with Professor S. Kret in the Institute of Physics of the Polish Academy of Sciences in Warsaw. As an example, the HRTEM images of the Cu/$\alpha - Al_2O_3$ interfaces in metal/ceramic composites and in hetero-structures obtained by the Pulsed Laser Deposi-

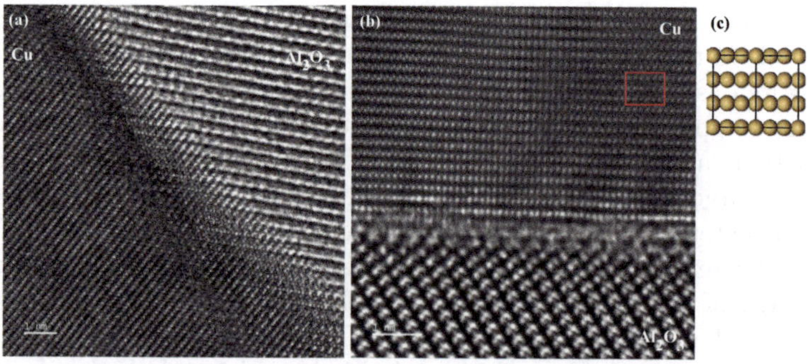

Figure 2. HRTEM images of Cu/$\alpha - Al_2O_3$ interfaces in a composite (a) and in a heterostructure obtained by the PLD (b) together with the identified Cu elementary cell (c).

tion (PLD) in the Laboratory of Professor Z. Szymański in the Institute of Fundamental Technological Research of the Polish Academy of Sciences in Warsaw, are presented (Fig.2a, b). In the case of systems formed by PLD, the Cu elementary cell (Fig. 2 b, c) is subjected not only to uniform stretching in the base plane but also contraction in the perpendicular direction. The compression effect comes from farther Cu layers which are deposited during the ablation process. If the strain process continues then the face centered cubic (fcc) structure transforms into the simple cubic (sc) one and subsequently into body centered cubic (bcc) one (see Fig. 3). The energy changes along this deformation path, so called trigonal one, are presented in Fig. 4. The bcc structure can be also obtained directly from fcc one when the same process of strain is applied to the cubic elementary cell (Figs. 5 and 6). Such a deformation path is called tetragonal one or the Bain path.

The limit states sc and bcc one arise at the definite changes in the energy per atom $\Delta E_{fcc\rightarrow sc}$ and $\Delta E_{fcc\rightarrow bcc}$ (see Figs. 4 and 6). Hence, the difference $E - E_{undef}$ becomes the natural measure of the material effort in the processes of the trigonal and tetragonal deformation.

Universal Binding Energy Relation (UBER) provides the physical explanation why most of considered materials reveal asymmetry of the elastic range, i.e. the limit of strength in compression is higher than the limit of strength in tension. In this way, the UBER justifies the assumptions of the material effort hypothesis by (Burzyński, 1928, 1929). In (Rose et al., 1984) it was observed on the basis of a vast set of ab initio data obtained for different metals that the total energy of a metal crystal per atom changes in the process of uniform volume expansion according to the following general formula:

$$E\left(r\right)=-E_{coh}\left[1+\eta\left(\left(\frac{V}{V_0}\right)^{1/3}-1\right)\right]e^{-\eta\left(\left(\frac{V}{V_0}\right)^{1/3}-1\right)} \tag{3}$$

where $\eta=\sqrt{9B_0V_0/E_{coh}}$ is the scaling parameter, while B_0, V_0 and E_{coh} are the equilibrium volume per atom, bulk modulus and the cohesion energy (in positive value), respectively. Additionally V constitutes the actual crystal volume per atom. For cubic metals the ratio $(V/V_0)^{1/3}$ reduces to the ratio of lattice constants a/a_0 . As an example, the UBER curve for copper reconstructed by a semi-empirical potential, (Nalepka, 2013) is presented (Fig. 7). The function matches to the ab initio data (Mishin et al., 2001) very well. In the final stage of the volume expansion, interactions between a central atom and its nearest neighbours become negligible, i.e. the interatomic distance exceeds the cutoff radius r_{ct}, ascribed to the semi-empirical potential. Then the Cu crystal loses the continuity (Fig. 8). In this way,

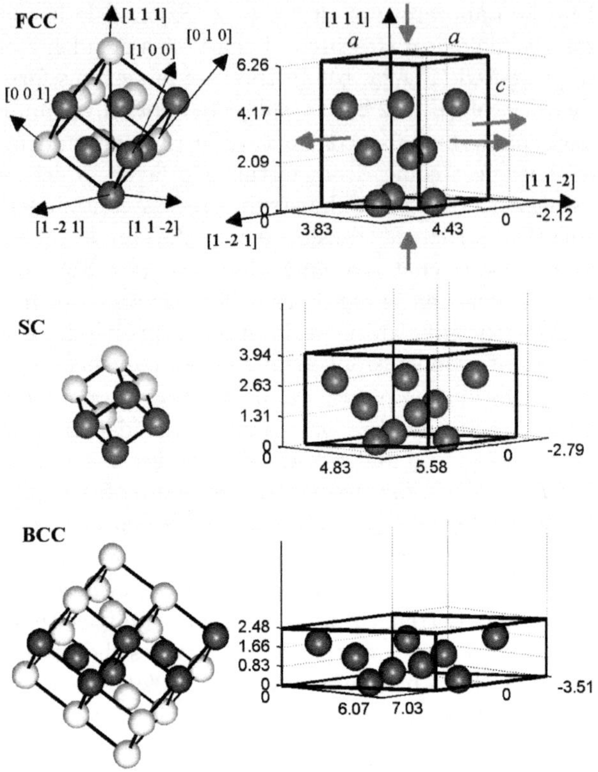

Figure 3. Trigonal deformation process during which the fcc structure transforms into sc and bcc structure (Nalepka, 2013).

the UBER identifies another limit state: the material strength. Using this relation we can register and visualize changes in the *material effort* of the Cu crystal. The relation (3) as well as Fig. 7 show that the process of the volume expansion differs from the isotropic compression. This asymmetry is visible at the continuum level in the processes of dynamic loadings which induce high pressures in material.

A certain *measure of material effort* is required to estimate the distance between the given stress state and the limit surface resulting from a considered failure criterion. This surface can be formed in the six-dimensional space of stresses evaluated by means of the reference system related with

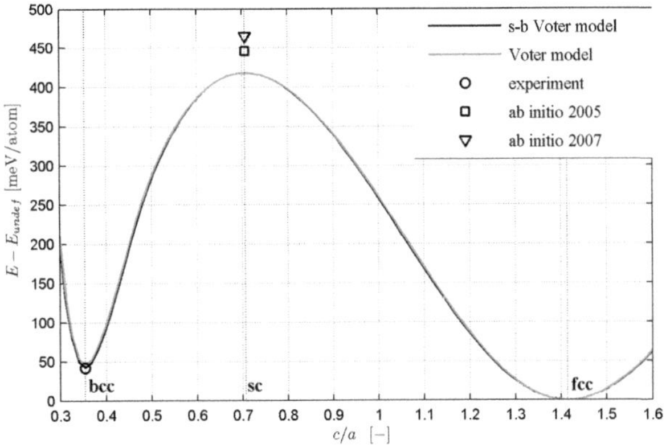

Figure 4. Change in the copper crystal energy per atom $E - E_{undef}$ along the trigonal path obtained by two semi-empirical potentials: original Voter model (Mishin et al., 2001) and Voter model specified by symmetry-based (s-b) conditions (Nalepka, 2012), (Nalepka, 2013). For comparison, the experimental and the ab initio data are included (Wang et al., 2004), (Černý et al., 2005), (Dai et al., 2007).

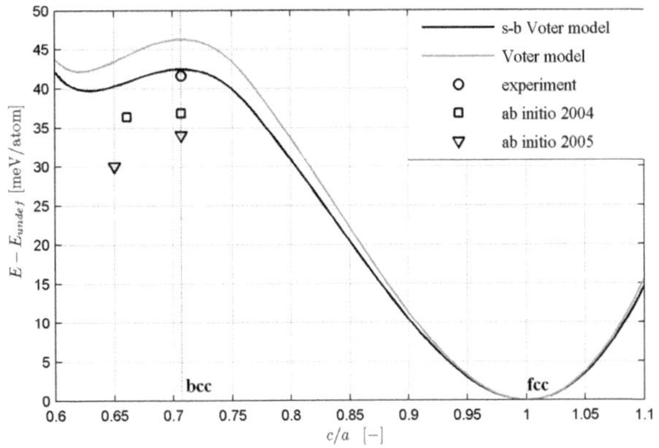

Figure 5. Change in the copper crystal energy per atom $E - E_{coh}$ along the Bain path obtained by two semi-empirical potentials (Nalepka, 2012). For comparison, the experimental and the ab initio data are included (Wang et al., 2004), (Černý et al., 2005), (Mehl et al., 2004).

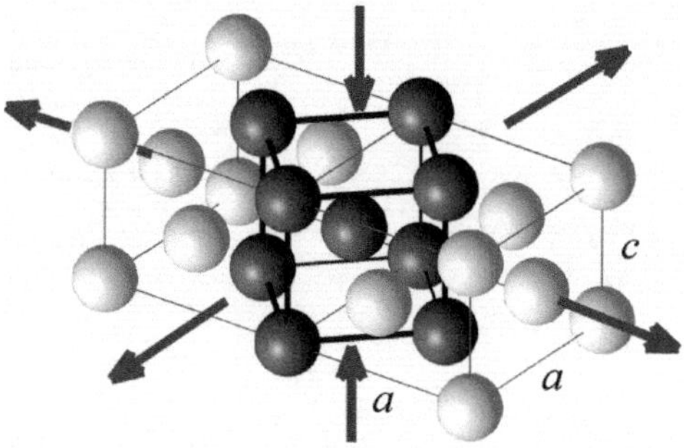

Figure 6. Bcc structure obtained along the Bain path.

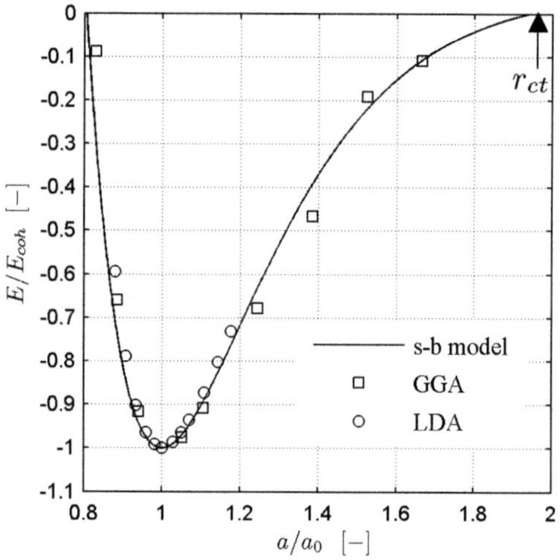

Figure 7. UBER curve for copper reconstructed by semi-empirical potential (Nalepka, 2013) compared with the ab initio data (Mishin et al., 2001).

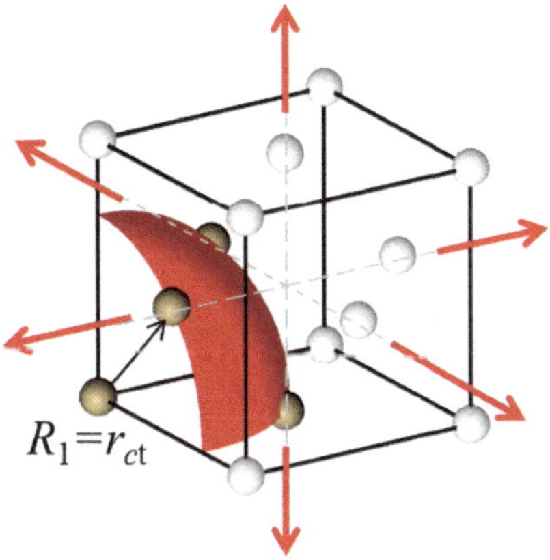

Figure 8. Loss of the continuity in copper crystal during volume expansion.

the considered material or in three-dimensional space of principal stresses, which is depicted in Fig. 9. The *measure of material effort* is defined in the frame of an assumed hypothesis. The analysis of deformation processes at the atomistic level shows that a natural measure of tension in material is change in the energy of interatomic interaction. The quantity of this change can be expressed in the continuum approach as the density of the elastic energy (1). Observations at the atomistic level enable further specification of the *measure of material effort.* The basis constitutes the identification of deformation processes which are crucial for the formation of new defects in material and / or the development of existing ones. In this way, a *measure of material effort* becomes a definite part of the elastic energy related with the critical strain processes.

Some of existing hypotheses of material effort were formulated on the basis of the reasoning similar to that presented above. Analyzing results of the experimental studies, (Burzyński, 1928, 1929) assumed the *measure of material effort* as the sum in which the first component constitutes a variable part of the energy density stored in volume deformation and the other one is the energy density of distortion. It is worth mentioning that James Clerk Maxwell (Maxwell, 1936) was the first, who proposed in the letter to William Thomson of 18 December 1856 the work of elastic distortion as a measure of material effort, and called it originally *resilience*:

"Now my opinion is that these two parts may be considered as independent U_1 being the work done in condensation and U_2 that done in distortion.

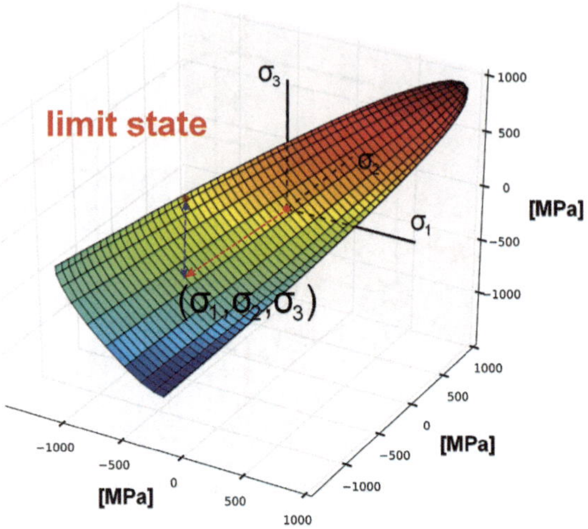

Figure 9. The schematic illustration of the measure of material effort with use of the yield surface picture of Inconel 718, discussed in (Pęcherski et al., 2011) and (Nowak et al., 2011).

Now I would use the old word "Resilience" to denote the work necessary to be done on a body to overcome its elastic forces.
The cubical resilience R is a measure of the work necessary to be expended in compression in order to increase the density permanently. This must increase rapidly as the body is condensed, whether it be wood or lead or iron. The resilience of rigidity R_2 (which is the converse of plasticity) is the work required to be expended in pure distortion in order to produce a permanent change of form in the element. I have strong reasons for believing that when... (U_2)... reaches a certain limit $= R_2$ then the element will begin to give way". p. 31, (Maxwell, 1936).

According to Maxwell, the term "resilience" can be then used as an equivalent for the measure of material effort understood as elastic energy density of distortion. Independently, Huber (Huber, 1904) considered, in a particular case, elastic energy of distortion as a measure of material effort. The limit criteria resulting from both formulations are identical. Discussion of more general statement of the Huber hypothesis is provided in (Burzyński, 1928). It is worth mentioning that Huber (Huber, 1904) used also simple physical reasoning of molecular interactions to justify the energy hypothesis, cf. also (Burzyński, 1928), (Pęcherski, 2008) for more detail discussion and further references. We owe to Heinrich Hencky, (Hencky, 1924), the proof that the result of rather formal, but brilliant, smooth approximation of corner-like Tresca criterion made by Richard von Mises, (Mises, 1913) is equivalent to the criterion of the limit energy of distortion considered earlier by Huber (Huber, 1904).

2.2 Experimental motivation

Recent experimental observations reveal that the sole elastic constants are not sufficient to assess mechanical behaviour of materials in the limit state of plastic yield initiation or fracture. It appears that these observations are in contradiction with common wisdom, which stems from the early search for simple empirical correlations between the mechanical hardness and plastic resistance versus elastic properties of crystals. For instance, (Gilman, 1960) derived an empirical relation showing that the yield stress is proportional solely to the Youngs modulus in fcc and bcc metals, cf. also (Gilman, 2003).

Many experimental investigations of mechanical properties of coarse-grained, ultra fine grained (ufg) and nanocrystalline (nc) metals reveal very strong increase of yield strength in decreasing grain size, while elastic constants are kept on the same level, (Nowak et al., 2007). For example, in (Valiev et al., 1994) deformation behaviour of ultra fine grained copper is investigated and the results of quasi-static compression tests for copper samples with mean grain sizes of 210 nm and 30 μm are confronted: the yield strength for nc Cu reached the value of 350 MPa, while in the latter case the yield strength is equal 50 MPa. The Young modulus in both cases was considered by the authors the same, E = 130 GPa. Also in (Wang et al., 2002) and (Lu et al., 2004) the typical tensile stress-strain curves for coarse-grained Cu as well as for pure nc Cu and nano-twin Cu are presented. The yield strength measured in the case of coarse-grained Cu is of the order of 50 MPa (Wang et al., 2002), while the yield strength of nc Cu is in excess of 400 MPa, (Wang et al., 2002), (Lu et al., 2004). The value of the tensile yield strength (at 0.2% offset) reached as high as 900 MPa for the nano-twin Cu, (Lu et al., 2004). The value of elastic Young modulus was not specified by the authors, but from the investigation of the tensile stress-strain curves for nc Cu and nano-twin Cu depicted in Fig. 2, (Lu et al., 2004), the conclusion can be drawn that in both cases the Young modulus is of the same order. Therefore, it is visible that the sole information about the elastic constants does not provide sufficient information about the behaviour of investigated materials. The coarse-grained Cu as well as nc Cu and nano-twin Cu are characterised with the yield limits, which differ by about two orders of magnitude. This is related with different capacity of accumulation of elastic energy density and different mechanisms responsible for plastic strain, dislocation mediated slip in coarse grained metals and micro-shear banding in the ufg or nc metals, (Nowak et al., 2007). It means that the considered materials: coarse-grained Cu, nc Cu and nano-twin Cu characterise with different values of *resilience*, as it was called by Maxwell or with different values of the *measure of material effort*, according to Huber and Burzyński. It is worthwhile mentioning that there are also publications, in which the semi empirical theory of mechanical hardness and related plastic resistance based on elastic constants and supplementary information on electronic structure of solids are investigated (Clerc and Ledbetter, 1998), (Clerc, 1999).

3 Measure of material effort as a definite part of the elastic energy density

3.1 Evolution of the energy-based measure of material effort

The aforementioned discussion leads to the conclusion that the concept of energy should be applied as universal measure of material effort. On the atomistic level certain part of energy of the system of atoms is related with the change of the strength of chemical bonds due the change in the energy of interatomic interactions. Similarly, on the macroscopic level some precisely defined contributions of the density of elastic energy accumulated in the strained body contribute to the measure of material effort. The symmetry of the elastic and strength properties controls the partition of the total elastic energy density, (Ostrowska-Maciejewska et al., 2013). The energy measure of material effort proposed by Burzyński, (Burzyński, 1928), (Olesiak and Pęcherski, 2013), reflects in an appropriate way the varying contributions of the volumetric and distortional parts of elastic energy density, in the course of deformation process. This new idea of energy-based approach found later confirmation by other authors. For instance, Freudenthal states: *"Correlation of behaviour of the different levels is possible only in terms of concept which on all levels has the same meaning in both Newtonian and statistical mechanics, the same dimension, and the same tensorial rank. This concept is energy. Being a scalar, that is, a tensor of rank zero, it is an algebraically additive quantity and has the same meaning on all levels of group of formation."* p. 20 in (Freudenthal, 1950). The need of proper account for the interplay between the both parts of elastic energy density: distortional and volumetric ones, was also underlined by (Christensen, 2004, 2006). This issue is discussed in more detail in (Pęcherski,

2008) and (Olesiak and Pęcherski, 2013).

3.2 Hypothesis of variable limit energy of volume change and distortion

Burzyński presented comprehensively in (Burzyński, 1928) the contemporary knowledge about limit criteria with the underlying hypotheses and furthermore proposed an innovative at that time approach determining the measure of material effort for isotropic solids with asymmetry of the elastic range. The original Burzyński's hypothesis of material effort states: *"The measure of local material effort in elastic and plastic ranges is the sum of density of quasi-energy of distortion and a certain part dependent on the state of stress and individual properties of a body of the density of the pseudo-energy of volume change"* – p. 141 (of English translation) (Burzyński, 1928). By adding "quasi" or "pseudo" it has been emphasized that the analytic expressions used in continuation do not mean for a certain group of bodies or relatively in certain experimental fields elastic energy in the sense of linear theory of elasticity.
Mathematical formulation of the hypothesis of variable limit energy of volume change and distortion is expressed as follows:

$$\Phi_f + \eta\Phi_v = k, \qquad \eta = \omega + \frac{\delta}{3p}, \qquad p = \frac{\sigma_1 + \sigma_2 + \sigma_3}{3} \tag{4}$$

where Φ_f denotes the density of elastic energy of distortion

$$\Phi_f = \frac{1}{12G}\left[(\sigma_1 - \sigma_2)^2 + (\sigma_1 - \sigma_3)^2 + (\sigma_2 - \sigma_3)^2\right] \tag{5}$$

and Φ_v is the elastic energy density of volume change

$$\Phi_v = \frac{1-2\nu}{6E}(\sigma_1 + \sigma_2 + \sigma_3)^2 = \frac{1-2\nu}{12G(1+\nu)^2}(\sigma_1 + \sigma_2 + \sigma_3)^2 \tag{6}$$

expressed in terms of principal stresses $\sigma_1 \geqslant \sigma_2 \geqslant \sigma_3$ and contributing to the total elastic energy for isotropic solid

$$\Phi = \Phi_f + \Phi_v \tag{7}$$

while ω and δ are material constants, which are to be specified and p denotes a mean stress expressed in principal stresses. Furthermore ν denotes Poisson's ratio and G is the Kirchhoff shear modulus. By introducing the function η Burzyński took into account the experimentally based information that the increase of mean stress p results in the diminishing contribution of the elastic energy density of volume change in the measure of material effort. Accounting for the above relations leads to the following final form of the Burzyński failure hypothesis (4):

$$\frac{1}{3}\sigma_f^2 + 3\frac{1-2\nu}{1+\nu}\omega p^2 + \frac{1-2\nu}{1+\nu}\delta p = 4GK \tag{8}$$

where $\sigma_f^2 = 12G\Phi_f$. The main idea of Burzyński's derivation lies in the conversion of material parameters (ω, δ, K) into the triplet (k_t, k_c, k_s) that can be obtained in the well known laboratory tests: elastic (plastic) limit obtained in uniaxial tension - k_t, in uniaxial compression - k_c and in torsion - k_s. The more detail discussion of this derivation and resulting failure criteria is given in (Burzyński, 1928) and in the recent papers (Frąś and Pęcherski, 2010), (Nowak et al., 2007) and (Frąś et al., 2010), (Pęcherski et al., 2011).
The discussion presented in the point 2 leads to the conclusion that material parameters (ω, δ, K) appearing in (5) can be determined also in a straightforward way basing on solid state physics instead of applying the results of the aforementioned mechanical experiments to determine the constants (k_t, k_c, k_s). First attempt of such an approach was made by (Zawadzki, 1957) as early as in the 1950s. More recent investigations concerning the semi empirical theories of hardness and plastic resistance in correlation with elastic constants and electronic properties of solids can be also applied, (Gilman, 2003), (Clerc and Ledbetter, 1998), (Clerc, 1999).

3.3 Specification of the Burzyński hypothesis

The general formulation of the hypothesis of variable limit energy of volume change and distortion presented in (5) can be transformed replacing the material parameters (ω, δ, K) with the material constants (k_t, k_c, k_s), obtained in the laboratory tensile test, compression test and in torsion or shear test, respectively (Burzyński, 1928):

$$\frac{k_c k_t}{3k_s^2}\sigma_e^2 + \left(9 - \frac{3k_c k_t}{k_s^2}\right)p^2 + 3\left(k_c - k_t\right)p - k_c k_t = 0, \qquad p = \frac{\sigma_1 + \sigma_2 + \sigma_3}{3} \tag{9}$$

where

$$\sigma_e = \frac{1}{\sqrt{2}}\sqrt{\left(\sigma_1 - \sigma_2\right)^2 + \left(\sigma_1 - \sigma_3\right)^2 + \left(\sigma_2 - \sigma_3\right)^2} \tag{10}$$

is an equivalent stress used in theory of plasticity. According to the detail study presented in (Burzyński, 1928) the equation (6) in the space of principal stresses $\sigma_1 \geqslant \sigma_2 \geqslant \sigma_3$ represents particular cases of quadric surfaces depending on the interplay between three material constants k_t, k_c, k_s. In [3] the following basic modes of stress are considered:

I. Uniaxial tension: $\sigma_1 = k_t, \quad \sigma_2 = 0, \quad \sigma_3 = 0$
II. Uniaxial compression: $\sigma_1 = 0, \quad \sigma_2 = 0, \quad \sigma_3 = -k_c$
III. Torsion (shear): $\sigma_1 = k_s, \quad \sigma_2 = 0, \quad \sigma_3 = -k_s$
IV. Biaxial uniform tension: $\sigma_1 = k_{tt}, \quad \sigma_2 = k_{tt}, \quad \sigma_3 = 0$
V. Biaxial uniform compression: $\sigma_1 = 0, \quad \sigma_2 = -k_{cc}, \quad \sigma_3 = -k_{cc}$
VI. Triaxial uniform tension: $\sigma_1 = k_{ttt}, \quad \sigma_2 = k_{ttt}, \quad \sigma_3 = k_{ttt}$
VII. Triaxial uniform compression: $\sigma_1 = -k_{ccc}, \sigma_2 = -k_{ccc}, \quad \sigma_3 = -k_{ccc}$

The first case of quadric, which can be obtained from general relation (9) corresponds, in the three dimensional system of principal stress coordinates, to ellipsoid of revolution with respect to the symmetry axis $\sigma_1 = \sigma_2 = \sigma_3$ for $k_s > \sqrt{\frac{k_c k_t}{3}}$. In the plane coordinate system (σ_f, p), where $\sigma_f = \sqrt{2}\sigma_e$ is the Burzyński‘s stress invariant (Burzyński, 1928), the criterion is depicted by an ellipse or a circle, Fig. 10. The discussed above basic modes of stress, from I to VII, are illustrated in Fig. 10 and the subsequent Figs. 11 and 12, with use of broken straight lines.

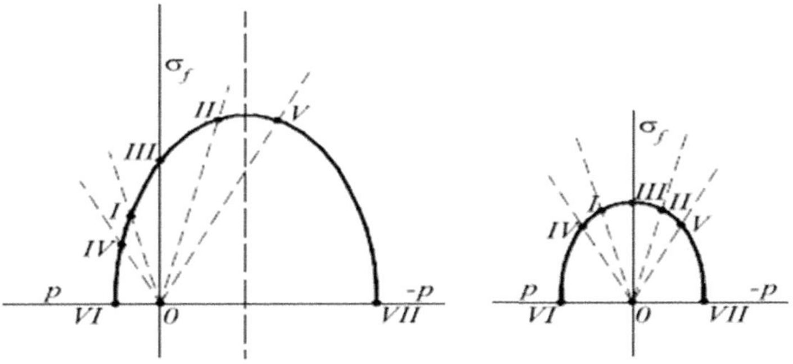

Figure 10. The original picture of the limit criteria with schematic illustration of the basic modes of stress published in (Burzyński, 1928). On the left hand side the plot of a half ellipse, for $k_c \neq k_t$ and the plot of a half circle for $k_c = k_t$ on the right hand side.

The depicted above ellipsoidal criteria can be applied for different kinds of cellular materials, e.g. metal, polyurethane and ceramic foams as well as for some metallic materials in which mechanism of twinning prevails at the limit state. In such a case we have $k_c < k_t$. The application of the Burzyński criterion for Alporas closed cell foams investigated by (Blazy et al., 2004) was studied in (Stręk, 2012). The similar elliptic shape of limit curves for metal foams was presented also in (Deshpande and Fleck, 2000) on the basis of certain empirical formula. Additional investigations are necessary, for other basic modes of stress to validate the hypothesis that the Burzyński limit criterion can appear useful for the analysis of limit states of cellular materials.

The second case corresponds, in the three dimensional system of principal stress coordinates, to paraboloid of revolution with respect to the symmetry axis $\sigma_1 = \sigma_2 = \sigma_3$, which can be obtained for $k_s = \sqrt{\frac{k_c k_t}{3}}$ from the general formula (6) in the following form:

$$\sigma_e^2 + 3\,(k_c - k_t)\,p - k_c k_t = 0 \tag{11}$$

In the plane coordinate system (σ_f, p), the criterion is depicted by a parabola, Fig. 11.

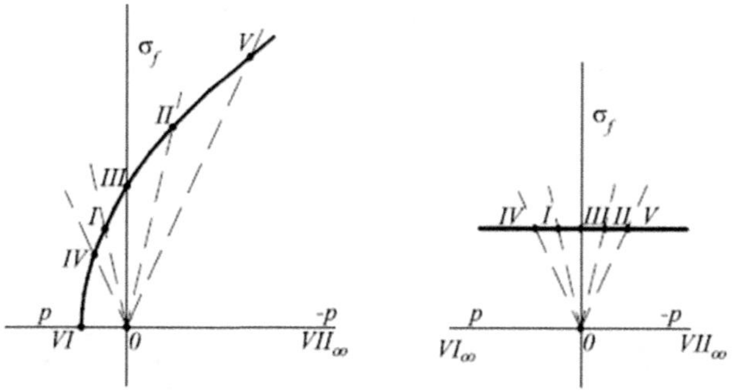

Figure 11. The original picture of the limit criteria with schematic illustration of the basic modes of stress published in (Burzyński, 1928). On the left hand side a half parabola, for $k_c \neq k_t$ and a line parallel to the axis p for $k_c = k_t$ on the right hand side corresponding to Huber-Mises condition.

The third case corresponds, in the three dimensional system of principal stress coordinates, to hyperboloid of revolution with respect to the symmetry axis $\sigma_1 = \sigma_2 = \sigma_3$, which can be obtained from the general formula (6) for $\frac{2}{\sqrt{3}}\frac{k_c k_t}{k_c + k_t} < k_s < \sqrt{\frac{k_c k_t}{3}}$. Generally, the surface has two separate sheets, but from practical point of view only one sheet of the hyperboloid comes into play. In the plane coordinate system (σ_f, p), the criterion is depicted by a hyperbola, Fig. 12. There is a special case: $k_s = \frac{2}{\sqrt{3}}\frac{k_c k_t}{k_c + k_t}$, then the hyperboloid degenerates to rotationally symmetric conical surface. One sheet of the surface is accounted for the description of the limit criterion, which in this case transforms from (9) into the following formula:

$$\sigma_e^2 + 3\frac{k_c - k_t}{k_c + k_t} - 2\frac{k_c k_t}{k_c + k_t} = 0 \tag{12}$$

In the plane coordinate system (σ_f, p), the criterion is depicted by two intersecting straight lines, the right hand side of Fig. 12.

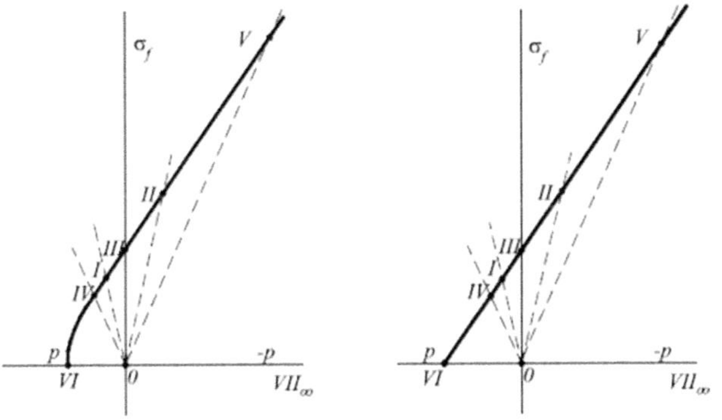

Figure 12. The original picture of the limit criteria with schematic illustration of the basic modes of stress published in (Burzyński, 1928). On the left hand side a half hyperbola and a straight line on the right hand side.

The yield condition of the form given in (12) is widely known in the literature since 1952 as Drucker-Prager yield criterion (Drucker and Prager, 1952). It is worthwhile to note, however, that some authors have recognised the earlier contribution of (Burzyński, 1928) for instance: (Sendeckyj, 1972), (Nardin et al., 2003) as well as (Pisarenko and Lebedev, 1973), (Jirasek and Bazant, 2002), (Yu, 2004) (Yu et al., 2006). The discussed above yield or strength criteria, which are obtained in the rigorous way from the energy-based material effort hypothesis proposed originally by (Burzyński, 1928) in the year 1928 were later re-discovered again and again for different ranges of empirical parameters independently by many researchers. The part of these findings is described by (Życzkowski, 1981, 1999) and (Skrzypek, 1993). These authors are presenting at the same time the original achievements of Burzyński. His yield condition in confrontation with the works of other authors was also discussed in (Pęcherski, 2008).

4 Examples and visualizations of the particular Burzyński failure criteria

In this paragraph some examples of the visualizations of the discussed above limit criteria are displayed. They are based on experimental investigations performed mostly in the plane state of stress. In such a case the yield surface in the three dimensional space of principal stresses is reconstructed with use of (Wolfram Mathematica, 6.0). The mentioned experimental tests are related usually with the uniaxial modes of stress I and II, the mode of pure shear III as well as with the biaxial ones IV and V for the plane states of stress inscribed in certain hypothetical yield limit of material with the thought strength differential ratio $\kappa = k_c/k_t = 1.3$, Fig. 13. The two

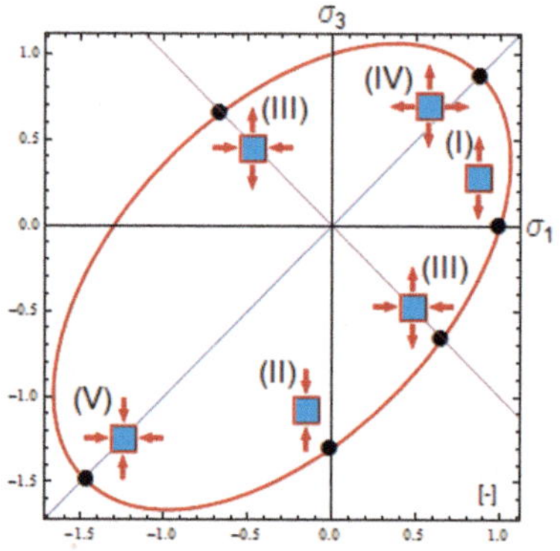

Figure 13. The hypothetical yield curve in the plane $\sigma_2 = 0$ for the case of strength differential ratio $\kappa = 1.3$ with schematically indicated points corresponding to the results of particular laboratory tests.

discussed above cases of particular specifications of the Burzyński hypothesis of variable limit energy of volume change and distortion, which is related with the ellipsoidal and paraboloidal surfaces will be illustrated for a chosen materials.

4.1 Ellipsoidal failure surface

An example of the material revealing the before discussed inequality $k_c < k_t$ is the magnesium alloy AZ31, for which according to the data published in (Yoshikawa et al., 2008): the yield strength in tension $k_t = 200MPa$, and the yield strength in compression $k_c = 120MPa$, while the yield strength in shear $k_s = 120MPa$. The data were obtained in the quasi-static test conditions with the assumed offset strain 0.002. According to these data the yield surface can be visualized as an ellipsoid in the principal stress coordinates as well as in the plane $\sigma_2 = 0$ and in the plane coordinate system (σ_e, p) as is depicted in Fig. 14, (Frąś, 2013).

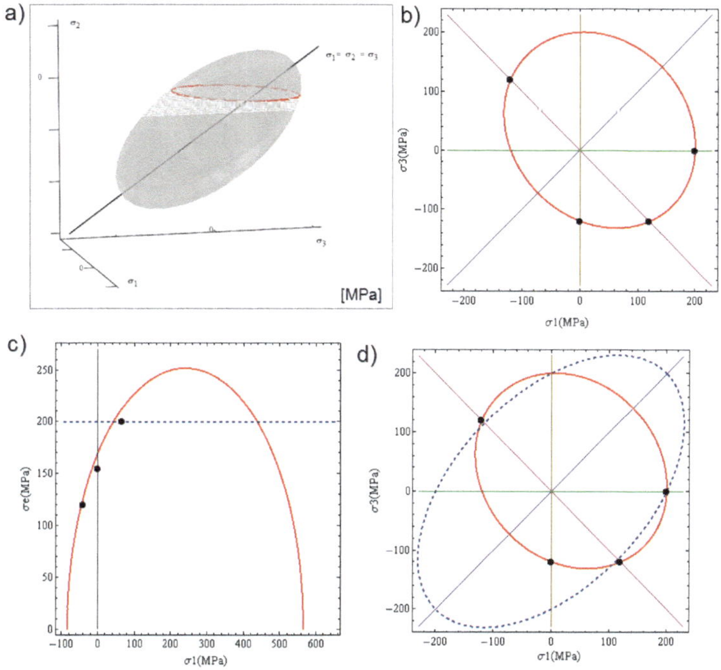

Figure 14. Presentation of Burzyński ellipsoid criterion, for $k_s > \sqrt{\frac{k_c k_r}{3}}$, applied for Mg alloy investigated in (Yoshikawa et al., 2008) confronted with the Huber-Mises criterion depicted in the plane $\sigma_2 = 0$ with blue line, (Frąś, 2013).

4.2 Paraboloid failure surfaces

The classical results of experimental investigations collected and elaborated on pure geometrically basis by (Theocaris, 1995), and discussed also within the context of paraboloidal failure criterion in (Frąś and Pęcherski, 2010), appear to be appropriate data for visualizing paraboloidal failure surfaces. The first results revealing very distinct strength differential effect measured by the ratio $\kappa = 3$ are related with investigations of gray cast-iron by (Grassi and Cornet, 1949) and (Coffin, 1950).

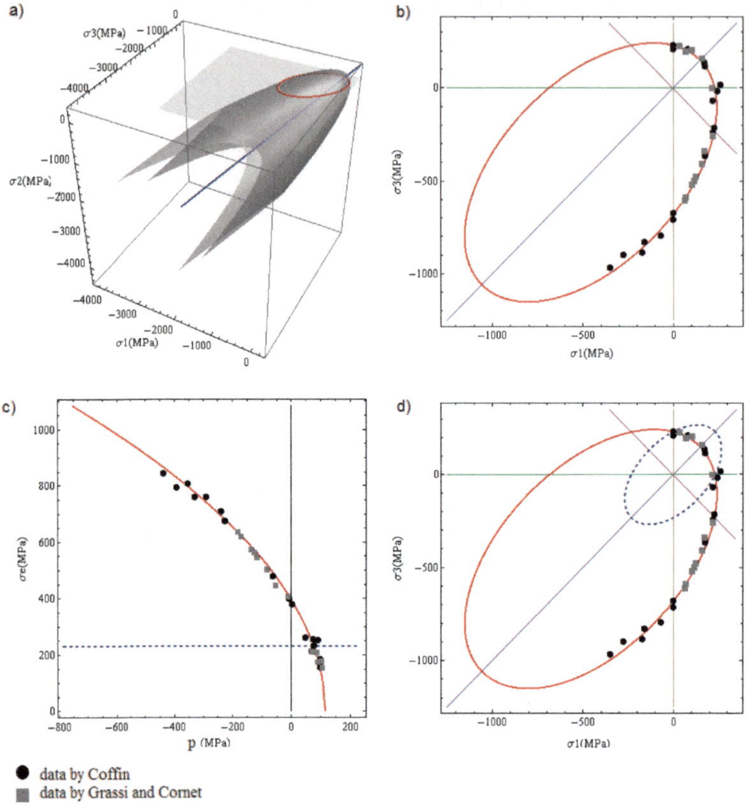

Figure 15. Presentation of the paraboloid failure surface according to the data of (Grassi and Cornet, 1949) and (Coffin, 1950) for gray cast - iron a) in the space of principal stresses, b) in the plane $\sigma_2 = 0$ c) in the coordinate system (σ_e, p), d) comparison with the Huber-Mises criterion presented in the plane $\sigma_2 = 0$, (Frąś and Pęcherski, 2010), (Frąś, 2013).

The another example is related with the first investigations of the yield curves in the complex states of stress performed by (Lode, 1926), as well as, by (Taylor and Quinney, 1931), which were collected and expressed in the non-dimensional form by (Theocaris, 1995), who suggested that the deviation of experimental results from the prediction according to Huber-Mises yield condition might be accounted for the strength differential effect measured by the ratio $\kappa = 1.3$. The mentioned deviation can be also explained as a result of slight initial anisotropy produced by manufacturing of material elements, e.g. drawing of rods and machining of samples. Another results are related with the collected by (Theocaris, 1995) historical data of experimental investigations for different polymers, behaviour of which apparently is pressure sensitive and the strength differential ratio for such materials equals about 1.3. The displayed experimental data show very good correlation with the Burzyński paraboloidal failure condition.

The last example is related with the innovative nc metals investigated in numerical simulations by (Lund and Schuh, 2003), (Schuh and Lund, 2003). The authors performed molecular simulations of multiaxial deformation of metallic glass using an energy minimization technique. The results reveal the asymmetry between the magnitudes of the yield strength in tension and compression, with the strength difference ratio$\kappa = 1.24$. At ambient temperature, metallic glasses deform by a process of shear banding, where plastic strain is highly localized into strips of nanometer-thickness. The mechanism of shear banding on the nanometer-scale produces unique mechanical properties at macroscopic scale. For example, the very low values of tensile elongation recorded for amorphous metals are attributed to rapid failure along a single shear band. In constrained modes of loading like compression, plastic yielding is observed in a serrated way and the strain is produced by a sequence of single shear bands. These properties have been investigated in a variety of glassy alloys with different compositions, and the observation appears to be general to this class of materials. One important consequence of shear localization in amorphous metals is that the macroscopic yield criterion may exhibit the dependence not only upon the maximum shear stress, but also upon the hydrostatic pressure or the normal stress acting on the shear plane. The discussion and proposal of a model accounting for shear banding mechanism in ufg and nc metals is given in (Nowak et al., 2007) and (Frąś et al., 2011). The results displayed in Fig. 13 show that the Burzyński paraboloidal yield condition provides adequate correlation with the discussed in (Lund and Schuh, 2003), (Schuh and Lund, 2003) data.

4.3 Concluding remarks

It is appealing to see that the universal energy-based approach provides the criteria, which accurately describe the discussed above experimental data for a wide range of different materials. It would be interesting to study the possibilities of applications of the mentioned above hyperboloid

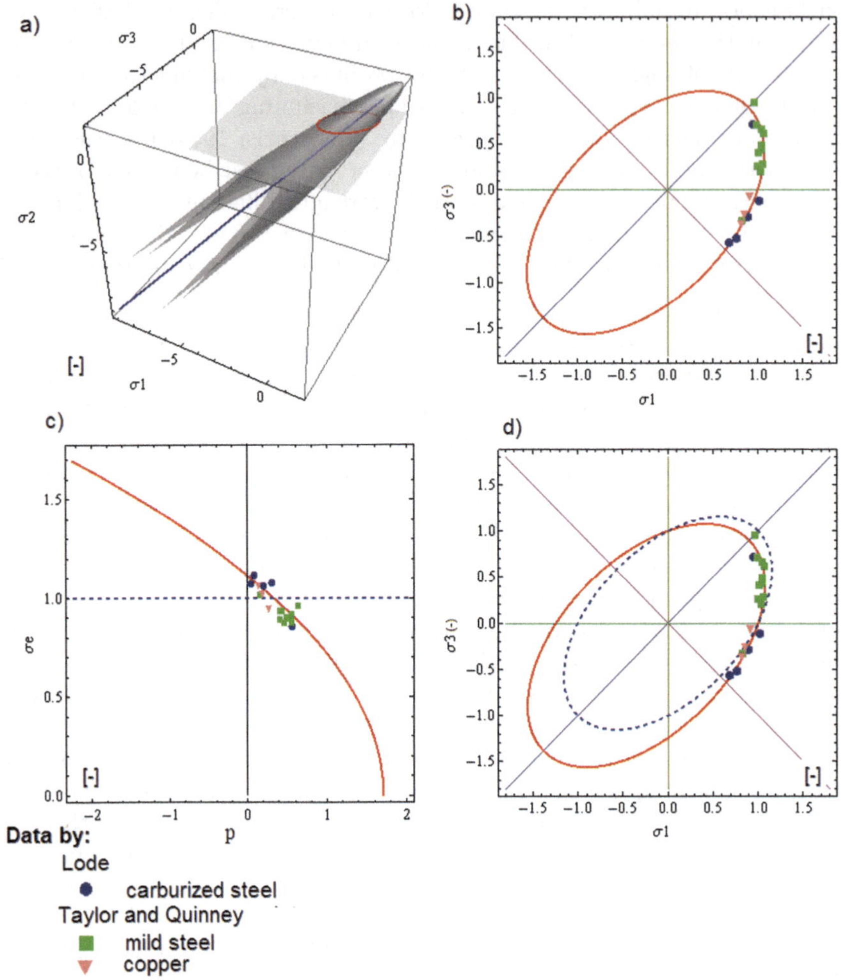

Figure 16. a) Presentation of the paraboloid yield surface according to the data of (Lode, 1926) and (Taylor and Quinney, 1931), collected by (Theocaris, 1995), in the space of principal stresses, b) in the plane $\sigma_2 = 0$ c) in the coordinate system (σ_e, p), d) comparison with the Huber-Mises criterion presented in the plane $\sigma_2 = 0$ (Frąś and Pęcherski, 2010), (Frąś, 2013).

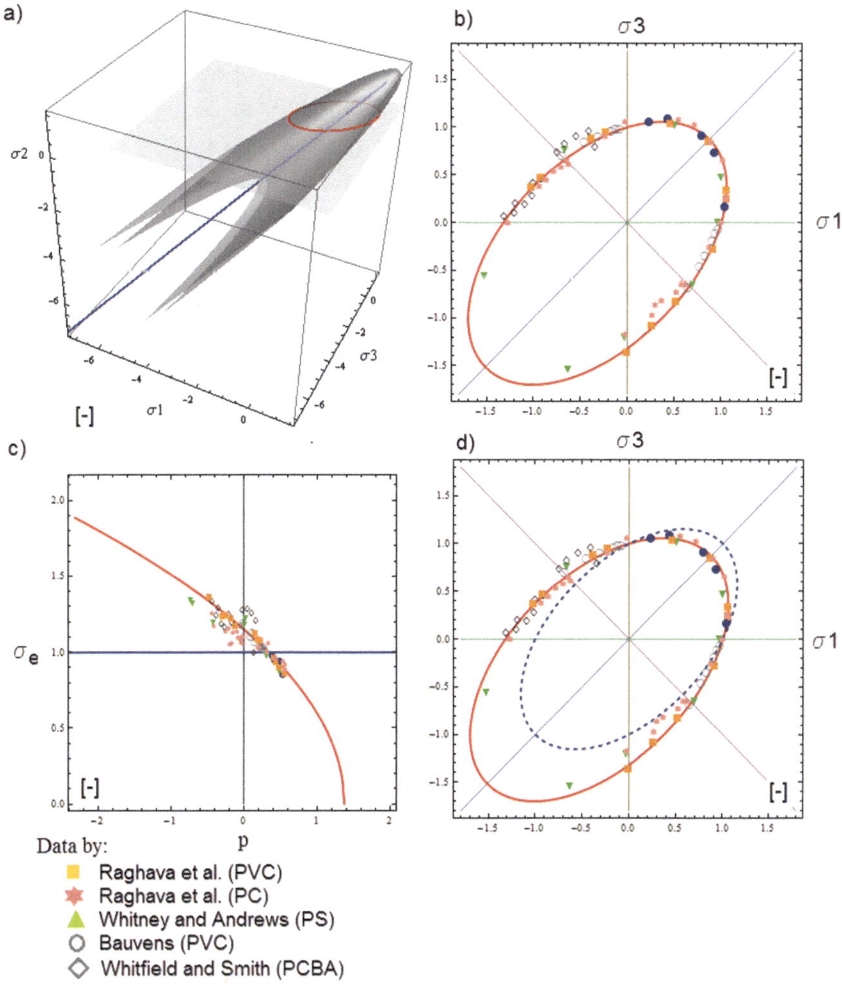

Figure 17. Presentation of the paraboloid failure surface according to the data collected by (Theocaris, 1995) for polymers in the space of principal stresses, b) in the plane $\sigma_2 = 0$ c) in the coordinate system (σ_e, p), d) comparison with the Huber-Mises criterion presented in the plane $\sigma_2 = 0$, (Frąś and Pęcherski, 2010), (Frąś, 2013).

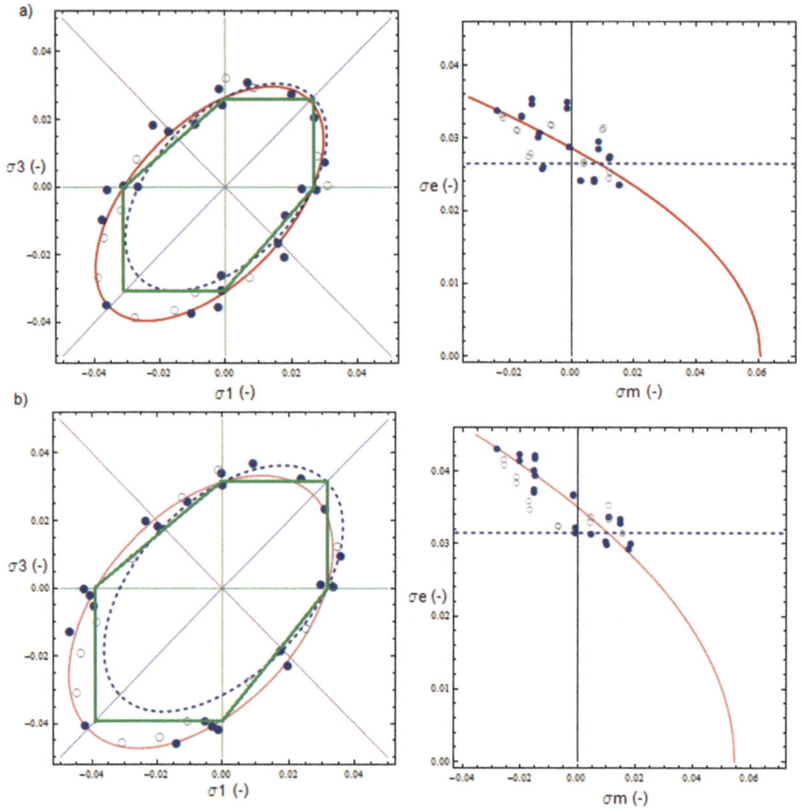

Figure 18. Presentation of the paraboloid failure surface according to the data collected by (Lund and Schuh, 2003), (Schuh and Lund, 2003), for metallic glass and comparison with the Huber-Mises yield condition and Coulomb-Mohr criterion, a) in the plane $\sigma_2 = 0$ and in the coordinate system (σ_e, p), for the offset strain 0.002, b) in the plane $\sigma_2 = 0$ and in the coordinate system (σ_e, p), for the offset strain 0.0005, (Frąś and Pęcherski, 2010), (Frąś, 2013).

criterion. Implementation of this criterion still remains an open question. The program of experimental investigations applied for new materials is necessary. Careful measurements of the limit values (k_t, k_c, k_s) should enable checking if the inequality $\frac{2}{\sqrt{3}}\frac{k_c k_t}{k_c+k_t} < k_s < \sqrt{\frac{k_c k_t}{3}}$ holds. The nanomaterials and foams of metallic, polymer or ceramic skeleton should be taken into consideration by that.

On the other hand, in commercial finite element programs a kind of hyperboloid yield function is considered, which is defined rather artificially as an "extended Drucker-Prager condition", cf. e.g. (Abaqus Analysis User's Manual, 6.12). Such an approach is implemented in the program within the framework of non-associated flow law, what loses its physical foundations and seems to be rather confusing.

5 The numerical implementation of elasto-plasticity equations associated with paraboloid yield condition

The paraboloid yield criterion resulting from the Burzyński hypothesis of variable limit energy of volume change and distortion (3) was applied in the numerical scheme of the integration of elasto-plasticity equations in (Vadillo et al., 2011) and independently by the authors of this paragraph. For metallic materials elastic strains are assumed infinitesimally small and therefore the additive form of the total strain rate distribution is justified:

$$\dot{\boldsymbol{\varepsilon}} = \dot{\boldsymbol{\varepsilon}}^e + \dot{\boldsymbol{\varepsilon}}^p \tag{13}$$

The Hooke's law reads in the case of isotropic material

$$\boldsymbol{\sigma} = \boldsymbol{C} : (\dot{\boldsymbol{\varepsilon}} - \dot{\boldsymbol{\varepsilon}}^p) \tag{14}$$

The elastic stiffness is defined by:

$$\boldsymbol{C} = 2G\boldsymbol{I} + K\mathbf{1}\otimes\mathbf{1}, \qquad G = \frac{E}{2(1+\nu)}, \qquad K = \frac{E}{3(1-2\nu)} \tag{15}$$

where E and ν is Young modulus and Poisson's ratio, respectively, whereas $\mathbf{1}$ is the second-order unit tensor and $\boldsymbol{I}$ is the unit deviatoric fourth-order tensor. The plastic part of the strain rate $\dot{\boldsymbol{\varepsilon}}^p$ is, according to the assumed associative flow law, normal to the plastic potential surface, which at the same time is the paraboloid yield surface:

$$\dot{\boldsymbol{\varepsilon}}^p = \dot{\lambda}\frac{\partial\Phi}{\partial\boldsymbol{\sigma}} \tag{16}$$

where λ is the plastic proportionality factor, and Φ corresponds to the paraboloid yield surface:

$$\Phi(p, \sigma_e, k_t) = \frac{1}{2\kappa}\left[3(\kappa - 1)p + \sqrt{9(\kappa - 1)^2 p^2 + 4\kappa\sigma_e^2}\right] - k_t \tag{17}$$

The detail formulation of the radial return algorithm and the consistent tangent operator for the elasto-plasticity equations with the associated plastic flow law (16) and paraboloid yield potential (17) was given for the first time, according to the authors knowledge, in (Vadillo et al., 2011), where also the integration algorithms programmed and implemented in the commercial finite element program (Abaqus Analysis User's Manual, 6.12) were applied for the solution of particular examples validating the practical applications of paraboloid yield condition. The calculations were performed to simulate the tests on round notched tensile specimens of 2024-T351 aluminum alloy investigated by (Wilson, 2002). Good correlation of the numerical results with experimental data was shown in (Vadillo et al., 2011) proving that the plasticity model with the associated paraboloid yield surface provides the well founded theoretically and physically replacement of the model using the cone surface originating from Drucker-Prager condition applied in the analysis in (Wilson, 2002). More complex problem of the simulation of tests on the Inconel 718 thin-walled tubes subjected to torsion after tension or compression was studied as well (Vadillo et al., 2011). Also in this case the application of paraboloid yield condition provides good agreement of computational results with experimental data discussed in (Gil et al., 1999a) and (Gil et al., 1999b).

5.1 Examples

Basing on the results in (Vadillo et al., 2011) the own user-defined material model was developed and implemented in user subroutine (UMAT) for commercial finite element program Abaqus (Abaqus Analysis User's Manual, 6.12). One of the calculated examples deals with the problem of compression test of the cylindrical specimen with a slanting cut, the so called shear-compression specimen (SCS), (Rittel et al., 2002).

Theoretical analysis of deformation process during compression test with application of ideal plasticity model presented in (Vural et al., 2011) shows that it is possible to approximate the relations between measured experimentally load P and equivalent stress σ_e in the deformation zone within the slanting cut, as well as, the relation between measured displacement δ and equivalent plastic strain ε_e^p.

$$\varepsilon_e^p = \frac{1}{k_1(\theta)} \frac{\delta}{h} \tag{18}$$

$$\sigma_e = \sqrt{\frac{3}{2} \boldsymbol{S} : \boldsymbol{S}} = \sqrt{\frac{3}{4} 4\sigma_{xy}^2 + \sigma_{yy}^2} = k_1(\theta) \exp\left(-k_2(\theta)\varepsilon_e^p\right) \frac{P}{D \cdot t} \tag{19}$$

where $\boldsymbol{S}$ denotes deviatoric part of stress tensor and h, t is the height and respectively the width of shearing zone determined by slanting cut, D is the diameter of the specimen cross-section, while the functions $k_1(\theta)$ and $k_2(\theta)$ of the inclination angle θ of the slit with respect to vertical axis of the specimen were derived in (Vural et al., 2011):

$$k_1(\theta) = \frac{\sqrt{3}}{2} \cos(\theta) \sqrt{4 \sin^2(\theta) + \cos^2(\theta)} \tag{20}$$

$$k_2(\theta) = \frac{\sqrt{3}}{2} \frac{\cos(\theta)}{\sqrt{4 \sin^2(\theta) + \cos^2(\theta)}} \tag{21}$$

The considered material was a new composite AlMg5%SiC. Performing compression test of the cylindrical specimen of the SCS type, Fig. 19, and applying the approximate analytical relations the material characteristic was reconstructed and implemented in finite element computations with use of own user material element. The results of calculations are presented in Fig. 20 displaying the comparison of calculated load-displacement relations with experimental data. It is visible that application of standard model with Huber-Mises yield condition, the strength differential ratio $\kappa = 1$, does not conform experimental observations. The identified strength differential ratio equals $\kappa = 1.15$.

5.2 Conclusions

The presented above analysis leads to the conclusion that application of the model and corresponding UMAT with paraboloid yield surface provides satisfactory prediction of material behaviour. The presented example illustrates a method of identification of the stress-strain characteristic and strength differential ratio of new materials. More adequate analysis requires separate compression tests for smooth specimens in order to get independently the stress-strain characteristics.

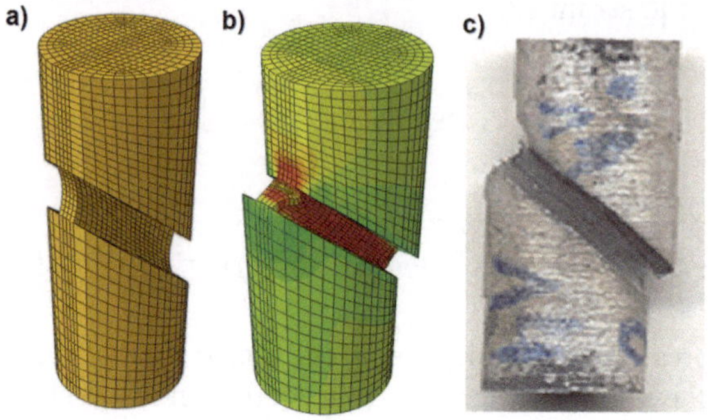

Figure 19. The finite element mesh of the specimen with slanting cut: element C3D8, number of finite elements 11 540, number of nodes 13871, b) the configuration obtained after the numerical simulation of compression process, c) picture of the deformed specimen of AlMg5%SiC composite.

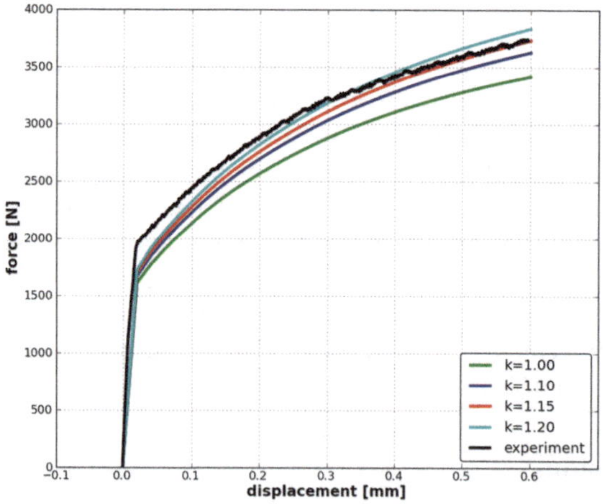

Figure 20. The comparison of calculated load-displacement relations with experimental data for different values of strength differential ratio.

6 Final remarks and further studies

The foregoing discussion has highlighted that energy-based Burzyński failure criteria, accounting in a natural way for strength differential effect and pressure sensitivity. It can be applied in the constitutive description of inelastic flow of materials and implemented into numerical simulations of impact problems and failure processes. At the same time, the following extensions and generalizations of the presented theory come into mind: an account for the third invariant of stress deviator in failure criteria for isotropic solids, for energy-based failure criteria of anisotropic solids and for the interplay of two basic mechanisms responsible for inelastic flow: crystallographic slip and shear banding.

6.1 An account for the influence of the third stress invariant

Experimental investigations related with failure processes and phase transformations in advanced metallic materials reveal that assumption about isotropic mechanical properties requires also accounting for the third invariant of stress tensor deviator, which corresponds to the Lode angle on the octahedral plane. The observations show that the effect of Lode angle transpires when shear deformation processes are taking place, (Pęcherski et al., 2011), (Nowak et al., 2011). According to the foregoing analysis in chapter 2, the shear process in a solid body is accompanied with such rearrangements of atoms that structure symmetry and the strength of chemical bonds is changing, mostly with downward tendency. Therefore, a degree of the domination of shear mode has essential influence on the energy-based measure of material effort. This leads to the conclusion that the contribution of shear modes of stress, quantified by means of the Lode angle, should be accounted for in the energetic criteria of failure. The main idea of the proposed new formulation is that the contribution of the density of elastic energy of distortion in the measure of material effort is controlled by the Lode angle (Pęcherski et al., 2011), (Nowak et al., 2011):

$$\eta_f(\theta)\Phi_f + \eta_\nu(p)\Phi_\nu = K \tag{22}$$

where the symbols η_f and η_ν denote respectively the Lode angle and pressure dependent influence functions and

$$\theta = \frac{1}{3}\arccos\left(\frac{3\sqrt{3}}{2}\frac{J_3}{J_2^{3/2}}\right) \tag{23}$$

is the Lode angle definition. More detail discussion on the different specifications of the influence functions known in the literature and the analysis of an example of yield surface for Inconel 718, shown in Fig. 9, is contained in (Pęcherski et al., 2011) and (Nowak et al., 2011). The specified relation for the yield surface can be applied in the formulation of inelastic flow law and implemented in the finite element simulations of deformation and failure processes.

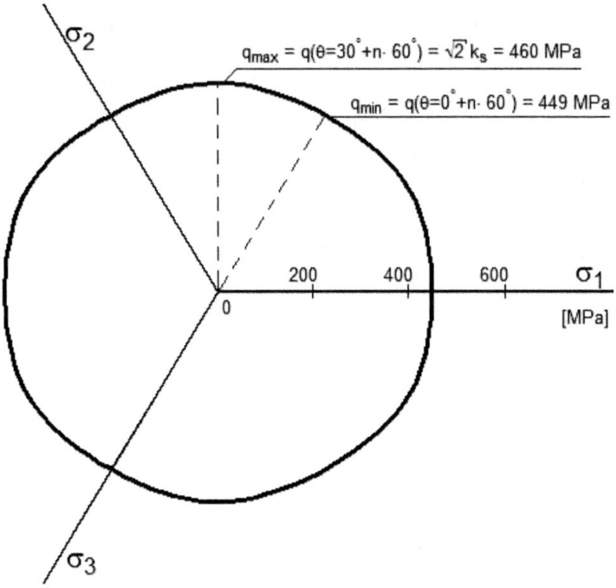

Figure 21. The effect of Lode angle illustrated on the octahedral cross-section of the yield surface for Inconel 718 shown in Fig. 9, (Nowak et al., 2011).

6.2 Energy-based hypothesis of material effort for anisotropic materials

First attempts of extending the energy-based hypothesis of material effort (4) for anisotropic materials belong to (Burzyński, 1928) and were continued in (Olszak and Urbanowski, 1956), and in (Olszak and Ostrowska-Maciejewska, 1985). The solution of this problem has been completed in the paper of Jan Rychlewski (Rychlewski, 1984), who proved that the original Mises criterion (Mises, 1928) of limit states for anisotropic solids expressed by means of stress deviators $\boldsymbol{S}$ (due to neglected effects of pressure):

$$\boldsymbol{S} \cdot \boldsymbol{H} \cdot \boldsymbol{S} \leqslant 1 \tag{24}$$

considerd in a more general form:

$$\boldsymbol{\sigma} \cdot \boldsymbol{H} \cdot \boldsymbol{\sigma} \leqslant 1 \tag{25}$$

can be expressed in terms of elastic energy density $\Phi(\sigma)$ partitioned over the energy densities $\Phi(\sigma_1), ..., \Phi(\sigma_\chi)$ defined for $\chi \leq 6$ eigen-states resulting from the symmetry of the fourth order limit tensor $\boldsymbol{H}$,(Rychlewski, 1984):

$$\boldsymbol{\sigma} \cdot \boldsymbol{H} \cdot \boldsymbol{\sigma} = \frac{1}{h_1}\Phi(\boldsymbol{\sigma}_1) + ... + \frac{1}{h_\chi}\Phi(\boldsymbol{\sigma}_\chi) \tag{26}$$

where h_χ denotes the limit values determined for a given material in the eigen-state $\chi \leq 6$ and the elastic energy density accumulated in the solid is the unique sum of the parts of energy pertinent to a particular eigen-state $\chi \leq 6$

$$\Phi(\boldsymbol{\sigma}) = \Phi(\boldsymbol{\sigma}_1) + ... + \Phi(\boldsymbol{\sigma}_\chi) \tag{27}$$

This is energy-based limit condition valid for anisotropic materials revealing symmetry of elastic range in the sense that the limit value in tension and compression are the same in each eigen-state $\chi \leq 6$. Some particular specifications of this general formulation were studied in (Ostrowska-Maciejewska amd Rychlewski, 1988), (Kowalczyk, 2003), (Kowalczyk-Gajewska and Ostrowska-Maciejewska, 2005). In (Kowalczyk-Gajewska and Ostrowska-Maciejewska, 2009) one can find a comprehensive review on spectral decomposition of Hooke's tensor for all symmetry groups of linear elastic solids with references to other authors dealing with the subject. A general formulation of an extension of the presented above criterion for anisotropic materials with asymmetry of elastic range, i.e. the limit values in tension and compression are different in at least one of eigen-states $\chi \leq 6$ was proposed in (Nowak et al., 2011), (Ostrowska-Maciejewska et al., 2012) and (Ostrowska-Maciejewska et al., 2013). However, experimental specification of particular criteria still remains unsolved and requires further studies.

6.3 Studies on inelastic flow and failure accounting for multiscale shear banding phenomena

The influence of impact loading on inelastic flow and failure, in particular on ductile fracture of metals, is one of most intensively studied phenomenon in thermomechanics of solids. Some issues of this problem are discussed in the accompanying chapters of the book. Having in mind a reach bibliography of the subject, only some particular points are mentioned. Phenomenological failure criteria are often considered, in which the critical failure strain depends on an accumulated equivalent strain and strain rate, which is usually averaged over a certain gage volume, (Arias et al., 2008). Sometimes, a stress triaxiality parameter and an internal variable representing microdamage are included, c.f. e.g. (Perzyna, 2005), (Perzyna, 2011) and (Rodríguez-Milán et al., 2012), where besides the stress triaxiality also the effect of Lode angle is accounted for. The recent studies show that an adequate prediction of dynamic fracture processes should take into considerations physical mechanisms activated on different levels of observation, (Perzyna, 2010) and (Perzyna, 2012). In particular, the multiscale mechanism of shear banding plays pivotal role in the evolution of ductile fracture. Although the effects of shear banding were implemented in the phenomenological models of plasticity of metallic solids (Pęcherski, 1992), (Pęcherski, 1997) as well as (Pęcherski, 1998) and (Kowalczyk-Gajewska and Pęcherski, 2009) and viscoplasticity of ultra fine grained (ufg) and nanocrystalline metals (nc metals), (Nowak et al., 2007) and (Frąś et al., 2011), the application of these mechanism for modelling and simulations of ductile fracture processes is rather scarce. However, some attempts of including effects of micro-shear bands in the analysis of ductile fracture processes were made in (Nowacki et al., 2010) and by (Perzyna, 2010) and (Perzyna, 2012). Future research should concentrate on phenomenological description of the effects of initiation and evolution of microdamage, which produces the strength differential effect on macroscopic level and should lead for application of paraboloid yield condition, (Korbel et al., 2006). Further studies are necessary to account for the effect of Lode angle related with multiscale shear banding in constitutive modelling and numerical simulations of inelastic flow processes and ductile fracture of metallic solids.

6.4 Conclusion

The novelty of our approach consists in a derivation of energy-based hypothesis of *material effort* well-founded on the multiscale analysis of deformation processes in solids: starting from electronic basis of chemical bonds, semi-empirical potentials of intermolecular interactions and assuming a *measure of material effort* as a definite part of elastic energy density on macroscopic level of continuous deformable body related with the critical strain processes.

Studying the bibliography of the subject one can observe that the yield or failure criteria, which are obtained in the rigorous way from the energy-based material effort hypothesis proposed originally by (Burzyński, 1928) were later rediscovered again and again for different ranges of empirical parameters independently by many researchers. More detail discussion of this issue is given in (Pęcherski, 2008). Also the mentioned above Mises criterion for anisotropic materials gained very efficient and physically clear specification for orthotropic materials in (Hill, 1948). Further studies in this field directed towards the applications, mainly in numerical simulations of sheet metal forming processes, led to abundance of modifications, mainly of empirical character, resulting in yield conditions without firm physical and theoretical foundations, cf. (Ostrowska-Maciejewska et al., 2013) for more detail discussion. The presented in this work approach based on the concept of energy measure of material effort may introduce some order and physical meaning in the field of constitutive description of inelastic flow and failure of solids. The efficient applications of the presented ideas require further studies, experimental, theoretical and computational ones to specify the particular yield and failure criteria.

Acknowledgement

The results presented in the paragraphs 2 and 5 have been obtained during the studies within the framework of the research program KomCerMet, *Composites and Nanocomposites for Aviation and Car Industry*, POIG 01.03.01-00-013/08. The studies related with the paragraph 2 were also partly supported by the Polish State Committee for Scientific Research within the framework of the Project N N501 15663.

Bibliography

J.C. Maxwell *Origins of Clerk Maxwell´s electric ideas as described in familiar letters to William Thomson*,Cambridge at the University Press 1937, Proc. Cambridge Phil. Soc., 32, 31-33, 1936.

M.T. Huber, *Specific work of strain as a measure of material effort*, Arch. Mech., 56, 173-190, 2004, English translation from the original edition in Polish, *Właściwa praca odkształcenia jako miara wytężenia materyału. Przyczynek do podstaw teoryi wytrzymałości*, Czasopismo Techniczne, XXII, 1904.

W. Burzyński, Selected passages from Wodzimierz Burzyński´s doctoral dissertation *Study on material effort hypotheses*, Engng. Trans., 57, 185-215(2009), English translation from the original edition in Polish: *Studium nad Hipotezami Wytężenia*, Akademia Nauk Technicznych, Lwów, 1928.

W. Burzyński, *Theoretical foundations of the hypotheses of material effort. Włodzimierz Burzyński (1900-1970)*, Engng. Trans., 56, 189-225(2008), English translation from the original edition in Polish, *Teoretyczne podstawy hipotez wytężenia*, Czasopismo Techniczne, Lwów, 47, 1-41, 1929.

R.L. Hayes, M. Ortiz, E.A. Carter, *Universal-binding energy relation for crystals that accounts for surface relaxation*, Physical Review, B69, 172104 2004.

M. He, S. Li, *An embedded atom hyperelastic constitutive model and multiscale cohesive finite element method*,Comput. Mech., 49, 337-355, 2012.

R. Phillips, *Crystals, Defects and Microstructures. Modelling Across Scales*, Cambridge University Press, Cambridge UK, 2001.

M. Buehler, *Atomistic Modelling of Materials Failure*, Springer Sciences + Business Media LLC, New York, 2008.

J.H. Rose, J.R. Smith, F. Guinea, J. Ferante, *Universal features of the equation of state of metals*, Physical review, B29, 2963 (1984).

A. Banerjea, J.R. Smith, *Origins of the universal binding-energy relation*, Physical Review B 37, 6632-6645 (????).

T. Frąś, R.B. Pęcherski, *Applications of the Burzyński hypothesis of material effort for isotropic solids*, Mechanics and Control, 29, 45-50, 2010.

R.B. Pęcherski, *Relation of microscopic observations to constitutive modelling of advanced deformations and fracture initiation of viscoplastic materials*, Arch. Mech., 35, 257-277, 1983.

R.B. Pęcherski, *Macroscopic effects of micro-shear banding in plasticity of metals*, Acta Mechanica, 131, 203-224, 1998.

J. Gilman, *Electronic Basis of Strength of Materials*, Cambridge University Press, Cambridge UK, 2003.

M Finnis, *Interatomic Forces in Condensed Matter*, Oxford University Press, Oxford UK, 2003.

K. Burke, *Perspective on density functional theory*, The Journal of Chemical Physics 136, 150901, 2012.

D. Wolf and S. Yip (eds.), *Material interfaces. Atomic-level structure and properties*, Chapman & Hall, London, 1992.

K. Nalepka, R.B. Pęcherski, *Modelling of the interatomic interactions in the Copper crystal applied in the structure* $(111)Cu \,||\, (0001)Al_2O_3$, Archives of Metallurgy and Materials., 54, 511-522, 2009.

K. Nalepka, *Efficient approach to metal/metal oxide interfaces within variable chargemodel*, The European Physical Journal B 85, 45, 2012.

K. Nalepka, *Material symmetry: a key to specification of interatomic potentials*, Bulletin of the Polish Academy of Sciences: Technical Sciences, 61, 1-10, 2013.

Y. Mishin, M.J. Mehl, D.A. Papaconstantopoulos, A.F. Voter, and J.D. Kress, *Structural stability and lattice defects in copper: Ab initio, tight-binding, and embedded-atom calculations*, Phys. Rev. B 63, 224106, 2001.

K. Nalepka, *Symmetry-based approach to parametrization of embedded-atom-method interatomic potentials*, Computational Materials Sciences, 56, 100-107, 2012.

Y. Wang, S. Curtarolo, C. Jiang, R. Arroyave, T. Wang, G. Ceder, L.-Q. Chen, and Z.-K. Liu, *Ab initio lattice stability in comparison with CALPHAD lattice stability*, Computer Coupling of Phase Diagrams and Thermochemistry 28, 79-90, 2004.

M. Černý, R. Boyer, M. Šob, and S. Yip, *Higher-energy structures and stability of Cu and Al crystals along displacive transformation paths*, Journal of Computer-Aided Materials Design 12, 161-173, 2005.

X.D. Dai, Y. Kong, and J.H. Li, *Long-range empirical potential model: Application to fcc transition metals and alloys*, Phys. Rev. B 75, 104101, 2007.

M.J. Mehl, A. Aguayo, and L.L. Boyer, *Absence of metastable states in strained monatomic cubic crystals*, Phys. Rev. B 70, 014105, 2004.

K. Kowalczyk, J. Ostrowska-Maciejewska, R.B. Pęcherski, *An energy-based yield criterion for solids of cubic elasticity and orthotropic limit state*, Archives of Mechanics, 55, 431-448, 2003.

J.J. Gilman, *The plastic resistance of crystals*, Aust. J. Phys., 13, 327-346, 1960.

Z. Nowak, P. Perzyna, R.B. Pęcherski, *Description of viscoplastic flow accounting for shear banding, Archives of Metallurgy and Materials*, 52, 217-222, 2007.

R. Z. Valiev, E. V. Kozlov, Yu. F. Ivanov, J. Liani, A. A. Nazarov, B. Baudelet, *Deformation behaviour of ultra-fine-grained Copper*, ActaMetall, Mater, 42, 2467-2475, 1994.

Y. Wang, M. Chen, F. Zhou, E. Ma, *High tensile ductility in a nanostructured metal*, Nature, 912-915, 2002.

L. Lu, Y. Shen, X. Chen, L. Qian, K. Lu, *Ultrahigh strength and high electrical conductivity in Copper*, Science, 304, 422-426, 2004.

D.G. Clerc, H.M. Ledbetter, *mechanical hardness: a semiempirical theory based on screened electrostatics and elastic shear*, J. Phys. Chem. Solids, 59, 1071-1095, 1998.

D.G. Clerc, *Mechanical hardness and elastic stiffness of alloys: semiempirical models*, J. Phys. Chem. Solids, 60, 83-102, 1999.

J. Zawadzki, *Ciśnienie zredukowane jako jeden z parametrów wytężenia (Przyrost waściwej energii swobodnej jako miara wytężenia) Reduced pressure as a strength parameter (The increase of unit free energy as a measure of effective stress)* in Polish (original English translation of the title), Rozprawy Inżynierskie (Engng. Trans.), 5, 359-398, 1957.

T. Frąś, Z.L. Kowalewski, R.B. Pęcherski, A. Rusinek, *application of Burzyski failure criteria. Part. I. Isotropic materials with asymmetry of elastic range*, Engng. Trans., 58, 3-13, 2010.

R.B. Pęcherski, M. Nowak, P. Szeptyński, *An extension of Burzyński hypothesis of material ef fort accounting for the third inwariant of stress tensor*, Archives of Metallurgy and Materials, 56, 503-508, 2011.

R. von Mises, *Mechanik der festenKrperimplastisch-deformablenZustand, Nachrichten von der Kngl. Gesellschaft der WissenschaftenzuGttingen*, Mathematisch-PhysikalischeKlasse, Z. 4 (1), 582-592, 1913.

H. Hencky, *Zur Theorie plastischer Deformationen und der hierdurchim Material hervorgerufenen Nach spannungen*, ZAMM, 4, 323-334, 1924.

R.B. Pęcherski, *Burzyński yield condition vis-a-vis the related studies reported in the literature*, Engng. Trans., 56, 311-324, 2008.

M. Nowak, J. Ostrowska-Maciejewska, R.B. Pęcherski, P. Szeptyński, *Yield criterion accounting for the third inwariant of stress tensor dewiator. Part. I. Proposition of the yield criterion based on the concept of influence function*, Engng. Trans. 59, 273-281, 2011.

J. Ostrowska-Maciejewska, P. Szeptyński, R.B. Pęcherski, *Mathematical foundations of limit criterion for anisotropic materials*, Archives of Metallurgy and Materials, 58, No. 3, 751-764, 2013.

Z.S. Olesiak, R.B. Pęcherski, *Włodzimierz Burzyński biography and work on material effort*, Bulletin of the Polish Society of Theoretical and Applied Mechanics, Yearbook 2012, PTMTS, Warszawa, 2013 (in Polish) available in print form oron the website: http://www.ptmts.org.pl/biul-2012-www.pdf.

A.M. Freudenthal, *The Inelastic Behaviorof Engineering Materials and Structures*, Wiley, N. York, 1950.

R.M. Christensen, *A two-property yield, failure (fracture) criterion for homogeneous, isotropic materials*, J. Eng. Mater. Technol., 126, 45-52, 2004.

R.M. Christensen, *A comparative evaluation of three isotropic, two property failure theories*, J. Appl. Mech., 73, 852-859, 2006.

J.S. Blazy, A. Marie-Louise, S. Forest, Y. Chastel, A. Pineau, A. Awade, C. Grolleron, F. Moussy, *Deformation and fracture of aluminium foams under proportional and non proportionalmultiaxial loading: statistical analysis and size effect*, Int. J. Mech. Sci., 46, 217-244, 2004.

A. Stręk, *Yield criteria and their verification for metal foams*, presentation on the 38th Solid Mechanics Conference, SolMech2012, August 27-31, 2012, Warsaw, 148-149, Book of Abstracts, R. Pęcherski, J. Rojek, P. Kowalczyk (eds.), Institute of Fundamental Technological Research, Polish Academy of Sciences, Warsaw, 2012.

V.S. Deshpande, N.A. Fleck, *Isotropic constitutive models for metallic foams*, Journal Mech. Phys. Solids, 48, 1253-1283, 2000.

T. Yoshikawa, M. Tokuda, T. Inaba, H. Iwasaki, K. Machino, N. Deguchi, *Plastic deformation of AZ31 magnesium Allom under various temperature conditions*, Jnl. Soc. Mat. Sci. Japan, 57(7), 688-695, 2008.

D.C. Drucker, W. Prager, *Soil mechanicsand plastic analysis for limit design*, Quart. Appl. Math., 10, 157-165, 1952.

G.P. Sendeckyj, *Empirical Strength Theories. Testing for Production of Material Performance in Structures and Components*, ASTM STP 515, 171-179, American Society for Testing and Materials, 1972.

A. Nardin, G. Zawarise, B.A. Schrefler, *Modelling of cutting tool soil interaction. Part. I: contact behaviour*, Computational Mechanics, 31, 327-339, 2003.

G.S. Pisarenko, A.A. Lebedev, *The Deformation and strength of Materials under Complex State of Stress*, NaukowaDumka, Kiev, 1973 (in Russian).

M. Jirasek, Z.P. Bazant, *Inelastic Analysis of Structures*, J. Wiley, Chichester, 2002.

M.-H. Yu, *Unified Strength Theory and its Applications*, Springer-Verlag, Berlin, Heidelberg, 2004.

M.-H. Yu, G.-W. Qiang, Y.-Q. Zhang, *Generalized Plasticity*, Springer-Verlag, Berlin, Heidelberg, 2006.

M. Życzkowski, *Combined Loadings in the Theory of Plasticity*, PWN-Polish ScientficPuublishers, Warszawa, 1981.

M. Życzkowski, *Discontinuous bifurcation in th case of the Burzyński Torre yield condition*, ActaMechanica, 132, 19-35, 1999.

J.J. Skrzypek, *Plasticity and Creep. Theory, Examples, and Problems*, R.B. Hetnarski (English Edition Editor), CRC Press, Boca Roca Raton, 1993.

T. Frąś, *Modelling of plastic yield surface of materials accounting for initial anisotropy and strength differential effect on the basis of experimentsand numerical simulation*, PhD Thesis, Université de Lorraine & AGH University of Science and Technology, Metz & Krakow, 2013.

Wolfram Mathematica, 6.0 Copyright 1988-2007, Wolfram Research Inc., Oxfordshire, U.K.

P. S. Theocaris, *Failure criteria for isotropic bodies revisited*, Engng. Fract. Mech., 51:239264, 1995.

R. Grassi, I. Cornet, *Fracture of gray cast-iron tubes under biaxial stresses*, Journal ofApplied Mechanics. Transactions of the ASME, 16, 178182, 1949.

L. F. Coffin, *The flow and fracture of brittle materials*, Journal of Applied Mechanics. Transactions of the ASME, 17, 223248, 1950.

W. Lode, *Versucheber den Einfluss der mittlerenHauptspannung auf das Fliessen der Metalle*, Z. der Physic, 36, 913919, 1926.

G. Taylor and H. Quinney, *The plastic distortion of metals*, Phil. Trans. Roy. Soc., A230, 323362, 1931.

A. C. Lund and C. A. Schuh, *Yield surface of a simulated metallic glass*, ActaMaterialia, 51, 53995411, 2003.
T. Frąś, Z. Nowak, P. Perzyna, R.B. Pęcherski, *Identification of the model describing viscoplasticbehaviour of high strength metals*, Inverse Problems in Science and Engineering, 19, 17-30, 2011.
G. Vadillo, J. Fernandez-Sáez, R.B. Pęcherski, *Some applications of Burzyński yield condition in metal plasticity*, Materials and Design, 32, 628-635, 2011.
ABAQUS (Abaqus Analysis User's Manual 6.12)
W.K. Nowacki, Z. Nowak, P. Perzyna, R.B. Pęcherski, *Effect of strain rate on ductile fracture. A new methodology*, Journal of Theoretical and Applied Mechanics, 48, 1003-1026, 2010
Á. Arias, J.A. Rodríguez-Martínez, J.A. Rusinek, *Numerical simulations of impact behaviour of thin steel plates subjected to cylindrical, conical and hemispherical non-deformable projectiles*, Engineering Fracture Mechanics, 75, 1635-1656, 2008.
P. Perzyna, *The thermodynamical theory of elasto-viscoplasticity*, Review paper, Engineering Transactions, 53, 235-316, 2005.
P. Perzyna, *Micromechanics of localized fracture phenomena in inelasticsolids generated by impact-loaded adiabatic processes*, Engineering Transactions, 59, 299-348, 2011.
M. Rodríguez-Milán, Á. Vaz-Romero, J.A. Rodríguez-Martínez, A. Rusinek, Á. Arias, *Influence of failure strain of different aluminium alloys on dynamic problems*, Engineering Transactions, 60, 345-353, 2012.
P. Perzyna, *The thermodynamic theory of elasto-viscoplasticity for description of nanocrystalline metals*, Engineering Transactions, 58, 15-74, 2010.
P. Perzyna, *Multiscale constitutive modelling of the influence of anisotropy effects on fracture phenomena in inelastic solids*, Engineering Transactions, 60, 25-284, 2012.
R.B. Pęcherski, *Modelling of large plastic deformations based on the mechanism of micro-shearbanding. Physical foundations and theoretical description in plane strain*, Arch. Mech. 44, 563- 584, 1992.
R.B. Pęcherski, *Macroscopic measure of the rate of deformation produced by micro-shear banding*, Arch. Mech. 49, 385-40, 1997.
K. Kowalczyk-Gajewska, R.B. Pęcherski, *Phenomenological description of the effect of micro-shear banding in micromechanical modelling of polycrystal plasticity*, Archives of Metallurgy and Materials, 54, 1145-1156, 2009.
K. Korbel, Z. Nowak, P. Perzyna, R.B. Pęcherski, *Viscoplasticity of nanometals based on Burzyński yield condition*, Proc. of the 35th Solid Mechanics Conference SOLMECH 2006, Kraków, September 4-8, Polish Academy of Sciences, Institute of Fundamental Technological Research, 145-146, 2006.

W. Olszak, W. Urbanowski, *The plastic potential and the generalized distortion energy in the theory of non-homogeneous anisotropic elastic-plastic bodies*, Arch. Mech. Stos., 8, 671-694, 1956.

W. Olszak, J. Ostrowska-Maciejewska, *The plastic potential in the theory of anisotropic elastic-plastic solids*, Engng. Fracture Mech., 21, 625-632, 1985.

R. Von Mises, *Mechanik der plastischen Formnderung von Kristallen*, ZAMM, 8, 161-185, 1928.

J. Rychlewski, *Elastic energy decomposition and limit criteria*, Engineering Transactions, 59, 31-63, 2011; English translation of the original paper in Russian from Advances in Mechanics (Uspekhi Mekhaniki), 7, 51-80, 1984.

J. Ostrowska-Maciejewska, J. Rychlewski, *Plane elastic limit states in anisotropic solids*, Arch. Mech.,40, 379-386, 1988.

K. Kowalczyk, J. Ostrowska-Maciejewska, *Energy-based limit conditions for transversally isotropic solids*, Arch. Mech., 54, 497-523, 2000.

K. Kowalczyk-Gajewska, J. Ostrowska-Maciejewska, *Energy-based limit criteria for anisotropic elastic materials with constraints*, Arch. Mech., 57, 133-155, 2005.

K. Kowalczyk-Gajewska, J. Ostrowska-Maciejewska, *Review on spectral decomposition of Hooke's tensor for all symmetry groups of linear elastic materials*, Engineering Transactions, 57, 145-183, 2009.

J. Ostrowska-Maciejewska, R.B. Pęcherski, P. Szeptyński, *Limit condition for anisotropic materials with asymmetric elastic range*, Engineering Transactions, 60, 125-139, 2012.

R. Hill, *A theory of the yielding and plastic flow of anisotropic metals*, Proc. Roy. Soc. (London), A 193, 281-297, 1948.

D. Rittel, S. Lee, G. Ravichandran, *A shear-compression specimen for large strain testing*, Experimental Mechanics, 42, 58-64, 2002.

C. Schuh, A.C. Lund, *Atomistic baris for the plastic yield criterion of metallic glass*, Nature Materials, 2, 449-452, 2003.

M. Vural, A.Molinari, N. Bhattacharya, *Analysis of slot orientation in shear-compression specimens (SCS)*, Experimental Mechanics, 51, 263-273, 2011.

C.D. Wilson, *A critical re-examination of classical metalplasticity*, J. Appl. Mech., 69, 63-68, 2002.

C.M. Gil, C.J. Lissenden, B.A. Lerch, *Yield of Inconle 718 by axial-torsional loading at temperatures up to 649 C.* J Test Eval, 27, 327-336, 1999.

C.M. Gil, C.J. Lissenden, B.A. Lerch, *Unusual nonlinear response of some metallic materials.* Mech Mater, 31, 565-577, 1999.

GPSR Compliance

The European Union's (EU) General Product Safety Regulation (GPSR) is a set of rules that requires consumer products to be safe and our obligations to ensure this.

If you have any concerns about our products, you can contact us on ProductSafety@springernature.com

In case Publisher is established outside the EU, the EU authorized representative is:

Springer Nature Customer Service Center GmbH
Europaplatz 3
69115 Heidelberg, Germany

Batch number: 10732220

Printed by Printforce, the Netherlands